Security and Privacy Preserving in Social Networks

Lecture Notes in Social Networks (LNSN)

For further volumes:
www.springer.com/series/8768

Richard Chbeir • Bechara Al Bouna
Editors

Security and Privacy Preserving in Social Networks

 Springer

Editors
Richard Chbeir
University of Pau and Adour Countries
LIUPPA Laboratory
Anglet
France

Bechara Al Bouna
Ticket Laboratory
Antonine University
Baabda
Lebanon

ISSN 2190-5428
ISSN 2190-5436 (electronic)
ISBN 978-3-7091-0893-2
ISBN 978-3-7091-0894-9 (eBook)
DOI 10.1007/978-3-7091-0894-9
Springer Wien Heidelberg New York Dordrecht London

Library of Congress Control Number: 2013949177

Printed on acid-free paper

Springer is part of Springer Science+Business Media (www.springer.com)

Preface

Richard Chbeir and Bechara Al Bouna

Abstract With the increasing success in social networks, the various facets of our "digital" lives are directly and indirectly accessible on the web with public and private details. To secure our digital footprints and protect our privacy on social networks, dedicated techniques and methods are required. The aim of this book is to assess the current approaches and technologies related to the security and privacy protection of social networks. It provides main concepts, an overview of the state of the art, latest techniques, studies, and approaches as well as future directions in this field, by including a wide range of contributions from various studies and research groups.

1 Introduction

Social networks provide users, within their platforms, powerful ways to interact with other users through different forms and modalities. Consequently, each user can search and check the profiles of her social network members (for various reasons), exchange messages with some of them, publish some videos, and post comments on shared photos, etc. Although such tools are attracting continuously more and more users, several security and privacy problems related to their usage have emerged. For instance, how to provide users with an easy way to protect their shared data? How to protect a data repository (e.g., photo album) while several related information about the same content is already published (by some user

R. Chbeir (✉)
University of Pau and Adour Countries, LIUPPA Laboratory, 64200 Anglet Cedex, France
e-mail: richard.chbeir@univ-pau.fr

B.A. Bouna
UPA University, Ticket Laboratory, Beirut, Lebanon
e-mail: bechara.albouna@upa.edu.lb

friends) on the same or other blog/wiki/social network? etc. Several privacy settings are provided on social networks to help users secure their profiles and data. Those privacy settings are indeed useful for basic cases. However, they become relatively complicated to be correctly tuned by (non-expert) users and also inappropriate for complex cases (particularly when multimedia data come to play). In addition, social networks suffer regularly from security breaches and sometimes leak intentionally or accidentally users' information to unauthorized entities or third parties.

The book explores emerging and new problems that everyone is facing in his/her daily digital life and it contributes to the awareness campaign shedding the light on the risk of publishing private information particularly on social networks. Thus, it is a valuable companion and comprehensive reference for:

- *Social network users* who would like to be aware of the risks and possible (current and future) solutions.
- *Undergraduate and postgraduate students* who are taking a course in security and protection and need to study related current concepts in social networks.
- *Developers* who need to implement advanced security and protection techniques and make benefit of existent technologies.
- *Researchers* who desire to understand exiting approaches, pin down their limits, and identify several future directions.

Why This Book Is Interesting?

As mentioned earlier, the primary target audience for this book includes researchers, scholars, postgraduate students, developers, as well as end users who are interested in social network security and protection. To achieve this:

- The book is organized in self-contained chapters to provide greatest reading flexibility.
- The book covers introductory material suitable for various reader profiles.
- Basic concepts will be explained in simple language with illustrations and examples, which will help readers new to this field understand concepts at play regarding the subject matter.
- There will be plenty of questions in the last chapter, which will be helpful for both lecturers and students to check the understanding of the book concepts.

2 Book Organization

This edited volume of the book aims to gather the latest advances and innovative solutions in security and privacy preserving in social networks. It assesses the current technologies, as well as major challenges and future perspectives in this field. In its current form, this book covers interesting works gathered from experts working in heterogeneous domains related to security and privacy. It is devoted to help the readers have a glance at the research priorities to be explored in social network throughout a security perspective.

The book contains eleven self-contained chapters organized into four parts as follows:

- Online Social Networks: Analysis, Privacy, and Terrorism
- Access Control, Reputation, and Semantic policies in social networks
- Security and Privacy in Mobile and P2P social networks
- Multimedia-based Authentication and Access Control Models for social networks

The last chapter contains a battery of problems and questions so as to allow the reader to better identify the main concepts in each chapter.

Part I deals with *Online Social Networks Analysis*, *Privacy*, and *Terrorism* and consists of three chapters.

In Chap. 1, "*Privacy in Online Social Networks*," Elie Raad and Richard Chbeir address privacy-related issues by resorting to social network analysis and link mining techniques. They first describe the fundamental of social networks, their common representations, and the main motivations associated with their use. Afterwards, they particularly show how privacy attacks can build on social network analysis and link mining techniques to reveal user-sensitive information. The chapter concludes with a discussion of some open challenges to address in future privacy-related works.

In Chap. 2, "*Online Social Networks: Privacy Threats and Defenses*", Shah Mahmood provides a brief overview of some threats to users' privacy. He classifies these threats as users' limitations, design flaws and limitations, implicit flows of information, and clash of incentives. He also discusses two defense mechanisms which deploy usable privacy through a visual and interactive flow of information and a rational privacy vulnerability scanner.

Shah Mahmood in Chap. 3, titled "*Online Social Networks and Terrorism: Threats and Defenses*," discusses how some users are misusing social networks for terrorism. To do so, he first provides the background, definition, and classification of terrorism. Second, he discusses how some terrorists may be using online social networks to (1) recruit new members to a terrorist organization and maintain the loyalty of their existing sympathizers; (2) plan attacks and share information about them; (3) gather intelligence; (4) train recruits for specific attacks; (5) raise funds for their causes; (6) propagate fear amongst the enemy population; and (7) engage in counterintelligence to uncover undercover agents. Third, he explores several mechanisms to detect terrorists using online social networks, including (1) keyword-based flagging; (2) sentiment analysis; (3) honeypots; (4) social network analysis; (5) facial recognition; and (6) view escalation. He shows that the keyword-only flagging mechanism used by US Department of Homeland Security to detect terrorists is potentially effective but certainly produces a large number of false positives, making it possibly less efficient in practice. Finally, he proposes the use of targeted advertisements to rehabilitate possible radicals using the online social networks.

Part II consists of three chapters and deals with *Access Control, Reputation*, and *Semantic Policies* in social networks.

In Chap. 4, "*User-Managed Access Control in Web Based Social Networks*," Gonzalez-Manzano et al. focus on the design and implementation of user-managed access control systems Web Based Social Networks (WBSNs). In this regard, there exists a well-known set of requirements (relationship-based, fine-grained, interoperability, sticky-policies, and data exposure minimization) that have been identified in order to provide a user-managed access control for WBSNs. These requirements, partially addressed by the works proposed in the literature, represent "building blocks" for a well-defined user-managed access control model. In this chapter, the authors first provide a conceptualization of a WBSN to propose an access control model, called $SoNeUCON_{ABC}$, and a mechanism that implements it. A set of mechanisms among the recently proposed in the literature are selected such that, when deployed over $SoNeUCON_{ABC}$, the whole set of user-managed requirements can be fulfilled.

Chapter 5, "*UPP+: A Flexible User Privacy Policy for Social Networking Services*" by Ramzi A. Haraty and Sally Massalkhy, presents a new privacy policy model, called UPP+, for enhancing privacy and security for ordinary users. To do so, they use the Alloy language to formalize the model and the Alloy analyzer to check for any inconsistencies.

In Chap. 6, "*Social Semantic Network-based Access Control*" by Villata et al., the authors address the privacy problem within the Social Semantic Web umbrella where social networks are integrated and enhanced by the use of semantic conceptual models, e.g., ontologies, where social information and links among users become semantic information and links. In this chapter, the authors discuss the benefits of introducing semantics in social network-based access control. In particular, they analyze and detail two approaches to manage the access rights of the social network users relying on Semantic Web languages only, and they highlight, thanks to these two proposals, what are the pros and cons of adopting semantic models and languages in social networks access control. Finally, they report on the other existing approaches coupling semantics and access control in the context of social networks.

Part III consists of three chapters and focuses on Security and Privacy in Distributed (*Mobile* and *P2P*) social networks.

In Chap. 7, "*Supporting Data Privacy in P2P Systems*" by Jawad et al., the authors explore large-scale data sharing in Peer-to-Peer (P2P) systems. To provide appropriate access control to sensitive data, they extend Hippocratic DataBases (HDB) approach to provide an effective solution to this problem. In essence, the use of HDB has been restricted to centralized systems. This chapter gives an overview of current solutions for supporting data privacy in P2P systems and develops in more detail a complete solution based on HDB.

In Chap. 8, "*Privacy Preserving Reputation Management in Social Networks*" by Omar Hasan and Lionel Brunie, the authors address the reputation management to help establish the trustworthiness of users in social networks to enforce various security requirements. For example, a reputation system can help filter fake user profiles. However, a major challenge in developing reputation systems for social networks is that users often hesitate to publicly rate fellow users or friends due to the fear of retaliation. This trend prevents a reputation system from accurately computing reputation scores. Privacy preserving reputation systems hide the individual ratings of users about others and only reveal the aggregated community reputation score thus allowing users to rate without the fear of retaliation. In this chapter, the authors describe privacy preserving reputation management in social networks and the associated challenges. In particular, they look at privacy preserving reputation management in decentralized social networks, where there is no central authority or trusted third parties, thus making the task of preserving privacy particularly challenging.

In Chap. 9, "*Security and Privacy Issues in Mobile Social Networks*," Teles et al. present an interesting chapter related to Mobile Social Networks (MSNs), since the use of contextual information gathered from mobile devices sensors, which in many cases occurs without the user's awareness, leads to new privacy issues. Here, the authors give a general overview on the challenges and the main concepts that affect privacy and security of MSNs. They show how privacy problems migrate from social networks to MSNs and provide some eventual solutions.

Part IV consists of two chapters and explores in a nice way *Multimedia-based* Authentication and Access Control Models for social networks.

Chapter 10, *"Avatar Facial Biometric Authentication Using Wavelet Local Binary Patterns"* by Abdallah A. Mohamed and Roman V. Yampolskiy, focuses on virtual worlds (e.g., Second Life) since they are populated by different types of people, businesses, and organizations. Users of virtual worlds, either individuals or organizations, might abuse the flexibility and adaptability offered by virtual environment by engaging in criminal activities. Even terrorist organizations have been active in virtual worlds. These organizations recently have used the virtual worlds for recruitment and to train their new members in an environment that is very similar to the real one. Since avatars are not just virtual creations as they have a great social and psychological correspondence with their creators, applying biometric techniques on avatars can give the law enforcement agencies and security experts the ability to identify who the actual users behind these avatars are. There is a mounting pressure to have techniques for verifying the real identities of the inhabitants of virtual worlds to secure cyberworld from incessant criminal activities (e.g., verbal harassment, fraud, money laundering, data or identity theft). In order to reduce the gap between our ability to recognizing human faces and avatar faces and to develop reliable tools for protecting virtual environments, the authors discuss in this chapter how one can use different versions of Local Binary Pattern (LBP) operators (traditional LBP, Multi-scale LBP, and Hierarchical multi-scale LBP) to recognize avatar faces from two different virtual worlds (Second Life and Entropia Universe). This chapter includes a definition of discrete wavelet transform from a face recognition research perspective, a summary of previous work done on this topic, characteristics of the data sets used in the experiments, as well as some suggestions for future work.

In Chap. 11, *"A Flexible Image-Based Access Control Model for Social Networks,"* Al Bouna et al. tackle the problems of publishing images and photos. It is true that social networks are tremendously spreading their tentacles over the web community providing appropriate and well-adapted tools for sharing images and photos. A fundamental glitch to consider is their ability to provide suitable techniques to preserve individuals' privacy. Indeed, there is an urgent need to guarantee privacy by making available to end users tools to enforce their privacy constraints. This cannot be done over images as simple as it has been designed so far for textual data. In fact, images, as all other multimedia objects, are of complex structure due to the gap between their raw data and their actual semantic descriptions. Without these descriptions, protecting their content is a difficult matter. In this chapter, the authors present a novel security model for image content protection. In the proposed model, the authors provide dynamic security rules based on first-order logic to express constraints that can be applied to contextual information as well as low-level features of images. They finally discuss a set of experiments and studies carried out to evaluate the proposed approach.

Acknowledgments

We would like to thank the authors who provided excellent chapters and timely revisions. We are also grateful for their trust in us and patience during the review process. We would like to express our sincere thanks to the reviewers (Prof. Bill Grosky, Prof. Alban Gabillon, Prof. Yudith Cardinale, Prof. Hiroshi Ishikawa, Prof. Kokou Yetongnon, Dr. Solomon Atnafu, Dr. Fekade Getahun, Dr. Joe Tekli, Dr. Youssef Bou Issa, Dr. Mohamad Nassar, Dr. Firas al Khalil, and Dr. Elie Raad) for their tremendous effort and challenging task of choosing high quality chapters and their valuable criticism that greatly improved the quality of final chapter versions. The editors also would like to thank, Professor Reda Alhajj, the editor-in-chief of the Lecture Notes in Social Networks series of Springer for editorial assistance and excellent cooperative collaboration to produce this important scientific work. We hope this volume motivates its readers to take the next steps beyond building models to implementing, evaluating, comparing, and extending proposed approaches and applications. We finally hope that readers will share our excitement to present this volume on *Security and Privacy Preserving in Social Networks* and find it useful.

About the Editors

Richard Chbeir received his Ph.D. in computer science from the University of INSA de Lyon-France in 2001 and then his Habilitation degree in 2010 from the University of Bourgogne. He is currently a full professor in the Computer Science Department in IUT de Bayonne in Anglet—France. His current research interests are in the areas of social networking, multimedia information retrieval, XML and RSS similarity, access control models, and digital ecosystems. He is member of IEEE and ACM since 1999. He has published (more than 40 peer-reviewed publications) in international journals and books (IEEE Transactions on SMC, Information Systems, Journal on Data Semantics, Journal of Systems Architecture, etc.), conferences (ER, WISE, SOFSEM, EDBT, ACM SAC, Visual, IEEE CIT, FLAIRS, PDCS, etc.) and has served on the program committees of several international conferences (SOFSEM, AINA, IEEE SITIS, ACM SAC, IEEE ISSPIT, EuroPar, SBBD, etc.). He has been organizing many international conferences and workshops (ACM SAC, ACM MEDES, CSTST, SITIS, etc.). He is currently the vice-chair of the ACM SIGAPP and the chair of its French Chapter.

Bechara Al Bouna received his B.S. degree and M.S. degree in networks and telecommunications engineering from the Antonine University—Lebanon in 2004. He received his Ph.D. degree from the LE2I laboratory of the Bourgogne University in France in 2009. From 2005 to 2008, he occupied teaching positions at the Antonine University and IUT of Dijon. Presently, he holds an assistant professor position at the Antonine University. His research interests include privacy preserving, data anonymization, multimedia access control models, and inference detection in multimedia environments.

Contents

Part I
Online Social Networks Analysis, Privacy and Terrorism

Chapter 1
Privacy in Online Social Networks

Elie Raad and Richard Chbeir

Abstract Online social networks have become an important part of the online activities on the web and one of the most influencing media. Unconstrained by physical spaces, online social networks offer to web users new interesting means to communicate, interact, and socialize. While these networks make frequent data sharing and inter-user communications instantly possible, privacy-related issues are their obvious much discussed immediate consequences. Although the notion of privacy may take different forms, the ultimate challenge is how to prevent privacy invasion when much personal information is available. In this context, we address privacy-related issues by resorting to social network analysis and link mining techniques. We first describe the fundamental of social networks, their common representations, and the main motivations associated with their use. Afterwards, we particularly show how privacy attacks can build on social network analysis and link mining techniques to reveal user-sensitive information. The chapter concludes with a discussion of some open challenges to address in future privacy-related works.

E. Raad (✉)
Memorial University of Newfoundland, St. John's, NL, Canada
e-mail: elie.raad@mun.ca

R. Chbeir
University of Pau and Adour Countries, Laboratoire LIUPPA, Anglet, France
e-mail: richard.chbeir@univ-pau.fr

R. Chbeir and B. Al Bouna (eds.), *Security and Privacy Preserving in Social Networks*,
Lecture Notes in Social Networks, DOI 10.1007/978-3-7091-0894-9_1,
© Springer-Verlag Wien 2013

1.1   Introduction

For the past few years, online social networks experienced an exponential growth in the number of their users and in the huge amount of available information. Many online social networks like Facebook,[1] LinkedIn,[2] Google+,[3] and Twitter[4] offer to web users new interesting means to communicate and interact. In reality, information available on these networks commonly describes persons along with their personal information (e.g., what they like, where do they live, and who they know) and interactions (e.g., with who they exchange messages, what comments they post, and how often they update their personal status).

With the proliferation of online social networks, information sharing on these networks is gaining an ever-increasing importance. Obviously, online social networks have found ingenious ways to collect data as users socialize. Not surprisingly, when socializing users communicate, interact, and tend to freely reveal personal information in line with their perceptions and preferences. To control the access to this personal information and to enforce its protection, online social networks promote the use of a number of built-in control mechanisms [1, 2]. However, social network users often fail to fully protect their profiles and personal data from undesirable forms of access as it has been revealed by previous studies [3–6]. This is due to the limited efficiency of the provided control mechanisms [1, 6, 7] and to the users' misconceptions about the networks' composition, the visibility of their profiles, and their misunderstandings of the privacy risks [3, 8].

As a result, more and more accessible personal information is available online, and yet, though the risk of security breaches and data exposures are manifold, adequate tools and efficient solutions are still missing. Social network users, overwhelmed with information, struggle to properly maintain privacy over their data to meet their actual expectations. However, social network users are not security experts and do not fully control their data. It is even hard to handle privacy threats as privacy breaches become numerous when dealing with personal information posted over years, across many online social networks, and shared with different types of contacts (e.g., colleagues, relatives, and friends). In addition, existing privacy settings are relatively complicated to be correctly managed by users [5, 6], online social networks may suffer from design conflicts issues (security and privacy vs. usability and sociability) [9] and may intentionally or accidentally leak users' information to unauthorized entities or third parties [10]. Consequently, it is vital to protect the tremendous amount of information from all sorts of attacks that may compromise users' privacy, invade their security, or disclose their data to unauthorized parties. Therefore, providing effortless mechanisms for social network

---

[1] http://www.facebook.com.

[2] http://www.linkedin.com.

[3] https://plus.google.com/.

[4] http://www.twitter.com.

users allowing them to control and reduce the potential exposure of their private information is of valuable importance.

With the huge number of social network users, it is therefore complex to delineate the concept of privacy. Privacy is a topic that received a lot of attention and has different facets [9, 11–13]. However, on online social networks some key characteristics that underlie privacy are commonly identified [13]. Among these concepts *anonymity*, *unlinkability*, and *unobservability* are the most interesting. Firstly, *anonymity* ensures that an attacker cannot sufficiently identify a user within a set of users. Secondly, *unlinkability* refers to the incapacity of an attacker to distinguish whether two or more items of interest are related or not. Thirdly, *unobservability* protects a user's activity so that an attacker or a third party cannot tell whether a resource or service is being used. Today, those are the most common users' privacy concerns.

In this chapter, we discuss privacy on social networks which is one of the most intriguing social networks' challenges. We argue that in order for a system to provide optimal privacy, its underlying algorithms must understand the characteristics of social networks and their associated analysis techniques. We also focus on the importance of social network data and explain how network analysis and data mining techniques [14, 15], useful in understanding users' behaviors and networks' characteristics, can become a source of privacy risk. On social networks, privacy concerns seem to be worldwide challenges for users, and thus novel privacy protection techniques must provide clear answers to a multitude of questions surrounding privacy:

– What are the most adequate analysis tools to use when dealing with specific privacy concerns?
– To which extent social networks' users behaviors are comparable to real-world persons' behavior and how to avoid that?
– How to prevent unwanted information leakage, data exploitation, and information linkage?
– What are the most important elements in order to protect users' privacy (e.g., type of data exchanged, relationship types, and networks structures)?

This chapter is structured as follows. We begin by defining the fundamental concepts of social networks in Sect. 1.2. We then focus on online social networks in Sect. 1.3 and we describe the main components of these networks as well as the main motivations associated with their use. This is followed by a description of the appropriate ways to represent social networks in Sect. 1.4 where we particularly highlight the graph-based representation. In Sect. 1.5, we discuss the challenges and opportunities related to the availability of social network data, its protection, its analysis and list some privacy protection techniques. We then present social network analysis in Sect. 1.6, a particularly important research area to study networks. We describe its most commonly used measures and their associated privacy threats. In Sect. 1.7, we detail link mining and its different tasks that are also used to analyze networks while emphasizing on social network links. We highlight the characteristics of the various link mining tasks and show their derived privacy threats. In Sect. 1.8, we present some open challenges yet to be addressed in future privacy-related systems before concluding this chapter in Sect. 1.9.

1.2   What Is a Social Network?

Networks have been used to model many systems of interest such as the World Wide Web [16], computer networks [17], biochemical networks [18], diffusion networks [19], and social networks [20]. Each of these networks is a structure that consists of a set of actors representing, for instance, web pages on the World Wide Web or persons in a social network, connected together by relations, representing links between web pages or friendships between persons. Besides these structural properties (actors and relations), Wasserman and Faust [14] identified a number of fundamental concepts like ties, dyads, triads, subgroups, and groups that characterize networks. For the purpose of this work, we start by detailing the concepts of actors, relations, and ties, the building blocks of social networks, before illustrating their use in online social networks (Sect. 1.3).

Definition 1. An actor is a social entity that interacts with other entities not only to maintain existing relations but also to establish new ones. On social networks, the concept of actors can refer to various types of entities such as persons, groups, and organizations.

Actors interact with each other through a variety of meaningful relations that denote different patterns of communication. Relations like friendship, collaboration, and alliance can vary across time, applications, or in terms of the involved actors [21]. Consequently, there are two main categories of networks that can be identified based on the type of actors, one-mode networks and two-mode networks [22]. While one-mode networks have a single type of actors, two-mode networks, also called bipartite, are networks with two types of actors. For instance, social networks modeling friendship between actors are an example of one-mode networks whereas those concerned with group memberships or attendance at events are two-mode networks.

Definition 2. A relation represents a connection from one actor to another one. A relation, also called relationship, plays an important role when studying the structure of social networks and the interactions among their actors. A relationship is characterized by various features such as its content, direction, and strength.

The relationship types have been addressed in several studies. Borgatti et al. [23] distinguished between four basic types of relationships: similarities, social relations, interactions, and flows. For instance, these relationships can express memberships (e.g., same club), kinships (e.g., mother of), affections (e.g., likes), interactions (e.g., talked to), and flows (e.g., flow of information), among others. Relationships on social networks can be directed or undirected. Depending on their content, relationships may (or may not) have a specific direction. While relationships such as "marriage" and "friendship" are undirected, other relationships such as "parent of" or "fan of" are directed. Social network relationships can also differ in strength. Usually, the strength can be estimated in a variety of ways using information about the actors, their interaction activities, or the correlation between them as the most common indicators [21, 24, 25].

Definition 3. A tie is the set of all relationships that exist between two actors. It is tightly connected to the concept of relationship as it aggregates the different types of relationships that exist between two actors. Just like relationships, ties also vary in terms of their content, direction, and strength.

Actors can be connected either with one relationship exclusively (e.g., employees of the same company) or with many relationships (e.g., employees of the same company and members of a sport club at the same time). Consequently, pairs of actors who maintain more than a single relationship are said to have a tie [26, 27]. While each individual relationship within a tie carries its own content and direction, the strength of a tie depends on many factors such as the number of relationships that actors maintain, the reciprocity of these relationships, and their duration. Granovetter [28] distinguished between strong and weak ties on the basis of the time actors spend together, their intimacy, and the emotional intensity of the existing relationships. Generally, weak ties are infrequently maintained with little interactions among actors (e.g., between distant acquaintances). Strong ties link similar actors, such as close friends, whose social circles tightly overlap with each other. Often, actors with strong ties that maintain many kinds of relations tend to communicate frequently with each other and use different channels of communication [29].

The previously defined concepts (actors, relations, and ties) are particularly important to understand and to study social networks. Besides the fact that social networks are made of several components, online social networks can hold different types of data and can have various representations as detailed in the next sections.

1.3 Online Social Networks

Interactions between actors and offline communications between persons have always been central in the study of social networks [30–32]. Many studies investigated ties between friends and relatives in order to understand why actors provide different types of social support [30], how social networks are formed, persist, and disappear [31], and what methods are appropriate in order to estimate the size of personal communication networks [32]. More recently, the impact of social-based technologies on users, and particularly the influence of online social networks, is becoming the major source of contemporary fascination and controversy [27, 33, 34]. A number of studies shed the light on different research directions like the implications of online social networks on individual connectivity [35], the capacity of technology to override cognitive limits in order to socialize with larger groups [36], and the challenge to maintain a balance between security, privacy, usability, and sociability on online social networks [9, 12]. Our focus in this chapter is on the privacy aspects of social networks, where research has primarily aimed to protect social network users with their profiles and relationships. In the following, we first highlight the main motivations associated with social networks' use and show some relevant statistics. We then describe the concepts of social network users, user profiles, and social relationships specifically in the context of online social networks. Note that in the following, we refer to social networks and online social networks interchangeably.

1.3.1 Motivations and Use

Social networks and content-sharing sites with social networking functionalities have become an important part of the online activities on the web and one of the most influencing media. Facebook, LinkedIn, Twitter, MySpace,[5] Flickr,[6] and Youtube[7] are among the most popular online social networks. These networks are attracting an ever-increasing number of users, many of whom are interested in establishing new connections, maintaining existing relations, and using the various social networks' services. Facebook, for instance, reported to have one billion monthly active users[8] that are uploading more than 250 million photos every day. On Twitter, 8 terabytes of data is generated on Twitter per day.[9] Another study published by Nielsen[10] on social networking reported that social networks and blogs dominate the time that users spend on the web and now account for nearly 20 % of the total time spent online on personal computers and 30 % of online time on mobile devices such as smartphones and tablets. With the huge number of users and the tremendous amount of shared data, such social networks will indisputably shape the future of online communication.

A large and growing body of literature has investigated the influence of using social networks on users' interactions [37, 38], gratifications [39, 40], self-estime [41, 42], and sharing practices [43, 44]. Recent studies examined the use of social networks, the behaviors that surround online interactions, and the benefits perceived by social network users [45–48]. The findings of these studies show that the motivations for using social networks are numerous. They indicate that the enjoyment is the most influential factor [49], followed by the users' interest to frequently interact with their real-world life friends [50], and the founded users' belief that social networks improve the efficiency of their shared information to enforce existing connections and to connect with new users [51].

---

[5]http://www.myspace.com/.

[6]http://www.flickr.com/.

[7]http://www.youtube.com/.

[8]https://www.facebook.com/press/info.php?statistics, accessed 01 October 2012.

[9]http://www.information-management.com/issues/21_5/big-data-is-scaling-bi-and-analytics-10021093-1.html, accessed 01 October 2012.

[10]http://blog.nielsen.com/nielsenwire/social/2012/.

1.3.2 Social Network Users

While many definitions exist for the term social network [52–54], all of them are centered around social network users. First, these users create a personal profile which usually contains identifying information (e.g., name, age, and photos) and captures users' interests (e.g., joining groups and liking brands). Afterwards, users start to socialize by interacting with other network members using a wide variety of communication tools offered by different social networks. In reality, each social network offers particular services and functionalities to target a well-defined community in the real world. Many of these available services are designed to help foster information sharing [55], bridge online and offline connections to enforce interactions [56], provide instant information help [46], and enable users to derive a variety of uses and gratifications from these sites [39]. To make use of the provided functionalities and to stay tuned with their related members, users create several accounts on various social networks where they disclose personal information with varying degrees of sensitivity [57]. Personal information available on these networks commonly describes users and their interactions, along with their published data.

1.3.3 User Profiles

Information about each social network user is maintained in a user profile which contains a number of attributes related to the demographics of users, their personal and professional addresses, their interests and preferences, as well as different types of user-generated contents (e.g., posts, photos, and videos) [58, 59]. Prior studies have noted the importance of user profiles to shape users' personalities, identities, and behaviors on social networks [7, 41, 42]. These studies showed that among the disclosed attributes such as personal information and user-generated contents, photos and status updates have higher preferences for users. User profiles also store the contact lists that consist of various interpersonal relationships as discussed in the following. Currently, social network sites do not all adopt the same user profile attributes' representation. Different technologies provide users with an extensive list of attributes to describe their profiles such as:

- *RDFa*[11] standing for *Resource Description Framework—in—attributes*, is a W3C recommendation used to embed semantic into XHTML. RDFa is a thin layer of markup that can be added to web pages and make them more understandable for machines as well as for persons. RDFa provides a consistent syntax and big expressivity by proposing an integration of the RDF triple concept (subject, predicate, attribute) with the flexible XHTML language, which is used by web browsers.

---

[11]http://www.w3.org/MarkUp/2009/rdfa-for-html-authors.

- *Microformats*[12] are little pieces of structured information embedded into XHTML documents. They transform documents to machine-readable semantic data such as contact details, social relationships, and event information. Currently, different microformats exist for different needs such as hCard used to describe persons, companies, and organizations with a limited set of elements representing business cards, calendars for events (e.g., hCalendar), decentralized tagging (e.g., rel-tag), etc.
- *XFN*[13] standing for *XHTML Friends Network*, represents 18 human relationships with a set of values and gives the possibility to authors, for example, to indicate which of the weblogs they read belong to friends they have met.
- *FOAF*[14] standing for *Friend Of A Friend*, is a machine-readable semantic vocabulary describing persons, their relationships, and activities. FOAF documents are written in XML syntax and adopt the conventions of the Resource Description Framework (RDF).[15] Among the many representations, FOAF is considered as the richest vocabulary to use in terms of describing users' profiles and has currently become a widely accepted standard [60]. FOAF defines a set of attributes, grouped into categories as shown in Fig. 1.1. A sample FOAF profile is illustrated in Fig. 1.2.

1.3.4 Social Relationships

While myriad social networks' services assist users to find new contacts and establish new connections (e.g., friend suggestion systems through locations [61], based on interactions [21]), users get connected to different types of contacts such as friends, relatives, colleagues, and strangers. Nevertheless, social relationship types between users and their contacts are rarely identified neither by the users nor by the existing social network sites [62–64]. This diversity, yet the different levels of social closeness between users and their contacts, entails an increasing need to analyze social interactions for better relationship (and consequently privacy) management. Currently, users are often provided with an exclusive and default relationship type connecting them to each of their contacts within a single social network site. However, it is common that social network users initiate connections with other contacts without any prior offline connection [65]. On Facebook, for instance, these contacts are known as *friends* even though social network users do not particularly know or trust them. Consequently, many privacy-related concerns are raised in terms of identity disclosure, information sharing, access control, etc. [9]. The default

---

[12]http://microformats.org.

[13]http://gmpg.org/xfn.

[14]http://xmlns.com/foaf/spec.

[15]http://www.w3.org/RDF.

FOAF Core	**Social Web**
Agent	nick
Person	mbox
name	homepage
title	weblog
img	openid
depiction (depicts)	jabberID
familyName	mbox_sha1sum
givenName	interest
knows	topic_interest
based_near	topic (page)
age	workplaceHomepage
made (maker)	workInfoHomepage
primaryTopic (primaryTopicOf)	schoolHomepage
Project	publications
Organization	currentProject
Group	pastProject
member	account
Document	OnlineAccount
Image	accountName
	accountServiceHomepage
	PersonalProfileDocument
	tipjar
	sha1
	thumbnail
	logo

Fig. 1.1 Main FOAF attributes grouped into categories: FOAF core and social web

```
<foaf:Person>
<foaf:name>Alexandre William</foaf:name>
<foaf:firstname>Alexandre</foaf:firstname>
<foaf:family_name>William</foaf:family_name>
<foaf:mbox rdf:resource=aw@somesite.com/>
<foaf:homepage rdf:resource=http://personalsite.com/aw/>
<foaf:workplaceHomepage rdf:resource=http://workaddress.com/>
<foaf:img>www.xyz.com/alex/photos/alex.jpg</foaf:img>
<foaf:interest>Paris, Software, Internet</foaf:interest>
<foaf:knows><foaf:Person>
<foaf:mbox rdf:resource=contact1@somesite.com />
<rdfs:seeAlso rdf:resource=http://contact1.net/foaf.rdf./>
</foaf:Person></foaf:knows>
<foaf:knows><foaf:Person>
<foaf:mbox rdf:resource=contact2@somesite.com />
</foaf:Person></foaf:knows>
</foaf:Person>
```

Fig. 1.2 Sample FOAF document

Table 1.1 Famous social networks with their main focus, default relationship(s), and the relationship's direction

Social Network	Focus	Default Relationship(s)	Relationship Direction
Facebook	General Use	Friendship	Symmetrical
Flickr	Photo-Sharing	Contact and optionally Friend or Family	Symmetrical
Google+	General Use	Friends, Family, Acquaintances and Following	Symmetrical
LinkedIn	Professional	Business	Symmetrical
MySpace	General Use	Friendship	Symmetrical
Twitter	Microblogging	Follower–Followee	Asymmetrical
Youtube	Video-Sharing	Subscribed-to	Asymmetrical

social relationship(s) among the users of a number of famous social networks, along with other information, can be found in Table 1.1. Given the diverse sources of social relationships, further research is needed to better understand the privacy needs of users as more friendships continue to be forged and maintained with online and offline contacts. The structure of the networks, the user-generated content, the level of interaction, as well as other dimensions, can also be used to analyze users' behaviors and understand their privacy needs. Next, we address in detail the structural representation of social networks.

1.4 How to Represent a Social Network?

Finding an appropriate representation that can facilitate efficient and accurate interpretation of network data is an important step in social network studies. Just as graphs are a set of interconnected nodes, social networks are built on the foundation of actors interconnected through relationships. The use of graphs is a powerful visual tool and a formal means to represent social networks as detailed in this section.

1.4.1 Why Graphs?

There are many notations to represent social networks: algebraic notations, matrices, and graphs. A sample algebraic notation, a matrix representation, and a graph are illustrated in Fig. 1.3a. Depending on the data to be processed, the notation whose representation best fits the social network to describe is typically selected. But, there are well-known limits to the extent to which social networks can be formalized using matrices or algebraic notations to be recalled here. First, social networks hold valued relations and user-related attributes that algebraic notations cannot handle. Second, matrices are mostly efficient for small networks. Consequently,

Table 1.2 Social network representations: advantages and drawbacks

Representation	Advantages	Drawbacks
Algebraic notations	– Useful for multi-relational networks as they can easily denote the combination of relations	– Cannot handle valued relations and user-related attributes
Matrices	– Efficient for small networks	– Not a best choice for large social networks
	– Easy to denotes ties between a set of actors (a matrix for each relationship)	– Difficult to use when network data contain information on attributes
Graphs	– Handle large social networks	– Scalable visualization techniques are needed
	– Provide a rich vocabulary to easily model social networks (labels, values, weights, etc.)	– Signed and valued graphs have to be used to represent valued relations
	– Provide mathematical operations that can be used to quantify structural properties and prove graph-based theorems	

due to the large size of social networks, matrices are not the most appropriate way to represent these networks. Note that to represent a social network using matrices, a two-way matrix, also called sociomatrix, can be used. A sociomatrix consists of rows and columns that denote social actors, and numbers or symbols in cells that denote existing relationships. Thus, graph-based representations are by far the most common form for modeling social networks [14, 66, 67]. Graphically representing social networks facilitates the understanding, labeling, and modeling of many properties of these networks (e.g., friendships networks with labeled actors and relationships). Hence, graphs can represent various social data properties and their attributes while handling large real-world networks. Beside an adequate vocabulary to denote structural properties, graph-based representations have shown their mathematical reliability as well as their capacity to prove theorems for different social structural properties [14]. More details about the advantages and drawbacks of each representation are provided in Table 1.2.

1.4.2 Graph Representation

Graphs are usually used to represent networks in different fields such as biology, sociology, and computer science [68]. Graphs consist of nodes to represent actors, and edges to represent relationships. The terms *nodes* and *objects* are usually used to denote *actors*. Likewise, *edges* may also be called *links*, or *relationships*. Nodes with multiple edges are used to represent *ties* related pairs of actors with more than one relationship.

More formally, a graph, $G = (V, E)$, consists of a set of nodes, V, and a set of edges, E. The number of elements in V and E are, respectively, denoted as $n = \|V\|$, the number of nodes, and $m = \|E\|$, the number of edges. The ith node, v_i, is usually referred to by its order i in the set V. Note that E consists of a finite set of relationships that is built from all relationships R_i, R_{i+1}, .., R_k, where k is the total number of relationships linking the pairs of actors. A subgraph $G' = (V', E')$ of $G = (V, E)$ is a graph such that $V' \subseteq V$ and $E' \subseteq E$. To represent different forms of data and to model the structural properties of social networks, graphs can have their edges and nodes labeled or unlabeled, directed or undirected, weighted or unweighted as explained in what follows.

Directed and Undirected Graphs

In an undirected graph, the order of the connected vertices of an edge is not important. We refer to each link by a couple of nodes i and j such as $e(i, j)$ or e_{ij}, i and j are the end-nodes of the link. A directed graph is defined by a set of nodes and a set of directed edges. The order of the two nodes is important: e_{ij} denotes the link from i to j, and $e_{ij} \neq e_{ji}$. To graphically indicate the direction of the links, directed edges are depicted by arrows. Depending on the nature of the relationship (asymmetric or symmetric), social network graphs can be undirected or directed. In fact, social networks can be modeled as undirected graphs when relationships between actors are mutual (e.g., symmetric relationships on Facebook where e_{ij} or e_{ji} both denote a *friendship* link between user i and user j). Social networks can also be modeled as directed graphs when relationships are not bidirectional (e.g., asymmetric relationships on Twitter where e_{ij} stands for user i is *following* user j). Figure 1.3b,c show, respectively, a representation of an undirected and a directed graph, both with $n = 5$ and $m = 6$. Directed links are important to evaluate the role of actors in a social network. They are key factors in measuring the centrality of actors in a social network. An interesting research work conducted by Brams et al. [69] described how to transform undirected graphs to directed ones in order to explore additional information about the networks' structure. This transformation is an important step in understanding the flow of influence in the context of terrorist networks. In another study, Morselli et al. [70] investigated and compared the structure of criminal and terrorist networks. The authors used links to compute a number of measures such as degree, betweenness, and centrality measures. These measures are used in order to discover the organizational hierarchy and to identify central and powerful criminal and terrorist actors. We detail these measures in Sect. 1.6.

Labeled and Unlabeled Graphs

Labels are important since they can identify the type of relationships between social network actors. When graphs are labeled, this means that a label is used to indicate the type of link that characterizes the relationship between the connected labeled

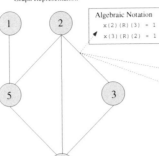

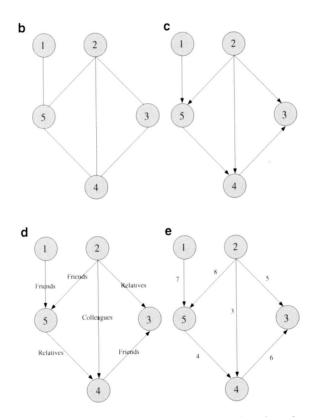

Fig. 1.3 A social network representation using a graph, its related matrix, and a sample algebraic notation (**a**), an undirected graph (**b**), a directed graph (**c**), a labeled directed graph (**d**), and a weighted directed graph (**e**) with $n = 5$ nodes and $m = 6$ links

nodes. Note that labeled graphs are considered to be signed graphs whenever their edges are labeled with either a + or a −. For example, a signed graph can be used to model the inferred trust or distrust relationships in online social networks [71]. Figure 1.3d shows a labeled graph where the relationship type between linked actors is indicated. On social networks, relationship can be used to organize contacts based on their relationship types. This is useful in different situations such as improving face clustering and annotation of personal photo collections [72], organizing friends into social circles [62, 64], and enforcing access control [73]. Relationship-based access control is highly interesting in order to enable users to manage and fine-tune their privacy settings.

Weighted and Unweighted Graphs

Weights represent the strength of relationships between social network actors. When graphs are weighted, this means that their edges are assigned with a numerical weight, w, that can provide various indications such as link capacity, link strength, level of interaction, or similarity between the connected nodes (e.g., the number of messages that actors have exchanged and the number of common friends). Figure 1.3e shows a weighted graph (on a scale of 0 to 10) where the numeric values are assigned to the links and indicate the level of interaction between social network's actors. One way to characterize relationships is by computing their strength. On social networks, link strength is highly correlated with the level of interaction between users. Link strength can be used to model different levels of friendship where high weights represent "close friends" and low weights represent "acquaintances." Xiang et al. [74] estimated the link strength from interaction activities (e.g., communication and tagging) and user similarities. Stutzman et al. [75] argued that link strength can be used to reduce the burden of manually specifying privacy settings for each contact within a user's social network. They proposed an automated grouping of users based on many criteria where link strength is highlighted as one of the most commonly considered factors. More recently, another research explored a more specific aspect related to the predictive capacity of link strength to generalize from one social network to another [25]. Typically, link strength is primarily used to build intelligent systems that can favor interactions with strong ties without missing interesting activities derived by weak ties. Specifically, this interesting study showed that the link strength model captured in one social network can be generalized to another network, one in which it did not train.

To sum up, structural characteristics of a graph are a key aspect for social networks as they can be used to analyze the activity and to understand the behaviors of social network users. In most cases, networks of interconnected users are mainly represented by graphs, while graphs resulting from users' activity are usually referred to as the activity graphs. The activity captured within social networks is between users (the nodes) sharing various directed or undirected relationships (the links) and different levels of interactions (strong and weak ties). In this regard, these characteristics can be used to identify well-connected, central, and influential users. This would give more visibility and understanding for the network analyzer but at the same time this can possibly reveal additional and sensitive information about the users, thus raising privacy concerns. In the next section, we discuss a number of challenges and opportunities related to the use of social networks from a user perspective.

1.5 Social Network Data: Opportunities and Challenges

Social networks have become an important platform for connecting users, sharing information, and a valuable source of social network data. Thus, the availability of such data represents an opportunity for people to study and analyze these networks. However, the various sources of data on social networks are not only perceived as sets of values and repositories of knowledge; rather their availability becomes a form of threat as they can be exploited by attackers to disclose various sensitive information (e.g., identities, attributes, and locations). In this section, we focus on social network data and address the challenges related to how data is collected, what data is collected, how data is protected, and we list a number of existing techniques used to protect the privacy of social networks users.

1.5.1 How Data Is Collected?

Traditionally, most of social network data were collected through questionnaires in order to study networks. These studies conducted face-to-face interviews [30], telephone surveys [76], or computer-based questionnaires [31]. To construct social networks using questionnaires, participants may spend a burdensome and unrealistic amount of time and effort in answering questions that can be difficult or repetitive. In addition, during the questionnaires participants may forget some relevant information or misinterpret questions. Consequently, such conventional methods have many limitations from different points of view related to scalability, subjectivity, inconsistency, error handling issues, etc. [77].

Today, the picture has changed. The use of electronic data extraction methods has been beneficial in collecting relevant network data, and their success spread to various domains such as hyperlink networks on the web [78], biochemical networks [18], and email messages archives [79]. In order to study social networks, novel techniques have been developed as well as adapted measures so to collect relevant data collections [24,80–82] . Marin et al. [82] highlighted the importance of the type of networks and the type of relationships in the process of collecting network data. In their study, the authors considered two important dimensions along which network data vary: whole vs. egocentric networks, and one-mode vs. two-mode networks. Note that the difference between one-mode networks and two-mode networks is explained in Sect. 1.2. As for the difference between whole and egocentric networks, it can be simply explained by noting that the egocentric networks privilege the study of one focal node (the ego) rather than considering all the nodes of the network as in the whole network analysis [14]. Many social network systems have been developed to collect, built, and analyze data from the web such as the Referral Web [80], Flink [81], and Polyphonet [24] or from online social networks such as Twitter [83], Flickr [84], and Facebook [85]. Currently, social networks allow users to exchange various types of information, including messages, photos, and comments. Many studies have shown that social network users are highly motivated to interact with their contacts and to share personal information [37, 38, 41–44]. As a matter of fact, social networks provide new possibilities to collect data more efficiently and cost-effectively.

1.5.2 What Data Is Collected?

There are many types of social network data that can be collected from various sources on the web (i.e., different social network sites) and extracted from the daily activities and interactions between users. In this context, Schneier [86] proposed a taxonomy of social data that we further develop into two main categories:

1. *Explicit data* is the set of explicit information that is provided by social network users or the data that is embedded in the provided information, i.e., metadata embedded in photos. Explicit information may include different forms of data such as text messages, photos, or videos. In this category, social network users actively participate in the creation of information.

   (a) *Service data* is the set of data that a user provides to the social network to create her account such as the user's name, date of birth, and country.
   (b) *Disclosed data* is what the user posts on her social network profile. This might include comments, posted photos, posted entries, captions, shared links, etc.
   (c) *Entrusted data* is what the user posts on other users' profiles. This might include comments, captions, shared links, etc.
   (d) *Incidental data* is what other social network users post about the user. It might include posted photos, comments, notes, etc.

2. *Implicit data* is the set of information that is not explicitly provided by social network users. However, social networks or third parties can use the set of explicit data to infer more information about the user. Inferring implicit data is founded on the analysis of the users' behaviors or derived from one or more user-provided information. For instance, it is possible to predict the characteristics of relationships between a number of users by examining the different aspects related to the patterns of communication between users (e.g., text messages, published photos, and number of common friends) [62, 87]. Consequently, in

this category social network users are considered to be passive since the inferred information is extracted from prior activities or previously posted data.

(a) *Behavioral data* is the data inferred from the user's behaviors. Social networks can collect information about the user's habits by tracking the patterns of activities of the user and consequently analyzing the user's behavior. Inferred behavioral data can reveal various information such as what the user usually do on the social networks, with whom the user usually interacts, and in what news topics the user is interested. Social networks collect such information by analyzing the articles that the user reads, the posts that the user publishes, the game that the user plays on social networks, etc.

(b) *Derived data* is the data about the user that can be inferred from all other data. It is not related to the habit of the user. For example, the IP address can be used to infer the users' actual location. The derived data can also be inferred from the combination of two (or more) information. For example, if a significant number of contacts live in one city, one can say that the social network user might live there as well. In this case, social networks or third parties must have access to two information in order to infer the derived data (the contacts of a user as the first information and their corresponding hometown as the second information).

1.5.3   *How Data Is Protected on Social Networks?*

Privacy on social networks is a complex concept which involves major challenges [2, 9]. A recent research in [1] addressed the topic of privacy settings on social networks and particularly investigated the privacy settings of Facebook. The results of this study show that privacy settings matched users' expectations in only 37 % of the time and up to 39 % when the users modified the default privacy settings [1]. The authors concluded that there is a big disparity between the desired and actual privacy settings and calls for new tools to manage privacy. This conclusion is in line with another study where social network users considered that the privacy settings are effective to manage threats coming from outside their social circles [6]. However, the same users experienced increasing concerns when it comes to sharing content with members of their social circle. Similar results were also identified when investigating privacy concerns and mechanisms surrounding tagged photos on social networks [7]. In the case of photos, the central point behind the users' priorities associated privacy concerns with identity and impression management. In most of the time, social network users were worried of seeing an unwanted photo of them online or being tagged on an unflattering photo. All of these studies indicate that the existing privacy systems as well as their designs must be improved to better address threats and meet users' expectations.

1.5.4   Are Existing Privacy Protection Techniques Useful?

Privacy is closely related to network anonymization [88–90], privacy preservation [91–94], and access control [73, 95, 96]. Nowadays, these conventional techniques have several disadvantages when it comes to protecting privacy of social network users.

Definition 4. Network anonymization consists to manipulate the network's information in order to make the process of nodes' identification difficult for attackers.

This can be achieved by modifying and removing all the attributes of the nodes in the network. However, anonymizing a network is often not enough to protect privacy since many attacks can apply de-anonymization techniques to re-identify nodes and their hidden attributes [88–90]. Attackers can use another release of the network to process the anonymized network in order to re-identify the protected nodes and consequently reveal sensitive information. As stated earlier, various sources of information are available on the web (e.g., users with accounts on different social networks and personal blogs) and thus many sources can be used as a background knowledge by the attackers. The structure of the social network and the background knowledge can be exploited by the attackers and consequently expose users to privacy issues. This is usually possible by using a variety of information such as the total number of contacts, the number of common contacts, and the relationship strength.

Definition 5. Privacy preservation focuses on protecting sensitive information primarily through using techniques such as hiding sensitive attributes, hiding users' identities, modifying data, and randomizing values.

Several efforts have been extensively investigating the protection of sensitive information using privacy preservation techniques [91, 92]. Besides the fact that privacy preservation has been most successful in dealing with relational data [93], privacy on social network is a confluence of several factors that when combined can lead to infer the original value of the sensitive information. All users' activities, relationships, and shared content can be potentially monitored, recorded, and analyzed by attackers. Consequently, a hidden attribute on a user's profile can still be inferred accurately. For instance, it is possible to predict the home address of a user by analyzing the geographical place of the most frequent updates posted at night or on the weekend [97]. Another study derived the user's location given the known location of the user's friends [98]. Similarly, the availability of metadata embedded within shared content (e.g., GPS location, date, time, and device name embedded in photos) as well as the use of location-based services (e.g., Foursquare[16] and Facebook) can significantly raise privacy concerns and complicate the task of protection [99]. Moreover, failing to provide an optimal identity protection can

---

[16]http://foursquare.com/.

lead to disclose other sensitive information such as the type of relationships among users [94].

Definition 6. Access control mechanisms seek to secure the access to sensitive information without explicit authorization by implementing appropriate access control mechanisms.

On social networks, there is a growing interest in implementing access control systems but very little work has been done in these directions. Notably, social network users strive to reduce the inefficiency of current privacy systems and look forward to enforce their privacy protection. Several access control systems have been proposed, including:

- Attribute-based access control [100]: the systems in this category grant or deny access based on the user's attributes. An attribute or a set of attributes form the digital credential and may contain attributes such as age, citizenship, employment, group membership, or credit status.
- Multimedia-based access control [101]: the multimedia-based systems tend to integrate multimedia objects in the decision process. Multimedia objects can yield valuable information sensed from multimedia devices about the users and their context (e.g., user's surrounding, moves, gestures, and people nearby).
- Purpose-based access control [102]: the purpose-based systems grant access to certain data with conditions. The notion of condition determines the access purpose in a dynamic manner. Such access control systems integrate the purpose of the access in the decision process and consequently dynamically associate the purpose with the requested data objects.
- Relationship-based access control [73, 95, 96]: the approaches in this category are designed to enforce users' privacy and enable users to tune their privacy settings by controlling access based on the type of relationship. For instance, the access to a specific content is authorized only for the user's colleagues, family members, etc. However, these existing works assume that relationship labels are provided with social networks. Consequently, relationship-based access control can be rarely implemented since relationship types are often missing [62–64].

1.5.5 Discussion

In the context of social networks, there are many challenges to overcome such as networks design and architecture, active user population and network dynamics, user interactions, user behavior, and most importantly privacy issues [103]. To improve users' experience while protecting their personal information is a challenge that requires adequate methods capable to analyze the different types of data, to explore the different components of social networks, and to understand the social interactions between users. As discussed in this section, the risk of disclosing private, personal, and potentially sensitive information is serious since social network users lack of appropriate means to efficiently control and easily protect their published data.

In the following, we show how attackers can resort to various techniques in order to reveal sensitive information. We mainly investigate privacy concerns that are derived from social network analysis [14] and link mining techniques [15]. Our classification inspiration draws from the literature of both techniques and illustrates the privacy threats associated with common social network analysis measures and link mining tasks. We start by introducing social network analysis and link mining before detailing their corresponding privacy threats in Sects. 1.6 and 1.7, respectively.

1.6 Social Network Analysis: Measures and Threats

Over the past years, there has been a surge of interest in social network analysis, with works ranging from exploiting networks' structures to examining actors' roles and their interaction patterns [14]. Understanding the characteristics of social networks, namely information related to structure, is of considerable importance to deal with privacy issues, and hence social network analysis presently attracting widespread interests. In this section, we briefly trace the history of social network analysis, provide an overview of its most common measures, and discuss where and how social network analysis has been used in the context of social networks' privacy.

1.6.1 Development and Measures

Social network analysis has a well-established tradition in psychology and in social sciences [66]. Since the beginning of the twentieth century, several social network analysis studies—primarily in educational and developmental psychology—have been conducted to study characteristics of groups (e.g., structure, formation, and behavior) and social ties (e.g., influence, interaction, and companionship) [104–106]. Drawing inspiration from these previous works, modern social network analysis emerged as interdisciplinary field with contributions from various areas of study such as sociology, anthropology, and mathematics [107, 108]. Concerned with the structural analysis of social interactions, modern social network analysis developed new models to study the fundamental properties of diverse theoretical and real-world networks [109]. The small-world model [110] and the scale-free model [111], useful in describing very large networks, are among the most important models that emerged from these efforts.

Social network analysis has been used in different application domains such as email communication networks [87], learning networks [29], epidemiology networks [112], terrorist networks [70], and online social networks [113]. These

Table 1.3 Main centrality measures and their characteristics

Centrality measure	Characteristic
Degree	Measures how much an actor is highly connected to other actors within a network
Closeness	Computes the length of paths from an actor to other actors in the network
Betweenness	Measures the extent to which an actor lies on the paths between other actors

works tried to answer a handful of questions such as how highly an actor is connected within a network? Who are the most influential actors in a network? How central is an actor within a network? To capture the importance of actors within a network, a number of measures have been proposed in the literature [114]. A commonly accepted measure is the centrality measure.

Definition 7. Centrality consists of giving an importance order to the actors of a graph by using their connectivity within the network.

Several structure-based metrics have been proposed to compute the centrality of an actor within a network, such as degree, closeness, and betweenness centrality [115]. Table 1.3 summarizes the characteristics of these structure-based centrality measures.

In what follows, we explain each of these metrics in details:

- *Degree Centrality:* Measures how much an actor is highly connected to other actors within a network. Degree centrality is a local measure since its value is computed by considering the number of links of an actor to other actors directly adjacent to it. A high degree centrality denotes the importance of an actor and gives an indication about potentially influential actors in the network. With a high degree of centrality, actors in social networks serve as hubs and as major channels of information in a network. Degree centrality, C_D, of an actor, v_i, can be computed as follows [115]:

$$C_D(v_i) = \sum_{i=1}^{n} a(v_i, v_j) \tag{1.1}$$

where n is the total number of actors in the social network, $a(v_i, v_j) = 1$ if and only if v_i and an actor, v_j, are connected by an edge; otherwise $a(v_i, v_j) = 0$.
- *Closeness Centrality:* Computes the length of paths from an actor to other actors in the network. By measuring how close an actor is to all other actors, closeness centrality is also known as the median problem or the service facility location problem. Actors with small length path are considered more important in the network than those with high length path. Closeness centrality, C_C, of an actor, v_i, can be computed as follows [115]:

$$C_C(v_i) = \frac{n-1}{\sum\limits_{i=1}^{n} d(v_i, v_j)} \qquad (1.2)$$

where n is the total number of actors in the social network, $d(v_i, v_j)$ is the geodesic distance from actor v_i to another actor v_j.

- *Betweenness Centrality:* Measures the extent to which an actor lies on the paths between other actors. It denotes the number of times an actor needs to pass via a given actor to reach another one, and thus represents the probability that an actor is involved into any communication between two other actors. Actors with high betweenness centrality facilitate the flow of information as they form critical bridges between other actors or groups of actors. Such central actors control the spread of information between groups of non-adjacent actors. Betweenness centrality, C_B, of an actor, v_i, can be computed as follows [115]:

$$C_B(v_i) = \sum_{j < k} \sum \frac{g_{jk}(n_i)}{g_{jk}} \qquad i \neq j \neq k \qquad (1.3)$$

where n is the total number of actors in the social network, $C_B(v_i)$ is the betweenness centrality for actor v_i, and g_{jk} is the number of geodesics linking actors v_j and v_k that also pass through actor v_i.

As shown in Fig. 1.4, different central actor(s) in a network can be identified using each of these structural measures (degree, closeness, and betweenness).

In the following, we present how social network analysis and its related structure-based measures have been used in the context of social networks' privacy and list some related privacy threats.

1.6.2 Privacy Threats

The availability of social network data has attracted the interest of the academic community, third-party advertisers, and governmental services for the purpose of data analysis. Anonymizing these networks before their release is important to enforce privacy but only hiding the identity of the users or removing all their attributes from their profiles does not always guarantee privacy. An attacker can potentially infer the true identities of the targeted users by referring to the structure of the network and by using a background knowledge [93].

Exploiting structural information with anonymized networks adds a new privacy-related dimension to consider and a large number of theoretical investigations and practical applications have been conducted on social networks [89, 90, 116, 117]. Anonymizing social network data is much more challenging than anonymizing relational data [93]. As stated earlier, social networks can be represented as graphs

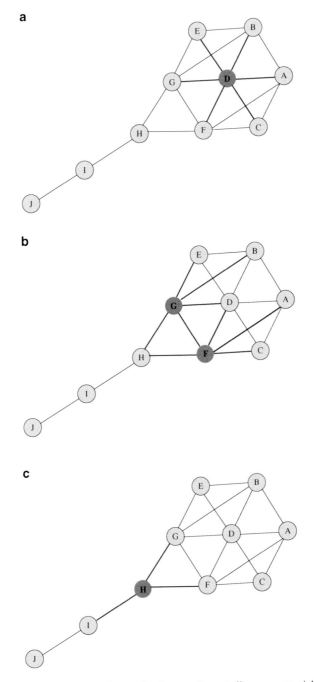

Fig. 1.4 A network shaped as a kite graph where each centrality measure yields a different central actor: degree centrality (D), closeness centrality (F and G), and betweenness centrality (H). (**a**) Centrality degree (**b**) Closeness degree (**c**) Betweenness degree

and thus the social network data can then be pre-processed and analyzed via social network analysis measures.

There has been much recent interest to study anonymized networks when an attacker has background information about the network structure [88, 118, 119]. The authors in [118] described a family of attacks based on the structural information of the network. In the first type of attacks, the active attacks, attackers were able to modify the network prior to its release and can potentially construct highly distinguishable subgraphs by inserting nodes and edges to the network. Passive attacks, the other type of attacks, are launched after the anonymized network is published and without inserting new nodes or edges. As observed in [119], the extent to which an individual can be distinguished using graphical position depends on the structural similarity of actors in a network and the background information an attacker can obtain. More specifically, the structural information is closely related to the degrees of the nodes and their neighbors in a network. In the same line of research, the authors in [88] pointed out that the degree of a node in a graph, among other structural characteristics, can, to a large extent, distinguish the node from other nodes. Consequently, attackers can greatly benefit from structural characteristics of networks that become identifying attributes. An interesting work presented in [120] where the attacks aimed to re-built the network from disparate pieces of information in order to gain more information and better visibility before launching the attacks. The attack consisted in acquiring information about local neighborhoods of different users in the network. Such attack is feasible and its effectiveness depends on the underlying social graph and the degree distribution of its nodes as detailed in the study [120]. The authors concluded that any social network that wishes to enforce the privacy of its users should take great care in decreasing the vulnerability of its interface, i.e. by not displaying the exact number of connections that each users has.

To sum up, social network analysis provides a set of measures that are used extensively to study networks' characteristics and users' behaviors. Though the study of social network is valuable to lot of people, there is serious risk of privacy concerns. Privacy attacks can take advantage of social network analysis to infer further knowledge about social network users using structural information. In addition, the proliferation of online social networks has resulted in huge amounts of available network data. It is quite challenging to protect users' privacy as many sources of background knowledge are widely available on the web. This is particularly true since the majority of social network users are not privacy experts. It is therefore a necessity that social networks enforce users privacy by protecting not only their data but also their established relationships. Moreover, it is highly desirable that social networks implement more flexible and more secure web interfaces. This would greatly benefit users to intuitively configure their privacy settings. At the same time, this would make more difficult for attackers to use search interfaces, Application Programming Interface (API), or users' connections in order to easily collect user-related information.

1.7   Link Mining: Tasks and Threats

Owing to the popularity of World Wide Web, the increase of computational power and performance, and the higher capacity to gather and analyze data, large-scale social networks studies are flourishing, spilling over all traditional disciplinary boundaries for social networks. Link mining studies, with their objective to efficiently discover valuable and inherent information from large databases, are highly related to privacy preservation [121]. While centrality measures are widely used in social network analysis [14], link mining techniques rely on recent advances in data mining and often put emphasis on the links between social network actors [15]. In the following, we start by presenting the various link mining tasks before describing relevant privacy threats related to each task.

1.7.1   Development and Tasks

Taking into account the links between social network actors, various data mining techniques [122] have contributed in the emergence of a new area commonly named *link mining*, where links that exhibit rich patterns are central in extracting hidden knowledge from available data. Link analysis, relational learning, web mining, and graph mining are among the most widely used techniques in link mining [15, 122]. By building predictive models (predicting attributes' values) or descriptive models (extracting interesting patterns), link mining can be regarded as data mining applied on social networks where links play a key role. Not only networks' links can be used to discover prominent actors within a network but also to reveal uncovered information related to identities, classes, and relationships between actors. In the following, we detail all the tasks that link mining embodies as presented in [15]:

1. *Node-related Approaches:*

    (a) *Link-based Node Ranking [123–126]:* The objective of link-based node ranking is to prioritize corresponding nodes based on their measured importance. In link mining, centrality measures (e.g., degree, closeness, and betweenness) are used to rank the nodes by exploiting the network structure.

    (b) *Link-based Node Classification [127–130]:* The link-based node classification task classifies the nodes of a network to a finite set of categories. This type of classification is based not only on nodes' attributes but also on their links to other nodes and on the attributes of these linked nodes.

    (c) *Link-based Node Clustering [131, 132]:* Node clustering, also called group detection, is another well-studied link mining task. Its objective is to identify similar nodes and group them together without predefining the clusters. Any two nodes, members of the same cluster, are more similar to each other than to any other node in a different cluster. They represent communities where the level of interaction or communication (emails, messages, collaborations, etc.)

between nodes of the same cluster is higher than with any node in another cluster.

(d) *Link-based Node Identification [57, 133–135]:* Link-based node identification, or entity resolution, aims at finding correspondences between nodes of distinct networks given that the nodes that have different identifiers may refer to the same real-world entity. In this case, these nodes form a matched entity pair.

2. *Link-related Approaches:*

(a) *Link Prediction [136–140]:* Link prediction, or link existence prediction, is the task of inferring the existence of a link between two nodes, based on the properties of the nodes. While link prediction in static networks aims at inferring missing links and facilitating the task of link formation, link prediction in dynamic networks consists of predicting the snapshot of links at a future time.

(b) *Link Type Prediction [62, 141]:* Unlike link prediction, where the aim is to predict the existence of a link between two nodes at a particular time, link type prediction aims to identify the type of an existing link (e.g., the type of relationship between two actors). In this task, it is assumed that the existence of a link between the nodes is already confirmed.

3. *Graph-related Approaches:*

(a) *Subgraph Discovery [142, 143]:* Subgraph discovery is a link mining task that detects similar substructures in pairs of graphs. Its aim is to find the set of subgraphs that are similar among the underlying graphs.

(b) *Graph Classification [144, 145]:* Graph classification aims at classifying an entire graph with respect to a specific category. Independently classifying each node in a large graph is a tedious task, sometimes infeasible, and may ignore useful information available from other nodes. Rather than trying to label each node within a graph, collective classification learns and infers labels of linked nodes together.

(c) *Graph-based Generative Models [146, 147]:* Generative models for graphs try to understand the characteristics of networks. Given an input network, generative models can produce a new network similar to the input one. They can model the structure similarities and the data distribution correctly. They are used to model the mechanisms of networks growth and evolution and to generate networks with realistic properties given few parameters. Studying generative models for graphs is becoming increasingly important, in particular when temporal metrics are considered (e.g., a social network that evolves over time).

As shown in Table 1.4, link mining techniques have been applied on various networks. Primarily the focus of most link mining approaches has been directed towards the bibliographic [123, 127, 136], biological [129, 135, 138, 143, 144, 147], and social networks [57, 62, 126, 132, 139, 140, 145, 146]. On these networks, a

Table 1.4 A summary of link mining tasks applied on different types of networks

Network type	Node-related	Link-related	Graph-related
Criminal [131]	✓		
Epidemiology [130]	✓		
Financial [124]	✓		
Linked data [125, 128, 141]	✓		
Database management [133]		✓	
Digital libraries [137]		✓	
Lexical [134]		✓	
Software behavior [142]			✓
Bibliographic [123, 127, 136]	✓	✓	
Biological [129, 135, 138, 143, 144, 147]	✓	✓	✓
Social [57, 62, 126, 132, 139, 140, 145, 146]	✓	✓	✓

number of node-related, link-related, and graph-related link mining tasks have been used. Meanwhile, on criminal [131], epidemiology [130], financial [124], and linked data networks [125, 128, 141], node-related techniques have been used. As for link-related approaches, they also examined the data management [133], digital libraries [137], and lexical networks [134]. Besides the biological [143, 144, 147] and social networks [145, 146], graph-related tasks have been applied also on software behavior networks [142]. In the following, we detail a number of privacy threats related to each of the mentioned link mining task.

1.7.2 Privacy Threats

Users concerns regarding privacy of personal information are rising in the light of the recent link mining advances. We describe in the following how each link mining task can be exploited by attackers or malicious users. Table 1.5 lists a number of approaches that may use link mining to compromise users' privacy.

Node-Related Threats

Social network users have strong expectations of privacy [118, 148]. Tracing users' interactions and reconstructing details of their behaviors are commonly unappreciated. However, *link-based node ranking* can be used to measure the influence and the importance of social network users. Exploiting the structure of the network makes it possible to infer meaningful relationships and quantify the interactions between social network users. Identifying influential users, who are capable of stimulating other users, is of considerable importance in many scenarios [149]. With the huge number of social network users, link-based node ranking has been applied in many areas such as marketing [150], diffusion of information [151], and governmental intelligence-gathering tasks [152].

Table 1.5 Link mining tasks and their corresponding implications

Technique	Application and threat
Link-based node ranking	[149–152]
Link-based node classification	[156, 157]
Link-based node clustering	[158–160]
Link-based node identification	[57, 88, 161]
Link prediction	[94, 162, 163]
Link type prediction	[121, 164]
Subgraph discovery	[12, 89, 90, 118]
Graph classification	[159, 165, 166]
Generative models	[64, 168]

The privacy implications of *link-based node classification* and *link-based node clustering* can reveal sensitive information such as membership to a particular group or a political party. Node classification and node clustering have been mainly used in the area of computer security. Their extend goes far beyond the simple case of social networks and targeted advertising to reach critical applications such as terrorist networks. Besides the typical profile attributes, users' activities and interactions over time are among the most important sources of information [153]. These online activities and interactions come in many guises such as establishing connections, exchanging messages, and publishing photos.

Link-based node classification can be a source of privacy threat. Prior studies have shown that communities are usually formed around users who share certain interests [154, 155]. Mislove et al. [156] have also reported that social network users are often friends with users who share their attributes. By combining the network graph structure with the fraction of available information, it is possible to infer the value of missing or hidden attributes. For instance, in many situations, some social network users would like to keep private their political affiliations because of privacy concerns. However, node classification techniques can easily infer a user's political affiliation by referring to her contacts and by using, for instance, available information revealing that the user participated in events hosted by a particular political party (e.g., electoral campaigns and debates) [157].

Link-based node clustering techniques are also a potential source of privacy threat. They are used to group users having the same type of activities, interested in the same hobbies, or seek the same kind of services. Although the user's personal interests are hidden, node clustering techniques can be used to reveal these information using, for example, some undirect data such as group memberships. In fact, it is possible to extract general groups' interests and then attribute them to users that are members of these groups. In addition, social network groups can be described based on their members' profiles [158]. This example of node clustering is one illustrative example among many that raise users' privacy concerns in social network [159, 160].

Social network users create several accounts on various sites where they disclose personal and professional information [57]. *Link-based node identification* can be used to associate a user profile with a real-world entity (person). Such profile-entity mapping leads to an identity disclosure problem whenever the user would like to keep her social network profile private or hidden. Furthermore, identity disclosure often causes attribute disclosure. Attribute disclosure occurs when sensitive information such as real name, address, and sexual orientation is revealed. In these types of privacy threat, the attacker might have access to an external knowledge and can use explicit identifiers or quasi-identifiers to reveal the identity of an anonymized user [88, 161].

Link-Related Threats

Link prediction can raise privacy concerns when the predicted link is between users who would like to keep their relationship private. In many cases, the link can be considered as the sensitive information to keep protected. Hiding its existence can be valuable in many real life situations to prevent user-associated sensitive information from being disclosed to third parties [94], to recommend accurately social links without disclosing sensitive information about users' contacts [162], to ensure web browsing anonymity [163]. More interestingly, a recent work shows the possibility to infer whether two non-members friends of a social network member (user) are friends themselves. The obtained results show a high rate of prediction success and is based only on information extracted from the friendship and email contact information of the social network members.

Link type prediction attacks can reveal the sensitive type of an existence relationship between two users although these users would like to keep this information private. Unlike link prediction, link type prediction is concerned in keeping private the type of the relationship (not the existence) between two users. This type of attack is also known as *link re-identification*. This occurs when an attacker is able to identity the type of a sensitive relationship or communication between two users [121]. The type of link between two users can be used to reveal much more information than just the existence of a relationship. Heatherly et al. used link types to classify unknown nodes as terrorist or non-terrorist [164]. In addition, link types can improve the classification accuracy when an attacker attempt to identify information related to personal interests, physical location, political affiliations, etc. For instance, friendship links are more important than professional links to infer personal interests (e.g., political affiliations and religious beliefs) particularly if a significant number of friends publicly display on their profiles such personal or sensitive information.

Graph-Related Threats

Attacks based on *subgraph discovery* attempt to acquire new information about an anonymized network using structural subgraph queries. Subgraph queries are useful to efficiently find similar structures in large social networks using two main types of attacks: passive and active attacks [89, 90]. In both types of attacks, structural information is used to reveal the true identities of the targeted users [118]. Consequently subgraph discovery attacks can be used to compromise the privacy of users and their contacts by raising identity and social link disclosure problems [12].

Graph classification which consists at classifying jointly a large number of interconnected nodes in a graph is one interesting aspect of collective classification [15]. In the context of social networks, a number of studies have revealed that users tend to establish friendship ties with other users who have similar interests [154, 155]. This tendency can cover a wide number of sensitive information, such as race and ethnicity, age, religion, education, and occupation, which are all personal attributes. Social network users can set the visibility of their attribute profiles, but this may not be enough to keep private their sensitive information because group memberships and friendship relationships in many cases remain visible and hence can be used to infer private information [159, 165]. In addition, information from the users' contacts can also be extracted from contacts who are not much concerned about securing their personal information or interact. For instance, it is possible to construct automatically users' profiles or infer the values of missing attributes as it is shown in [166] where the authors describe a user profiling approach in social networks. Likewise, other privacy attacks are described in [159] where a mixture of public and private user profiles are used to predict the private attributes of users by applying collective classification which aims at learning and inferring class labels of linked nodes together.

Understanding the structure and evolution of social networks over time has gained much attention recently, especially with many *graph-based generative models* [167–169]. The potential of these approaches to study network evolution and group formation is appealing and, as a consequence, the insights provided by these methods are highly interesting: How networks and group are formed? Why and when users join groups? Which groups will grow rapidly? At the same time, the consequences are rather severe for users' privacy when these approaches are misused. For instance, the authors in [168] address the problem of modeling social network generation and demonstrate the capacity of their generative model to reveal that users are joining groups for various reasons and that friendship with other group members is only one of these reasons. In another work [64], the author propose a generative model for friendships in social circles using a combination of both network and profile information. This model can automatically identify users' social circles (e.g., contacts who are friends, contacts from the same hometown, contacts from the same college) and predict to which circles a new contact should be assigned.

To conclude, link mining techniques can be used to infer new information about the users and consequently pose serious privacy concerns. This can take many forms within the different link mining tasks. In node-related tasks attackers can compromise users' privacy by exploring the structure of social networks, identifying key users, combining information about the network structure with the fraction of publicly available information, investigating group memberships, and associating users' profiles with their corresponding real-world entities. In link-related tasks, attackers can invade users' privacy by predicting link existence between users who would like to keep their relationship private, and revealing the type of a sensitive relationship between users. In graph-related tasks, attackers can violate users' privacy by re-identifying users of two similar structures (an anonymized network and a background of knowledge), classifying jointly a large number of interconnected nodes, extracting information from contacts with public profiles, and seeking to retrieve insights from the study of network evolution and group formation. Concerned about privacy implications, social network users are increasingly interested in solutions that enforce individual privacy and protect sensitive information. Next, we discuss some open challenges that future social network solutions must take into account.

1.8   Open Challenges

As information sharing is gaining a great deal of attention among social networks users, privacy on these sites is one of the most intriguing social networks' challenges [9]. Data on social networks usually open up questions related to users' privacy and data management. Privacy concerns vary significantly across these networks due to the open nature of how data is displayed and controlled by each site. In the following, we present some of the key challenges to consider in future privacy protection approaches.

1.8.1   Relationship Discovery and Management

Currently, social networks let users manage and control their privacy settings. However, they typically do so in terms of contact identities or manually grouped lists of contacts. Controlling access to own resources is driven more by relationships that social network users share with their contacts such as colleagues, relatives, and friends. Treating all their contacts in the same way, without differentiating one user from another, is an unsafe and restrictive practice. For instance, a user might want to:

1. Prevent her relatives from viewing content posted by her friends,
2. Prevent her relatives from connecting, posting comments, and communicating with her colleagues,
3. Share some data and interact with some members of their social network but not with all of them.

Unfortunately, relationship types between a user and her contacts are often missing [36]. Although some social network sites provide the possibility to define manually how a user knows each of her contacts, most of the time this option is skipped by social network users and only the link existence is indicated [170].

Relationship discovery and management is one of the key challenges that are closely related to privacy topics. On social networks where relationships multiply rapidly and evolve over time, relationship-based management mechanisms play a major role in enforcing and facilitating privacy-related settings. This involves mechanisms that let social network users protect their personal information by only granting access to some contacts (e.g., friends) while denying the access to other contacts (e.g., colleagues). Social network users are facing different kinds of misuse cases regarding their privacy due to the lack of efficient access control models. While relationship-based access control mechanisms seems highly interesting, such approaches are rarely implemented since relationship types on social networks remain unlabeled. Recently, the problem of missing relationship types in social networks has been investigated in [62,64] where a number of relationship discovery approaches are described. Such relationship-related capacities seem to be promising when approaching privacy issues, and social networks should explicitly consider them in their privacy settings far more than they currently do.

1.8.2   Multimedia Content and Metadata

Sharing and controlling access to published information has become an integral part of users' every day lives. Apparently, users are cautious about the information they reveal online in terms of [171]:

1. Degree of identifiability (pseudonyms/real names),
2. Type of information (school name, hobbies, interests, etc.),
3. Visibility of information (who can view user's profile).

At the same time, the number of created accounts on various sites and the volume of data available on social networks are exponentially growing at an incredible rate, going beyond social network users' ability to easily interact and manage their data. In addition, users are unaware of the risks associated with the disclosed information since they cannot fully control who can use these information and for what purposes [172].

Managing multimedia content and their metadata is another challenge for social network users. This particular privacy challenge is raised by the increasingly growing number of multimedia objects uploaded on social networks. For instance, 250 million photos are uploaded on Facebook[17] every day. However, photos can be used to identify persons (e.g., facial recognition), as well as to infer additional information through available metadata (e.g., GPS location, data and time, and device name). Personal photos published on social networks may be used for inappropriate purposes related to users' private life, friends, work information, habits, etc. For example, employers who want to justify a decision to fire an

---

[17]https://www.facebook.com/press/info.php?statistics, accessed 01 October 2012.

employee or to check the backgrounds of potential employee. Similarly, users' friends or the friends of their friends who are interested to know more about the past of a person or infer new information more than what the users chose to share with them (list of friends, physical location, political affiliation, sexual orientation, etc.). More importantly, such available information when associated with other profile content may be used by family members, friends, or colleagues to check information related to sexuality, relationship status, location, or details of personal problems that owners might consider embarrassing if it was widely known [173].

It is therefore obvious that unauthorized users or parties must not be able to link between the various user's activities in order to infer further information. Hiding the link between multiple actions is essential to prevent attackers to reconstruct user's profile. This aspect is essentially related to the unlinkability privacy requirement. Novel approaches must ensure that private data must be protected and no useful information can be leaked through the analysis of user's activities.

1.8.3 Social Media Preservation

There is no doubt that social networks sites can yield much more information from users' data such as selling insights from mined data for targeted advertising purposes [174], predicting relationships from social behavior data useful in understanding topics discussed between users and their personalities [153], and falsifying identities for criminal and terrorist groups [175]. At any time and often without users' knowledge, social networks can mine, copy, or archive personal information. Personal data can be mined and used to reveal information about users' private lives, their social relationships, or additional information that users would like to keep private. Unaware of an old posted information or a previously added online contact, an uncountable number of social network users went through bad experiences that affected their life.

Dynamically protecting sensitive information of the archived users' data is another privacy challenge on social networks. Social network sites have complete control over users' data and may intentionally or accidentally leak its content to unauthorized entities or third parties. A fundamental privacy concern for users is that social network providers and third-parties may potentially access and aggregate personal information from the users' archived data [10, 176, 177]. However, the social network users' perception of privacy goes beyond hiding information from being viewed by their contacts. It also involves mechanisms that enforce privacy and ensure that personal archived information is not misused. And if information is leaked, even though appropriate mechanisms were implemented, an enforced privacy must be able to guarantee that no personal information is revealed to unauthorized entities. This privacy aspect underlies the unobservability aspect of privacy. While many social networks provide various forms of access control mechanisms to restrict who can view a published information, they do not provide any form of protection regarding the possibility that archived information can be misused and consequently analyzed to reveal sensitive information. To date, none of these concerns has been successfully integrated into users' privacy settings, and the privacy-related issues derived from social media preservation have been largely left unexplored.

1.8.4   Social Digital Ecosystem

Along with the previously mentioned privacy challenges, future social networks must shed the lights on the real needs and expectations of their users. So far, social network providers set the boundaries within which users can socialize, share information, accept data ownership rights, permission policies, as well as other critical issues. Current social networks dictate such rules rather than letting their users set up their own rules for data sharing and ownerships to promote data decentralization and to go beyond the walled garden of the current social network sites.

Shifting social networks toward an ecosystem model is the challenge for the new generation of social networks. In essence, an ecosystem is an environment made of entities that interact within the system, maintain the system stable, are committed to ensure mutual respect, and can benefit from each other's participation [178]. Shifting the future forms of social networks toward an ecosystem environment would enable users to set their own preferences and to use the system more effectively. These preferences are initially the same for all social network users. After joining an ecosystem, users are free to personalize these preferences as they wish. One aspect of these settings is related to the privacy of the users who can design their privacy strategies as they wish. This would ensure a better protection level since all the social ecosystem are enforced to accept the *gold rule* that englobes the system's stability, the mutual respect, and the positive participation of the users.

1.9   Conclusion

This chapter has presented a global overview about privacy on social networks. While it is complex to give a precise definition of privacy as it is engorged with various and distinct meanings, understanding the characteristics of social networks would certainly contribute positively to the design of more adequate privacy protection approaches.

As stated earlier, by analyzing social networks' structure and content it is possible to infer further knowledge which may go beyond what the users want to disclose. In light of this, we reviewed various social network analysis measures and link mining techniques then we derived their associated privacy threats. Particularly, we have shown the importance of taking into consideration these techniques and the usefulness to integrate them in future privacy-related systems. Novel approaches must be able to cope with the various privacy challenges such as trust and privacy management, risks and threats of social networking, traceability analysis, user profiling and related risks, ethical conflicts in social networks as well as the moral implications, relationship management and discovery, anonymity preserving, social terrorism, social network-based access control, and abnormal activities on social networks.

Of all social network challenges, privacy protections is crucial for both users and social networks. Failing to provide an optimal privacy protection may have undesirable consequences on the popularity of such social networks and on the amount of information that social network users are willing to share. It is therefore up to social network sites to provide their users with a variety of support tools that align with users' perceptions of privacy such as enhanced relationship management capacities, intuitive interfaces, fine-grained access control, secure online data storage, and media preservation solutions.

References

1. Liu, Y., Gummadi, K.P., Krishnamurthy, B., Mislove, A.: Analyzing facebook privacy settings: user expectations vs. reality. In: Proceedings of the 2011 ACM SIGCOMM Conference on Internet Measurement Conference. IMC '11, pp. 61–70. ACM, New York (2011)
2. Lipford, H.R., Besmer, A., Watson, J.: Understanding privacy settings in facebook with an audience view. In: Proceedings of the 1st Conference on Usability, Psychology, and Security, USENIX Association Berkeley, CA, USA, pp. 1–8 (2008)
3. Acquisti, A., Gross, R.: Imagined communities: awareness, information sharing, and privacy on the facebook. In: Danezis, G., Golle, P. (eds.) Privacy Enhancing Technologies. Lecture Notes in Computer Science, Vol. 4258, pp. 36–58. Springer, Berlin/Heidelberg (2006)
4. Awad, N., Krishnan, M.: The personalization privacy paradox: an empirical evaluation of information transparency and the willingness to be profiled online for personalization. MIS Quarterly 13–28 (2006)
5. Madejski, M., Johnson, M., Bellovin, S.: A study of privacy settings errors in an online social network. In: 2012 IEEE International Conference on Pervasive Computing and Communications Workshops (PERCOM Workshops), pp. 340–345 (2012)
6. Johnson, M., Egelman, S., Bellovin, S.M.: Facebook and privacy: it's complicated. In: Proceedings of the Eighth Symposium on Usable Privacy and Security. SOUPS '12, pp. 1–15. ACM, New York (2012)
7. Besmer, A., Richter Lipford, H.: Moving beyond untagging: photo privacy in a tagged world. In: Proceedings of the SIGCHI Conference on Human Factors in Computing Systems. CHI '10, pp. 1563–1572. ACM, New York (2010)
8. Grimmelmann, J.: Saving Facebook. Iowa Law Rev. **94**, 1137–1206 (2009)
9. Zhang, C., Sun, J., Zhu, X., Fang, Y.: Privacy and security for online social networks: challenges and opportunities. IEEE Netw. **24**(4), 13–18 (2010)
10. Viswanath, B., Kiciman, E., Saroiu, S.: Keeping information safe from social networking apps. In: Proceedings of the 2012 ACM Workshop on Workshop on Online Social Networks. WOSN '12, pp. 49–54. ACM, New York (2012)
11. Brey, P.: Ethical aspects of information security and privacy. In: Petkovic, M., Jonker, W., Carey, M.J., Ceri, S. (eds.) Security, Privacy, and Trust in Modern Data Management. Data-Centric Systems and Applications, pp. 21–36. Springer, Berlin/Heidelberg (2007)

12. Zheleva, E., Terzi, E., Getoor, L.: Privacy in social networks. Synth. Lect. Data Min. Knowl. Discov. **3**(1), 1–85 (2012)
13. Pfitzmann, A., Hansen, M.: A terminology for talking about privacy by data minimization: Anonymity, unlinkability, undetectability, unobservability, pseudonymity, and identity management (August 2010) v0.34
14. Wasserman, S., Faust, K.: Social Network Analysis: Methods and Applications. Cambridge University Press, Cambridge (1994)
15. Getoor, L., Diehl, C.P.: Link mining: a survey. SIGKDD Explor. Newsl. **7**(2), 3–12 (2005)
16. Broder, A., Kumar, R., Maghoul, F., Raghavan, P., Rajagopalan, S., Stata, R., Tomkins, A., Wiener, J.: Graph structure in the web. Comput. Network **33**(1), 309–320 (2000)
17. Wellman, B.: Computer networks as social networks. Science **293**(5537), 2031–2034 (2001)
18. Ito, T., Chiba, T., Ozawa, R., Yoshida, M., Hattori, M., Sakaki, Y.: A comprehensive two-hybrid analysis to explore the yeast protein interactome. Proc. Natl. Acad. Sci. **98**(8), 4569–4574 (2001)
19. Gomez-Rodriguez, M., Leskovec, J., Krause, A.: Inferring networks of diffusion and influence. ACM Trans. Knowl. Discov. Data **5**(4) (2012)
20. Adamic, L., Buyukkokten, O., Adar, E.: A social network caught in the web. First Monday **8**(6) (2003)
21. Wilson, C., Sala, A., Puttaswamy, K.P.N., Zhao, B.Y.: Beyond social graphs: User interactions in online social networks and their implications. ACM Trans. Web **6**(4), 17:1–17:31 (2012)
22. Faust, K.: Centrality in affiliation networks. Soc. Network **19**(2), 157–191 (1997)
23. Borgatti, S., Mehra, A., Brass, D., Labianca, G.: Network analysis in the social sciences. Science **323**(5916), 892–895 (2009)
24. Matsuo, Y., Mori, J., Hamasaki, M., Nishimura, T., Takeda, H., Hasida, K., Ishizuka, M.: Polyphonet: An advanced social network extraction system from the web. Web Semant. **5**(4), 262–278 (2007)
25. Gilbert, E.: Predicting tie strength in a new medium. In: Proceedings of the ACM 2012 Conference on Computer Supported Cooperative Work. CSCW '12, pp. 1047–1056. ACM, New York (2012)
26. Haythornthwaite, C.: Social network analysis: An approach and technique for the study of information exchange. Library Inform. Sci. Res. **18**(4), 323–342 (1996)
27. Musial, K., Kazienko, P.: Social networks on the internet. World Wide Web **16**(1), 31–72 (2013)
28. Granovetter, M.S.: The strength of weak ties. Am. J. Sociol. **78**(6), 1360–1380 (1973)
29. Haythornthwaite, C.: Social networks and internet connectivity effects. Inform. Comm. Soc. **8**(2), 125–147 (2005)
30. Wellman, B., Wortley, S.: Different strokes from different folks: community ties and social support. Am. J. Sociol. **96**(3), 558–588 (1990)
31. Bernard, H., Johnsen, E., Killworth, P., McCarty, C., Shelley, G., Robinson, S.: Comparing four different methods for measuring personal social networks. Soc. Network **12**(3), 179–215 (1990)
32. Killworth, P., Johnsen, E., Bernard, H., Ann Shelley, G., McCarty, C.: Estimating the size of personal networks. Soc. Network **12**(4), 289–312 (1990)
33. Steinfield, C., Ellison, N., Lampe, C.: Social capital, self-esteem, and use of online social network sites: A longitudinal analysis. J. Appl. Dev. Psychol. **29**(6), 434–445 (2008)
34. Heidemann, J., Klier, M., Probst, F.: Online social networks: A survey of a global phenomenon. Comput. Network **56**(18), 3866–3878 (2012)
35. Hua, W., Wellman, B.: Social connectivity in america: changes in adult friendship network size from 2002 to 2007. Am. Behav. Sci. **53**(8), 1148–1169 (2010)
36. Dunbar, R.I.M.: Social cognition on the internet: testing constraints on social network size. Phil. Trans. Roy. Soc. B Biol. Sci. **367**(1599), 2192–2201 (2012)
37. Subrahmanyam, K., Reich, S., Waechter, N., Espinoza, G.: Online and offline social networks: Use of social networking sites by emerging adults. J. Appl. Dev. Psychol. **29**(6), 420–433 (2008)

38. Correa, T., Hinsley, A., de Zúñiga, H.: Who interacts on the web? The intersection of users' personality and social media use. Comput. Hum. Behav. **26**(2), 247–253 (2010)
39. Joinson, A.N.: Looking at, looking up or keeping up with people?: motives and use of facebook. In: Proceedings of the SIGCHI Conference on Human Factors in Computing Systems. CHI '08, pp. 1027–1036. ACM, New York (2008)
40. Papacharissi, Z., Mendelson, A.: Toward a new (er) sociability: uses, gratifications and social capital on facebook. Media Perspectives for the 21st Century, pp. 212–230 (2011)
41. Ryan, T., Xenos, S.: Who uses facebook? an investigation into the relationship between the big five, shyness, narcissism, loneliness, and facebook usage. Comput. Hum. Behav. **27**(5), 1658–1664 (2011)
42. Gentile, B., Twenge, J., Freeman, E., Campbell, W.: The effect of social networking websites on positive self-views: An experimental investigation. Comput. Hum. Behav. **28**(5), 1929–1933 (2012)
43. Nosko, A., Wood, E., Molema, S.: All about me: Disclosure in online social networking profiles: The case of facebook. Comput. Hum. Behav. **26**(3), 406–418 (2010)
44. Krasnova, H., Spiekermann, S., Koroleva, K., Hildebrand, T.: Online social networks: Why we disclose. J. Inform. Tech. **25**(2), 109–125 (2010)
45. Ross, C., Orr, E., Sisic, M., Arseneault, J., Simmering, M., Orr, R.: Personality and motivations associated with facebook use. Comput. Hum. Behav. **25**(2), 578–586 (2009)
46. Morris, M.R., Teevan, J., Panovich, K.: What do people ask their social networks, and why?: a survey study of status message q&a behavior. In: Proceedings of the SIGCHI Conference on Human Factors in Computing Systems. CHI '10, pp. 1739–1748. ACM, New York (2010)
47. Lin, K.Y., Lu, H.P.: Why people use social networking sites: An empirical study integrating network externalities and motivation theory. Comput. Hum. Behav. **27**(3), 1152–1161 (2011)
48. Nosko, A., Wood, E., Kenney, M., Archer, K., De Pasquale, D., Molema, S., Zivcakova, L.: Examining priming and gender as a means to reduce risk in a social networking context: Can stories change disclosure and privacy setting use when personal profiles are constructed? Comput. Hum. Behav. **28**(6), 2067–2074 (2012)
49. Sledgianowski, D., Kulviwat, S.: Using social network sites: The effects of playfulness, critical mass and trust in a hedonic context. J. Comput. Inform. Syst. **49**(4), 74–83 (2009)
50. Pempek, T., Yermolayeva, Y., Calvert, S.: College students' social networking experiences on facebook. J. Appl. Dev. Psychol. **30**(3), 227–238 (2009)
51. Kwon, O., Wen, Y.: An empirical study of the factors affecting social network service use. Comput. Hum. Behav. **26**(2), 254–263 (2010)
52. Adamic, L., Adar, E.: How to search a social network. Soc. Network **27**(3), 187–203 (2005)
53. Boyd, D., Ellison, N.: Social network sites: Definition, history, and scholarship. J. Comput. Mediat. Comm. **13**(1), 210–230 (2007)
54. Schneider, F., Feldmann, A., Krishnamurthy, B., Willinger, W.: Understanding online social network usage from a network perspective. In: Proceedings of the 9th ACM SIGCOMM Conference on Internet Measurement Conference. IMC '09, pp. 35–48. ACM, New York (2009)
55. Bakshy, E., Rosenn, I., Marlow, C., Adamic, L.: The role of social networks in information diffusion. In: Proceedings of the 21st Annual Conference on World Wide Web. WWW'12, pp. 519–528 (2012)
56. Ellison, N., Steinfield, C., Lampe, C.: The benefits of facebook "friends:" social capital and college students' use of online social network sites. J. Comput. Mediat. Comm. **12**(4), 1143–1168 (2007)
57. Raad, E., Chbeir, R., Dipanda, A.: User profile matching in social networks. In: Proceedings - 13th International Conference on Network-Based Information Systems, NBiS 2010, pp. 297–304 (sept. 2010)
58. Thelwall, M.: Social networks, gender, and friending: An analysis of mySpace member profiles. J. Am. Soc. Inform. Sci. Tech. **59**(8), 1321–1330 (2008)
59. Abel, F., Henze, N., Herder, E., Krause, D.: Interweaving public user profiles on the web. In: Bra, P., Kobsa, A., Chin, D. (eds.) User Modeling, Adaptation, and Personalization.

Volume 6075 of Lecture Notes in Computer Science, pp. 16–27. Springer, Berlin/Heidelberg (2010)

60. Graves, M., Constabaris, A., Brickley, D.: FOAF: connecting people on the semantic web. Cataloging Classification Quarterly **43**(3–4), 191–202 (2007)
61. Cranshaw, J., Toch, E., Hong, J., Kittur, A., Sadeh, N.: Bridging the gap between physical location and online social networks. In: UbiComp'10 - Proceedings of the 2010 ACM Conference on Ubiquitous Computing, pp. 119–128 (2010)
62. Raad, E., Chbeir, R., Dipanda, A.: Discovering relationship types between users using profiles and shared photos in a social network. Multimed. Tool Appl. **64**(1), 141–170 (2013)
63. Tang, L., Liu, H.: Scalable learning of collective behavior based on sparse social dimensions. In: Proceedings of the 18th ACM Conference on Information and Knowledge Management. CIKM '09, pp. 1107–1116. ACM, New York (2009)
64. McAuley, J., Leskovec, J.: Learning to discover social circles in ego networks. Adv. Neural Inform. Process. Syst. **25**, 548–556 (2012)
65. Ellison, N., Steinfield, C., Lampe, C.: Connection strategies: Social capital implications of facebook-enabled communication practices. New Media Soc. **13**(6), 873–892 (2011)
66. Newman, M.: The structure and function of complex networks. SIAM Rev. **45**(2), 167–256 (2003)
67. Boccaletti, S., Latora, V., Moreno, Y., Chavez, M., Hwang, D.U.: Complex networks: structure and dynamics. Phys. Rep. **424**(4–5), 175–308 (2006)
68. Fortunato, S.: Community detection in graphs. Phys. Rep. **486**(3–5), 75–174 (2010)
69. Brams, S., Mutlu, H., Ramirez, S.: Influence in terrorist networks: From undirected to directed graphs. Stud. Conflict Terrorism **29**(7), 703–718 (2006)
70. Morselli, C., Giguère, C., Petit, K.: The efficiency/security trade-off in criminal networks. Soc. Network **29**(1), 143–153 (2007)
71. Bachi, G., Coscia, M., Monreale, A., Giannotti, F.: Classifying trust/distrust relationships in online social networks. In: Privacy, Security, Risk and Trust (PASSAT), 2012 International Conference on and 2012 International Confernece on Social Computing (SocialCom), pp. 552–557 (Sept. 2012)
72. Zhang, T., Chao, H., Tretter, D.: Dynamic estimation of family relations from photos. In: Lee, K.T., Tsai, W.H., Liao, H.Y., Chen, T., Hsieh, J.W., Tseng, C.C. (eds.) Advances in Multimedia Modeling. Volume 6524 of Lecture Notes in Computer Science, pp. 65–76. Springer, Berlin/Heidelberg (2011)
73. Carminati, B., Ferrari, E., Perego, A.: Enforcing access control in Web-based social networks. ACM Trans. Inf. Syst. Secur. **13**(1), 1–38 (2009)
74. Xiang, R., Neville, J., Rogati, M.: Modeling relationship strength in online social networks. In: Proceedings of the 19th International Conference on World Wide Web. WWW '10, pp. 981–990. ACM, New York (2010)
75. Stutzman, F., Kramer-Duffield, J.: Friends only: examining a privacy-enhancing behavior in facebook. In: Proceedings of the SIGCHI Conference on Human Factors in Computing Systems. CHI '10, pp. 1553–1562. ACM, New York (2010)
76. Kogovšek, T., Ferligoj, A., Coenders, G., Saris, W.: Estimating the reliability and validity of personal support measures: Full information ML estimation with planned incomplete data. Soc. Network **24**(1), 1–20 (2002)
77. Groves, R.: Survey Errors and Survey Costs, vol. 536. Wiley-Interscience, Hoboken, NJ (2004)
78. Gonzalez-Bailon, S.: Opening the black box of link formation: Social factors underlying the structure of the web. Soc. Network **31**(4), 271–280 (2009)
79. Tyler, J., Wilkinson, D., Huberman, B.: E-mail as spectroscopy: Automated discovery of community structure within organizations. Inform. Soc. **21**(2), 133–141 (2005)
80. Kautz, H., Selman, B., Shah, M.: Referral web: combining social networks and collaborative filtering. Comm. ACM **40**(3), 63–65 (1997)
81. Mika, P.: Flink: Semantic web technology for the extraction and analysis of social networks. Web Semant. **3**(2–3), 211–223 (2005)

82. Marin, A., Wellman, B.: Social network analysis: An introduction. Handbook Soc. Network Anal. **22**(January), 11–25 (2010)

83. Kwak, H., Lee, C., Park, H., Moon, S.: What is twitter, a social network or a news media? In: Proceedings of the 19th International Conference on World Wide Web. WWW '10, pp. 591–600. ACM, New York (2010)

84. Kazienko, P., Musial, K., Kajdanowicz, T.: Multidimensional social network in the social recommender system. IEEE Trans. Syst. Man Cybern. A Syst. Hum. **41**(4), 746–759 (2011)

85. Catanese, S., Meo, P., Ferrara, E., Fiumara, G., Provetti, A.: Extraction and analysis of facebook friendship relations. In Abraham, A. (ed.) Computational Social Networks, pp. 291–324. Springer, London (2012)

86. Schneier, B.: A taxonomy of social networking data. IEEE Secur. Privacy **8**(4), 88 (2010)

87. Diesner, J., Frantz, T., Carley, K.: Communication networks from the enron email corpus "it's always about the people. enron is no different". Comput. Math. Organ. Theor **11**(3), 201–228 (2005)

88. Liu, K., Terzi, E.: Towards identity anonymization on graphs. In: Proceedings of the 2008 ACM SIGMOD International Conference on Management of Data. SIGMOD '08, pp. 93–106. ACM, New York (2008)

89. Narayanan, A., Shmatikov, V.: De-anonymizing social networks. In: 2009 30th IEEE Symposium on Security and Privacy, pp. 173–187 (may 2009)

90. Hay, M., Miklau, G., Jensen, D., Towsley, D., Li, C.: Resisting structural re-identification in anonymized social networks. VLDB J. **19**(6), 797–823 (2010)

91. Agrawal, R., Srikant, R.: Privacy-preserving data mining. SIGMOD Rec. (ACM Special Interest Group on Management of Data) **29**(2), 439–450 (2000)

92. Verykios, V., Bertino, E., Fovino, I., Provenza, L., Saygin, Y., Theodoridis, Y.: State-of-the-art in privacy preserving data mining. SIGMOD Rec. **33**(1), 50–57 (2004)

93. Zhou, B., Pei, J.: The k-anonymity and l-diversity approaches for privacy preservation in social networks against neighborhood attacks. Knowl. Inform. Syst. **28**(1), 47–77 (2011)

94. Li, N., Zhang, N., Das, S.K.: Relationship privacy preservation in publishing online social networks. In: PASSAT/SocialCom., pp. 443–450 (oct. 2011)

95. Fong, P.W., Siahaan, I.: Relationship-based access control policies and their policy languages. In: Proceedings of the 16th ACM Symposium on Access Control Models and Technologies. SACMAT '11, pp. 51–60. ACM, New York (2011)

96. Cheek, G.P., Shehab, M.: Policy-by-example for online social networks. In: Proceedings of the 17th ACM Symposium on Access Control Models and Technologies. SACMAT '12, pp. 23–32. ACM, New York (2012)

97. Li, N., Chen, G.: Sharing location in online social networks. IEEE Network **24**(5), 20–25 (2010)

98. Backstrom, L., Sun, E., Marlow, C.: Find me if you can: improving geographical prediction with social and spatial proximity. In: Proceedings of the 19th International Conference on World Wide Web. WWW '10, pp. 61–70. ACM, New York (2010)

99. Cunningham, S., Masoodian, M., Adams, A.: Privacy issues for online personal photograph collections. J. Theor. Appl. Electron. Commerce Res. **5**(2), 26–40 (2010)

100. Frikken, K., Atallah, M., Li, J.: Attribute-based access control with hidden policies and hidden credentials. IEEE Trans. Comput. **55**(10), 1259–1270 (2006)

101. Bouna, B., Chbeir, R., Marrara, S.: Enforcing role based access control model with multimedia signatures. J. Syst. Architect. **55**(4), 264–274 (2009)

102. Peng, H., Gu, J., Ye, X.: Dynamic purpose-based access control. In: International Symposium on Parallel and Distributed Processing with Applications, 2008. ISPA '08, pp. 695–700 (2008)

103. Willinger, W., Rejaie, R., Torkjazi, M., Valafar, M., Maggioni, M.: Research on online social networks: time to face the real challenges. SIGMETRICS Perform. Eval. Rev. **37**(3), 49–54 (2010)

104. Wellman, B.: The school child's choice of companions. J. Educ. Res. **14**(2), 126–132 (1926)

105. Bott, H.: Observation of play activities in a nursery school. Genet. Psychol. Monogr. **4**(1), 44–88 (1928)

106. Moreno, J.: Who Shall Survive? vol. 58. Nervous and Mental Disease Publishing Company, Washington, DC (1934)
107. Wellman, B.: Network analysis: Some basic principles. Socio. Theor 1(1), 155–200 (1983)
108. Barnes, J.A.: Class and committees in a norwegian island parish. Hum. Relat. 7(1), 39–58 (1954)
109. Luke, D., Harris, J.: Network analysis in public health: History, methods, and applications. Annu. Rev. Publ. Health 28, 69–93 (2007)
110. Watts, D.: The "new" science of networks. Annu. Rev. Sociol. 30, 243–270 (2004)
111. Barabási, A.L., Bonabeau, E.: Scale-free networks. Sci. Am. 288(5), 60–69 (2003)
112. Christley, R., Pinchbeck, G., Bowers, R., Clancy, D., French, N., Bennett, R., Turner, J.: Infection in social networks: using network analysis to identify high-risk individuals. Am. J. Epidemiol. 162(10), 1024–1031 (2005)
113. Mislove, A., Marcon, M., Gummadi, K.P., Druschel, P., Bhattacharjee, B.: Measurement and analysis of online social networks. In: Proceedings of the 7th ACM SIGCOMM Conference on Internet Measurement. IMC '07, pp. 29–42. ACM, New York (2007)
114. Koschützki, D., Lehmann, K., Peeters, L., Richter, S., Tenfelde-Podehl, D., Zlotowski, O.: Centrality indices. In: Brandes, U., Erlebach, T. (eds.) Network Analysis. Volume 3418 of Lecture Notes in Computer Science, pp. 16–61. Springer, Berlin/Heidelberg (2005)
115. Freeman, L.: Centrality in social networks conceptual clarification. Soc. Network 1(3), 215–239 (1978)
116. He, X., Vaidya, J., Shafiq, B., Adam, N., Atluri, V.: Preserving privacy in social networks: A structure-aware approach. In: Web Intelligence and Intelligent Agent Technologies, vol. 1, pp. 647–654 (Sept. 2009)
117. Campan, A., Truta, T.: Data and structural k-anonymity in social networks. In: Bonchi, F., Ferrari, E., Jiang, W., Malin, B. (eds.) Privacy, Security, and Trust in KDD. Volume 5456 of Lecture Notes in Computer Science, pp. 33–54. Springer, Berlin/Heidelberg (2009)
118. Backstrom, L., Dwork, C., Kleinberg, J.: Wherefore art thou r3579x?: anonymized social networks, hidden patterns, and structural steganography. In: Proceedings of the 16th International Conference on World Wide Web. WWW '07, pp. 181–190. ACM, New York (2007)
119. Hay, M., Miklau, G., Jensen, D., Weis, P., Srivastava, S.: Anonymizing social networks. Technical report (2007)
120. Korolova, A., Motwani, R., Nabar, S.U., Xu, Y.: Link privacy in social networks. In: Proceedings of the 17th ACM Conference on Information and Knowledge Management. CIKM '08, pp. 289–298. ACM, New York (2008)
121. Zheleva, E., Getoor, L.: Preserving the privacy of sensitive relationships in graph data. In: Bonchi, F., Ferrari, E., Malin, B., Saygin, Y. (eds.) Privacy, Security, and Trust in KDD. Volume 4890 of Lecture Notes in Computer Science, pp. 53–171. Springer, Berlin/Heidelberg (2008)
122. Wu, X., Kumar, V., Ross, Q., Ghosh, J., Yang, Q., Motoda, H., McLachlan, G., Ng, A., Liu, B., Yu, P., Zhou, Z.H., Steinbach, M., Hand, D., Steinberg, D.: Top 10 algorithms in data mining. Knowl. Inform. Syst. 14(1), 1–37 (2008)
123. Liu, X., Bollen, J., Nelson, M., Van De Sompel, H.: Co-authorship networks in the digital library research community. Inform. Process. Manag. 41(6), 1462–1480 (2005)
124. Creamer, G., Stolfo, S.: A link mining algorithm for earnings forecast and trading. Data Min. Knowl. Discov. 18(3), 419–445 (2009)
125. Li, P., Li, Z., Liu, H., He, J., Du, X.: Using link-based content analysis to measure document similarity effectively. In: Li, Q., Feng, L., Pei, J., Wang, S., Zhou, X., Zhu, Q.M. (eds.) Advances in Data and Web Management. Volume 5446 of Lecture Notes in Computer Science, pp. 455–467. Springer, Berlin/Heidelberg (2009)
126. Lin, Z., Wang, L., Guo, S.: Recommendations on social network sites: From link mining perspective. In: International Conference on Management and Service Science, 2009. MASS '09, pp. 1–4 (Sept. 2009)

127. Karamon, J., Matsuo, Y., Yamamoto, H., Ishizuka, M.: Generating social network features for link-based classification. In: Kok, J., Koronacki, J., Lopez de Mantaras, R., Matwin, S., Mladenic, D., Skowron, A. (eds.) Knowledge Discovery in Databases: PKDD 2007. Volume 4702 of Lecture Notes in Computer Science, pp. 127–139. Springer, Berlin/Heidelberg (2007)

128. Chakrabarti, S., Dom, B., Indyk, P.: Enhanced hypertext categorization using hyperlinks. SIGMOD Rec. **27**(2), 307–318 (1998)

129. Segal, E., Wang, H., Koller, D.: Discovering molecular pathways from protein interaction and gene expression data. Bioinformatics **19**(SUPPL. 1), i264–i272 (2003)

130. Stattner, E., Vidot, N.: Social network analysis in epidemiology: current trends and perspectives. In: 2011 Fifth International Conference on Research Challenges in Information Science (RCIS), pp. 1–11 (may 2011)

131. Fard, A., Ester, M.: Collaborative mining in multiple social networks data for criminal group discovery. In: International Conference on Computational Science and Engineering, 2009. CSE '09, vol. 4, pp. 582–587 (Aug. 2009)

132. Barbier, G., Liu, H.: Data mining in social media. In: Aggarwal, C.C. (ed.) Social Network Data Analytics, pp. 327–352. Springer, New York (2011)

133. Bhattacharya, I., Getoor, L.: Iterative record linkage for cleaning and integration. In: Proceedings of the 9th ACM SIGMOD Workshop on Research Issues in Data Mining and Knowledge Discovery. DMKD '04, pp. 11–18. ACM, New York (2004)

134. Ponzetto, S.P., Strube, M.: Exploiting semantic role labeling, wordnet and wikipedia for coreference resolution. In: Proceedings of the Main Conference on Human Language Technology Conference of the North American Chapter of the Association of Computational Linguistics. HLT-NAACL '06, pp. 192–199. Association for Computational Linguistics (2006)

135. Stein, L.: Integrating biological databases. Nat. Rev. Genet. **4**(5), 337–345 (2003)

136. Taskar, B., Wong, M.F., Abbeel, P., Koller, D.: Link prediction in relational data. In: Advances in Neural Information Processing Systems (NIPS). MIT Press, Cambridge, MA (2003)

137. Huang, Z., Li, X., Chen, H.: Link prediction approach to collaborative filtering. In: Proceedings of the 5th ACM/IEEE-CS Joint Conference on Digital Libraries. JCDL '05, pp. 141–142. ACM, New York (2005)

138. Yu, H., Paccanaro, A., Trifonov, V., Gerstein, M.: Predicting interactions in protein networks by completing defective cliques. Bioinformatics **22**(7), 823–829 (2006)

139. Zheleva, E., Getoor, L., Golbeck, J., Kuter, U.: Using friendship ties and family circles for link prediction. Lect. Note Comput. Sci. **5498 LNAI**, 97–113 (2009)

140. Buccafurri, F., Lax, G., Nocera, A., Ursino, D.: Discovering links among social networks. In: Flach, P., Bie, T., Cristianini, N. (eds.) Machine Learning and Knowledge Discovery in Databases. Volume 7524 of Lecture Notes in Computer Science, pp. 467–482. Springer, Berlin/Heidelberg (2012)

141. Getoor, L.: Link mining: a new data mining challenge. SIGKDD Explor. Newsl. **5**(1), 84–89 (2003)

142. Cheng, H., Lo, D., Zhou, Y., Wang, X., Yan, X.: Identifying bug signatures using discriminative graph mining. In: Proceedings of the Eighteenth International Symposium on Software Testing and Analysis. ISSTA '09, pp. 141–152. ACM, New York (2009)

143. Ciriello, G., Guerra, C.: A review on models and algorithms for motif discovery in protein-protein interaction networks. Brief. Funct. Genomic. Proteomic. **7**(2), 147–156 (2008)

144. Borgwardt, K., Ong, C., Schönauer, S., Vishwanathan, S., Smola, A., Kriegel, H.P.: Protein function prediction via graph kernels. Bioinformatics **21**(SUPPL. 1), i47–i56 (2005)

145. Rabelo, J., Prudêncio, R., Barros, F.: Leveraging relationships in social networks for sentiment analysis. In: Proceedings of the 18th Brazilian Symposium on Multimedia and the Web. WebMedia '12, pp. 181–188. ACM, New York (2012)

146. Zhou, D., Manavoglu, E., Li, J., Giles, C.L., Zha, H.: Probabilistic models for discovering e-communities. In: Proceedings of the 15th International Conference on World Wide Web. WWW '06, pp. 173–182. ACM, New York (2006)

147. Nguyen, C., Mamitsuka, H.: Kernels for link prediction with latent feature models. In: Gunopulos, D., Hofmann, T., Malerba, D., Vazirgiannis, M. (eds.) Machine Learning and Knowledge Discovery in Databases. Volume 6912 of Lecture Notes in Computer Science, pp. 517–532. Springer, Berlin/Heidelberg (2011)

148. Zhou, B., Pei, J., Luk, W.: A brief survey on anonymization techniques for privacy preserving publishing of social network data. SIGKDD Explor. Newsl. **10**(2), 12–22 (2008)

149. Bakshy, E., Hofman, J.M., Mason, W.A., Watts, D.J.: Everyone's an influencer: quantifying influence on twitter. In: Proceedings of the Fourth ACM International Conference on Web Search and Data Mining. WSDM '11, pp. 65–74. ACM, New York (2011)

150. Chen, W., Wang, C., Wang, Y.: Scalable influence maximization for prevalent viral marketing in large-scale social networks. In: Proceedings of the 16th ACM SIGKDD International Conference on Knowledge Discovery and Data Mining. KDD '10, pp 1029–1038. ACM, New York (2010)

151. Kempe, D., Kleinberg, J., Tardos, E.: Maximizing the spread of influence through a social network. In: Proceedings of the Ninth ACM SIGKDD International Conference on Knowledge Discovery and Data Mining. KDD '03, pp. 137–146. ACM, New York (2003)

152. National Research Council (US). Committee on Technical and Privacy Dimensions of Information for Terrorism Prevention and Other National Goals: Protecting Individual Privacy in the Struggle Against Terrorists: A Framework for Program Assessment. National Academies Press (2008)

153. Adali, S., Sisenda, F., Magdon-Ismail, M.: Actions speak as loud as words: predicting relationships from social behavior data. In: Proceedings of the 21st International Conference on World Wide Web. WWW '12, pp. 689–698. ACM, New York (2012)

154. McPherson, M., Smith-Lovin, L., Cook, J.: Birds of a feather: Homophily in social networks. Annu. Rev. Sociol. **27**, 415–444 (2001)

155. Macskassy, S., Provost, F.: Classification in networked data: A toolkit and a univariate case study. J. Mach. Learn. Res. **8**, 935–983 (2007)

156. Mislove, A., Viswanath, B., Gummadi, K.P., Druschel, P.: You are who you know: inferring user profiles in online social networks. In: Proceedings of the Third ACM International Conference on Web Search and Data Mining. WSDM '10, pp. 251–260. ACM, New York (2010)

157. Heatherly, R., Kantarcioglu, M., Thuraisingham, B.: Preventing private information inference attacks on social networks. IEEE Trans. Knowl. Data Eng. **PP**(99), 1 (2012)

158. Baatarjav, E.A., Phithakkitnukoon, S., Dantu, R.: Group recommendation system for facebook. In: Meersman, R., Tari, Z., Herrero, P. (eds.) On the Move to Meaningful Internet Systems: OTM 2008 Workshops. Volume 5333 of Lecture Notes in Computer Science, pp. 211–219. Springer, Berlin/Heidelberg (2008)

159. Zheleva, E., Getoor, L.: To join or not to join: the illusion of privacy in social networks with mixed public and private user profiles. In: Proceedings of the 18th International Conference on World Wide Web. WWW '09, pp. 531–540. ACM, New York (2009)

160. Chaabane, A., Acs, G., Kaafar, M.A.: You are what you like! Information leakage through users' interests. In: Proc. Annual Network and Distributed System Security Symposium (NDSS) (2012)

161. Li, N., Li, T., Venkatasubramanian, S.: t-closeness: Privacy beyond k-anonymity and l-diversity. In: IEEE 23rd International Conference on Data Engineering, 2007. ICDE 2007, pp. 106–115 (April 2007)

162. Machanavajjhala, A., Korolova, A., Sarma, A.D.: Personalized social recommendations: accurate or private. Proc. VLDB Endow. **4**(7), 440–450 (2011)

163. Ying, X., Wu, X.: On link privacy in randomizing social networks. Knowl. Inform. Syst. **28**, 645–663 (2011)

164. Heatherly, R., Kantarcioglu, M., Thuraisingham, B.: Social network classification incorporating link type values. In: IEEE International Conference on Intelligence and Security Informatics, 2009. ISI '09, pp. 19–24 (June 2009)

165. Donath, J., Boyd, D.: Public displays of connection. BT Tech. J. **22**(4), 71–82 (2004)

166. Gayo-Avello, D.: All liaisons are dangerous when all your friends are known to us. In: Proceedings of the 22nd ACM Conference on Hypertext and Hypermedia. HT 2011, pp. 171–180 (2011)
167. Kubica, J., Moore, A., Schneider, J., Yang, Y.: Stochastic link and group detection. In: Proceedings of the National Conference on Artificial Intelligence, pp. 798–804 (2002)
168. Zheleva, E., Sharara, H., Getoor, L.: Co-evolution of social and affiliation networks. In: Proceedings of the 15th ACM SIGKDD International Conference on Knowledge Discovery and Data Mining. KDD '09, pp. 1007–1016. ACM, New York (2009)
169. Backstrom, L., Leskovec, J.: Supervised random walks: predicting and recommending links in social networks. In: Proceedings of the fourth ACM International Conference on Web Search and Data Mining. WSDM '11, pp. 635–644. ACM, New York (2011)
170. Hogg, T., Wilkinson, D., Szabo, G., Brzozowski, M.: Multiple relationship types in online communities and social networks. In: Proceedings of the AAAI Symposium on Social Information Processing, pp. 30–35. AAAI (2008)
171. Hart, J., Ridley, C., Taher, F., Sas, C., Dix, A.: Exploring the facebook experience: a new approach to usability. In: Proceedings of the 5th Nordic Conference on Human-Computer Interaction: Building Bridges. NordiCHI '08, pp. 471–474. ACM, New York (2008)
172. Krishnamurthy, B., Wills, C.E.: Characterizing privacy in online social networks. In: Proceedings of the First Workshop on Online Social Networks. WOSN '08, pp. 37–42. ACM, New York (2008)
173. Thelwall, M.: Social network sites: Users and uses. Adv. Comput. **76**, 19–73 (2009)
174. Clemons, E.: The complex problem of monetizing virtual electronic social networks. Decis. Support Syst. **48**(1), 46–56 (2009)
175. Boongoen, T., Shen, Q., Price, C.: Disclosing false identity through hybrid link analysis. Artif. Intell. Law **18**(1), 77–102 (2010)
176. Enamul Kabir, M., Wang, H., Bertino, E.: A conditional purpose-based access control model with dynamic roles. Expert Syst. Appl. **38**(3), 1482–1489 (2011)
177. McNealy, J.: The privacy implications of digital preservation: social media archives and the social networks theory of privacy. Elon Law Rev. **3**, 133 (2011)
178. Fisher, B., Turner, K., Morling, P.: Defining and classifying ecosystem services for decision making. Ecol. Econ. **68**(3), 643–653 (2009)

Chapter 2
Online Social Networks: Privacy Threats and Defenses

Shah Mahmood

Abstract With over 1 billion users connected through online social networks, user privacy is becoming ever more important and is widely discussed in the media and researched in academia. In this chapter we provide a brief overview of some threats to users' privacy. We classify these threats as: users' limitations, design flaws and limitations, implicit flows of information, and clash of incentives. We also discuss two defense mechanisms which deploy usable privacy through a visual and interactive flow of information and a rational privacy vulnerability scanner.

2.1 Introduction

The level of human connectivity has reached unprecedented levels with over 1 billion people using one or more online social networks including Facebook, Twitter, YouTube, and Google+.

The immense amount of data provided and shared on these social networks may include the following information about a user: date of birth, gender, sexual orientation, current address, hometown, email addresses, phone numbers, web sites, instant messenger usernames, activities, interests, favorite sports, favorite teams, favorite athletes, favorite music, television shows, games, languages, his religious views, political views, inspirations, favorite quotations, employment history, education history, relationship status, family members, and software applications. The user also provides updates in the form of status messages or Tweets, which could include: a thought, an act, a link they want to share, or a video. All these information reveal a lot about the user, which will be of interest to various groups including governments, advertisers, and criminals.

S. Mahmood (✉)
Department of Computer Science, University College London, London, UK
e-mail: shah.mahmood@cs.ucl.ac.uk

R. Chbeir and B. Al Bouna (eds.), *Security and Privacy Preserving in Social Networks*,
Lecture Notes in Social Networks, DOI 10.1007/978-3-7091-0894-9_2,
© Springer-Verlag Wien 2013

Employers have used these social networks to hire or fire employees on the basis of their behavior on social networks [33]. Universities can screen applicants or ensure discipline by monitoring their students on social networks [7]. According to a survey by Social Media Examiner, 92 % of marketers use Facebook as a tool [40]. Phishers have improved their techniques by personalizing their schemes based on the data they acquire from social networks and these have been shown to be more useful than traditional phishing schemes [21, 36]. A woman in Indiana (US) was robbed by a social network friend after she posted on her Facebook profile that she was going out for the night [32].

Social networks, due to many such unfavorable incidents, have been criticized for breaching the privacy of their users. Both in academia and in the media, the importance of a user's privacy has been repeatedly discussed. In addition to some proposed technical solutions, there have been a vast number of initiatives to educate users so that they do not provide an excessive amount of personal information online. Facebook privacy consciousness is displayed in this reply by President Obama when a school student who wanted to become the President of the United States asked him for advice. He replied [34],

> Be careful about what you post on Facebook, because in the YouTube age, whatever you do will be pulled up again later somewhere in your life ...

Nonetheless, despite increased user awareness, we still come across numerous stories in which a user's privacy has been breached, with unfortunate consequences.

Before further classifying the threat and its solutions, we think it is essential to define privacy.

2.2 Definitions of Privacy

There is no single agreed definition of privacy in academia or in government circles. Over the course of time several definitions have been proposed. In this section we look into some of those definitions.

One of the first definitions of privacy, by Aristotle, makes a distinction between political activity as public and family as private [37].

Implicit here are boundaries that might be suggested by the walls of a family house, an assumption which is made explicit, though also modified, in a far more recent definition, that of Associate Justice John Paul Steven of the US Supreme Court. For Steven [38]:

> The 4th Amendment protects the individual's privacy in a variety of settings. In none is the zone of privacy more clearly defined than when bounded by the unambiguous physical dimensions of an individual's home - a zone that finds its roots in clear and specific constitutional terms: the right of the people to be secure in their ... houses ... shall not be violated.

Here, the home is not the exclusive locus of privacy, but is, rather, the informing image or motif in light of which privacy in other contexts may be construed.

This is an interesting definition. The Internet has managed to blur the boundaries that would have been suggested by the walls of a house. Is the right of security in one's house also assured when users are connected through the Internet? Would this extend to usage outside the home, such as in a car?

A definition of privacy less beholden to images of hearth and home and perhaps clearer in consequence is that adopted by the Calcutt Committee, headed by Sir David Calcutt, in 1990. The committee defined privacy as:

> The right of the individual to be protected against intrusion into his personal life or affairs, or those of his family, by direct physical means or by publication of information.

However, privacy on the Internet is a more complex affair than physical metaphors of intrusion and exposure can capture alone. Defense against publication of private information can protect the exposure of that information, but what if it is used, rather to produce targeted advertisements, with no publication?

William Parent provides a definition of privacy which does not rest on an implicit physical dimension, as follows [35]:

> Privacy is the condition of not having undocumented personal knowledge about one possessed by others. A person's privacy is diminished exactly to the degree that others possess this kind of knowledge about him.

This definition rests on the notion of "informed consent" as defined by Aristotle [37]. On this view, for any information to be held about another, there must be documentary proof of consent. An idea of privacy breach understood in these terms thus remains very valid in the era of cloud computing. But the importance of issues of control and freedom from interference of all kinds is suggested by definition of privacy that is most widely used in the computer science community, derived from Samuel Warren's and Louis Brandies' 1890 paper [42], "The Right to Privacy," in which they refer to Judge Cooley summarizing it as consisting of the right "to be let alone." The current formulation casts this as a right to:

> Control over information about oneself.

It is in this tradition of thought that Alan Westin defined privacy as an individual right, held by all people, [45]:

> To control, edit, manage, and delete information about them[selves] and decide when, how, and to what extent information is communicated to others.

In the light of the concerns of this thesis a survey by Alessandro Acquisti and Jens Grossklags is salient. More than 90 % of users agreed with this definition of privacy as ownership and control of their personal information. What is more, they found that while 61.2 % of users defined privacy as a matter of personal dignity, a significant minority, 26.4 %, nonetheless thought of it as the "ability to assign monetary value to each flow of personal information" [1], suggesting personal autonomy and control, rather than inviolability, is at issue.

Fig. 2.1 Classification of the causes of uses' privacy leaks

2.3 Privacy Threats

In this section we discuss four causes of privacy leaks along with real online social network examples, where possible. These four causes include: user limitations, design flaws or limitations, implicit flows of information, and clash of interests, as shown in Fig. 2.1. Let us discuss each of them in detail now.

2.3.1 Users' Limitations

Users of online social networks are (mostly) human beings. Humans have inherent limitations and flaws. It is due to these flaws that Alexander, in Malaysia, shared his atheistic views on Facebook resulting in him being sent to jail. He acted in accord with his free will, in spite of awareness of the possible risk. Similarly, Congressman Wiener shared his inappropriate pictures with a female correspondent online, resulting in a controversy leading to his resignation [3]. On many occasions, we share a lot of content on social networks without thinking about the short-term or long-term consequences of such information flow.

Human rationality in decision making is limited by the amount of time we have to make a decision; the amount of data involved; the arrangement of this data; and other cognitive limitations of the mind [4]. Two reasons for bad privacy decisions by human beings are: (1) bounded rationality and (2) limited working memory. Human decision makers should not be expected to make optimal decisions in complex scenarios, in a limited time. Similarly, when a user logs onto a social network, they are not expected to spend hours reading the privacy policies, understand the technical jargon, be aware of all risks and latest attacks, etc. With the lack of all this required information and the limited decision time, are we doing justice to the user by expecting them to make the best decision?

Human working or operant memory, the part of memory which is available for use at a particular instant, is limited [2]. A user's memory cannot hold too much information at the same time. So, whatever decision is made, it is made on the basis of whatever is present in the working part of the memory. When users log into an online social network or provide information on any other site, they are psychologically distracted in several different ways. These distractions make the users forget about their privacy or cause them to make their privacy decision on the basis of very limited information. Spiekermann et al. [39] conducted an e-privacy experiment which suggested that people appreciate highly communicative environments and thus forget about their privacy concerns. One explanation of this could be that within this communicative environment, the working memory of the user is flooded with marketing information. Privacy consciousness is removed from operational memory, and the user ends up providing sensitive information without noticing it.

Humans are considered to be inherently wired to trust other human beings. This trusting factor enables an adversary to engage in social engineering attacks. Social engineering is the art of manipulating another person. Here, a victim allows the attacker into his sphere of trust and provides him with information or access that the attacker would not have otherwise been authorized to get. It is due to human trust and social engineering attacks that users have been found to add large numbers of strangers to their Facebook accounts, enabling these strangers to view their personal content, intended to be shared with friends only.

Boshmaf et al. [9] launched a social bot attack against users on Facebook. They created social bots[1] for this purpose, where each social bot sent 25 friend requests to randomly selected users per day. They limited the per day requests to avoid the social bots being detected by Facebook's Immune System (FIS) which protects Facebook from threats and prevents malicious activity in real time. In the first phase, they sent 5,053 friend requests over 2 days. 2,391 of the requests were sent from male social bots and 2,661 from female social bots. Over a period of 6 days, 976 of their requests were accepted, with an acceptance rate of 19.3 %. For the male social bots the acceptance rate was 15.3 % and for the female social bots it was 22.3 %. Then over the next 6 weeks they requested another 3,517 users from the extended neighborhood of those who already accepted their requests; 2,079 of their requests were accepted, putting the average acceptance rate at 59.1 %. They noted that the acceptance rate increased with an increased number of mutual friends. In total they added 3,055 Facebook users, using 102 social bots, over a period of eight weeks, to demonstrate that 35.6 % of their friend requests were accepted. Social bots have been previously used by criminals and are available online for a few dollars.

Mahmood and Desmedt [30] introduced the targeted friend attack. For this attack they created a pseudonymous profile. During the first phase, which lasted for 99, days, they sent 255 friend requests out of which 90 were accepted. They did not allow the receipt of any friend requests during this time. From Day 100 to Day 285 they started accepting friend requests. They sent a total of 595 friend requests in total, out of which 370 were accepted, resulting in an acceptance rate of 62 %, as shown in Fig. 2.2. The interesting part of their experiment was that they received another 3,969 friend requests, which is 6.67 times more than the number of users they requested. The received friend request distribution is shown in Fig. 2.3. In total, their single pseudonymous profile had access to the private data of a 4,339 users.

---

[1] Social bots are bot nets on social networks.

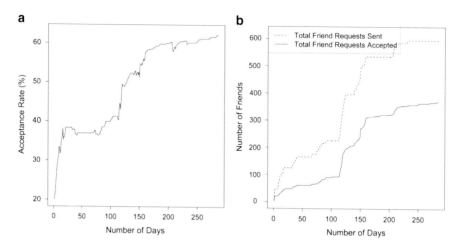

Fig. 2.2 (**a**) Probability of acceptance of our friend requests on Facebook, (**b**) Total number of friends requests sent and total accepted [30]

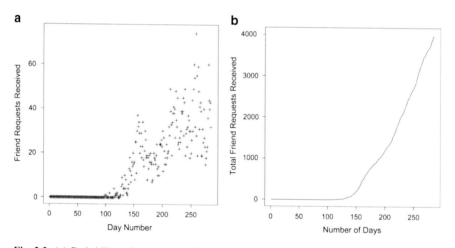

Fig. 2.3 (**a**) Probability of acceptance of our friend requests on Facebook, (**b**) Total number of friends requests sent and total accepted [30]

Similarly, an attacker could create profiles for a famous scientist. For example, someone has created a Facebook profile for Claude Shannon. It has got 180 friends.[2] Moreover, there are Facebook profiles made for animals.[3]

---

[2]http://www.facebook.com/claude.shannon—Checked on September 12, 2012, Shannon's profile may have been made by a fan to pay tribute but similar approaches can be used by attackers.

[3]The profile of a cat at the first author's previous residential hall has 170 friends. This could again be used to spy on users—http://www.facebook.com/CelebratingStarlight.

Human limitations in privacy decision making is not limited to social networks. Despite the awareness that loyalty cards by supermarkets can be used to track the behavioral attributes of a user and launch correlation attacks against them, users sign up for them even when the benefits are minimal. For example, the author signed up for a Tesco Club Card in the United Kingdom and regularly shopped at the store for his groceries for a period of thirteen months only to get £7 worth of cash coupons. In this little experiment the author was paid only £7 to allow the supermarket track his eating habits for a period of over 1 year. Users have also been found selling their DNA samples for a McDonald's Big Mac burger or writing their PINs on their credit/debit cards. In the latter case, when they lose the card, they have already attached the secret to it, making the job of the adversary much simpler.

Lack of privacy and over-sharing on social networks has also affected criminals. Grasso, a drug fugitive from Italy was caught in London after he posted photos about his life in London. He posted photos with the wax figures of President Obama and Prime Minister Cameron, taken at London's Madame Tussauds [13].

These are certainly not examples of rational decision making. However, they may be classified as rational within the inherent bounds and limits of humans at the instant of that decision.

2.3.2 Design Flaws and Limitations

Having looked at the limitations of users, let us now turn to the design flaws and limitations in some social networks. The design limitations and flaws include the weak privacy controls by social networks and the possibility of explicit attacks including cloning attacks.

Facebook Privacy Setting Evolution and Devolution

Facebook was initially launched as a student social network, with its approach to privacy being initially network-centric, which meant all data shared by users was visible to all the members of the network [48]. A network could have been based on an academic institution or a city. In the former case the profile of a user would be visible to all his fellow students within that institution, while in the latter case, his profile would be visible to everyone who selected that city as their city of residence. As the network grew and its users increased to millions, the privacy settings were changed several times until they reached their present form, in which by default different levels of user information are visible to "Friends," "Friends of Friends," and "Everyone." Today, Facebook does allow a user the option of sharing information only with himself through an "Only Me" option in the privacy settings. It also allows making an exception for specific groups of people, called "lists." These lists are very similar to Google+ circles [26]. It has been shown in the literature that users rarely change the default settings [8].

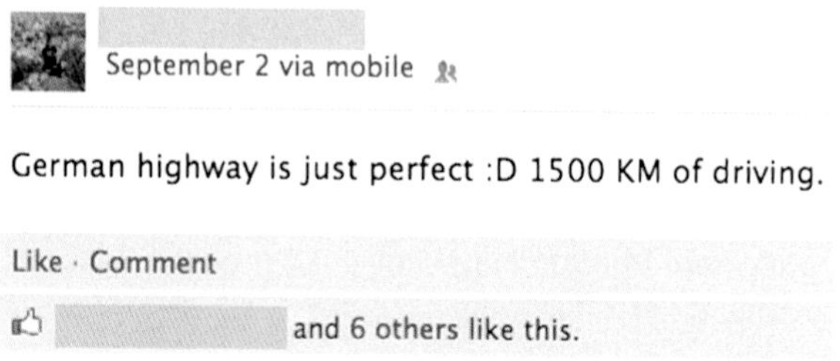

September 2 via mobile

German highway is just perfect :D 1500 KM of driving.

Like · Comment

and 6 others like this.

Fig. 2.4 Reconstructing friend-list on Facebook from wall posts

As Facebook's default settings evolved to a level where not all information was visible to a user's network, attackers and researchers tried to glean information from what was publicly available. A new area of social graph extraction, where by the communities and friends of users were extracted, emerged, e.g. [7,22,43,46]. Although the information this provides is limited when compared to all the information provided by users on their profile, it still reveals a lot. In 2009, Facebook showed a list of eight friends of a user when he was searched for either within the network or through an external search engine by someone who was not their friend. These eight friends were selected on the basis of some internal algorithm and would change every time the searchers' cache was reset. According to Facebook, revealing eight friends would help a searcher decide whether they have found the person they were looking for. If they were the person the searcher was looking for, the searcher could add them, and if they were not that person, then the searcher could look into the next user returned by the search. However, another consequence was that using this information, researchers were able to approximate the social graph of users on Facebook [8]. Facebook no longer displays the list of eight friends to a searcher.

For added privacy, Facebook users have the option to restrict who can view their friend-list, but, this does not mean a friend attacker[4] cannot reconstruct that user's friend-list [24]. For at least a partial reconstruction, a friend attacker can enumerate the names/user IDs of all the users who comment on posts visible to friends only. In Fig. 2.4, even though the user's friend-list is not visible to the author, we are able to find the names of at least seven friends of the victim. These friends liked the post. Any users who make comments on these posts will also be visible. By analyzing more posts, over a longer duration of time, an attacker can find the names and user IDs of more friends of the victim.

Similarly, when a user is tagged in a photo, we can see the name of the person who tagged the user by rolling the mouse over their name. It displays "Tagged by" and the tagger's name. As only a user's friends are allowed to tag them on Facebook, this also helps in reconstructing the friend-list. Moreover, Facebook does not allow users to hide their mutual friends. These can also be added to the reconstruction of

---

[4]A friend attacker is an attacker who is a friend on Facebook.

the victim's friend-list. This way the attacker can reconstruct a very significant part of a user's friend-list.

Timeline, a new virtual space in which all the content of Facebook users is organized and shown, was introduced on December 15, 2011 [19]. In addition to re-organization of users' content, Timeline comes with some default and unchangeable privacy settings. First, it is no longer possible for a Facebook user to hide their mutual friends, which was possible before Timeline. Second, it is not possible to limit the public view of "cover photos." These cover photos could be a user's personal pictures or political slogans and their widespread sharing may have various short-term and long-term consequences for that user. Third, with Timeline, depending on the user's privacy settings, if the likes and friend-list of a user are shared with a list of users, then that list of users can also see the month and the year when those friends were added or when the user liked those pages. This will allow an attacker to analyze the sentiments and opinions of a user, e.g. when did a user start liking more violent political figures and un-liking the non-violent ones. Finally, with Timeline, if a user makes a comment on a page or a group, he does not have the option to disable the probability of being traced back to the profile. Before Timeline, a user could make themselves searchable by a specific group (e.g., "Friends" or "Friends of friends") and even if they commented on pages and groups, people outside those allowed groups would not be able to link back to the commenters profile. Facebook can solve these problems by allowing users to change the settings to share their content with their desired audience.

Fake Accounts and Cloning Attacks

Social networks do not currently offer any certification or verification of the authenticity of user accounts. Thus, it is very easy for an attacker to register accounts in the name of someone else, although it is prohibited by the privacy policies of most service providers. The act of creating fake accounts is also known as a sybil attack.

An attacker can use personal information, e.g. pictures and videos of the victim, on the fake profile to win the trust of his friends and let them allow the fake account into their circle of trust. This way the attacker will have access to the information of the friends of his victim, which his friends have agreed to share with the victim and not necessarily the attacker.

The process of creating fake accounts, is called a "cloning attack," when the attacker clones (creates almost exact copies) of real social network accounts and then adds the same and/or other contacts as their victim. Bilge et al. [5] showed the ease of launching an automated identity theft attack against some popular social networks by sending friend requests to friends of hundreds of their cloned victims.

Permanent Takeover of a Facebook Account

Facebook allows a user to recover their compromised account using several verification mechanisms, but they all fail if the attacker disassociates the victim's current

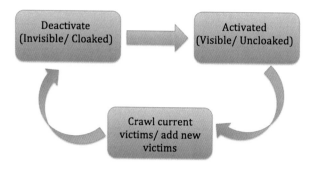

Fig. 2.5 Basic concept of the deactivated friend attack

login email address from the victim's account and associates it with a new dummy account [24]. The attacker can associate a new email address with the victim's original account and permanently take it over. As an example, let us suppose the victim is "Bob" and his login email address is "bob@victim.com." The attacker will associate the original account of Bob with a new email address "attacker@hacked.com" and associate a new account named "Dummy" with "bob@victim.com." Thus, when Bob recovers his account, he gets access to the Dummy account. Presently, there is no way to recover an account after such an attack.

Facebook should not allow the association of used email addresses with new accounts. This would prevent the "permanent takeover" attack.

Facebook's Deactivated Friend Attack

Deactivated friend attack[5] occurs when an attacker adds their victim on Facebook and then deactivates his own account. As deactivation is temporary in Facebook, the attacker can reactivate his account as he pleases and repeat the process of activating and deactivating for an unlimited number of times. While a friend is deactivated on Facebook, he becomes invisible. He cannot be unfriended (removed from a friend's list) or added to any specific list. The only privacy changes that may apply to him are those applied to *all* friends, or to the particular list of which he is already a member.

This deactivated friend, i.e. the attacker may later reactivate the account and crawl his victims profiles for any updated information. Once the crawling has finished, the attacker will deactivate again. While activated, the attacker is visible on the victim's friend list. The concept here is very similar to that of cloaking in Star Trek where Badass Blink or Jem'Hadar has to uncloak (be visible), even if only for a moment, to open fire. Figure 2.5 provides the abstract conceptual view of the attack. Facebook provides no notification of the activation or deactivation of friends to its users.

---

[5] Based on our paper [30].

The deactivated friend attack is only possible due to the presence of a cloaked channel, which is defined below.

Definition 1. *Cloaked Channel:* A channel is called a cloaked channel if, and only if, it is invisible when cloaked, and reappears when uncloaked [30].

This kind of attack is very serious for several reasons. First, it is very hard to detect this kind of attack. The attacker could activate his account at times when he is least likely to be detected and crawl his victims' profile for information, with which to update his records. Various groups of information aggregators could find this attractive as a permanent back door to the private information of Facebook users, including: marketers; background checking agencies; governments; hackers; spammers; stalkers; and criminals. Second, it continues to be useful even if a user becomes more security conscious. He may want to adjust his privacy settings but will not be able to affect his attackers access, unless he applies an update to all his friends. Third, by closely monitoring a few users on Facebook, an attacker can get a deeper insight into a large network. This possibility has been enhanced by Facebook's addition, last year, of a "browsing friendship" feature. This would help the attacker in analyzing the bond between two of his victims by browsing their friendship which provides information including: the month and year since which they have been Facebook friends; events they both attended; their mutual friends; things they both like; their photos; the messages they have written/write on each others' walls, etc. This would give a very deep insight into the level of their relationship, the intensity of their activity at a particular time, the degree of their interactivity, etc. This information could be used for several attacks including social engineering and social phishing attacks. This vulnerability was fixed by Facebook using one of our solutions [30]. Now the deactivated friends are visible to a user on their friend-list and a user can un-friend them or add them to another list, possibly with a limited view.

Google+ Photo Metadata

When a user uploads a photo on Google+, some metadata are made available to those with whom the photo is shared, including: the name of the photo owner; the date and time the photo was taken; the make and model of the camera, etc.[6] This set of information, in particular the date and time, may at first seem relatively innocent and trivial, but could in reality lead to some serious privacy concerns. On August 10, 2007, in Pennsylvania (USA), a divorce lawyer proved his client's spouse to have been unfaithful, when electronic toll records showed him in New Jersey (USA) on a particular night and not in a business meeting in Pennsylvania as he had claimed [16]. With the metadata revealed by Google+ a user might leak enough information to be legally held liable on a similar basis.

---

[6]Based on our paper [26].

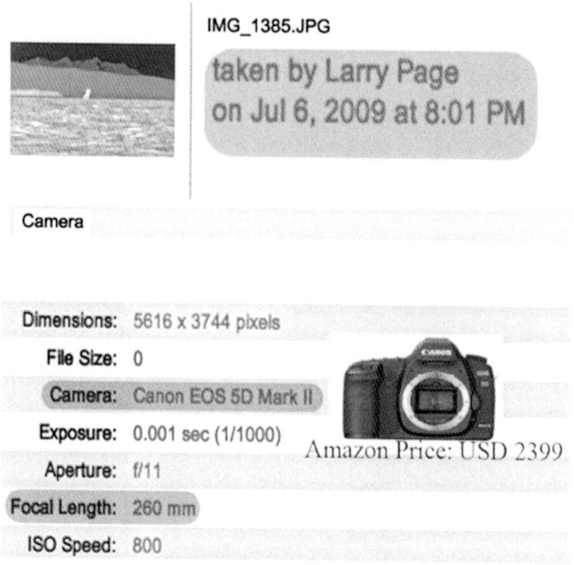

IMG_1385.JPG

taken by Larry Page
on Jul 6, 2009 at 8:01 PM

Camera

Dimensions:	5616 x 3744 pixels
File Size:	0
Camera:	Canon EOS 5D Mark II
Exposure:	0.001 sec (1/1000)
Aperture:	f/11
Focal Length:	260 mm
ISO Speed:	800

Amazon Price: USD 2399

Fig. 2.6 Metadata from a photo by Larry Page on Google+ [26]

Similarly, the make of the camera with which a photo was taken could be another concern for privacy. Higher end cameras cost thousands of dollars. There have been past incidents where victims were killed for their cameras. In May 2011, a Greek citizen, 44, was killed for his camera when taking his wife to hospital for the birth of their child [44].

Just to give an example of the level of information a picture exposes about the camera, look at the metadata of the publicly shared pictures (from his Google+ profile) of Google co-founder Larry Page, shown in Fig. 2.6.[7] It reveals that he used a Canon EOS 5D Mark II to shoot his vacation photographs. This camera is worth approximately USD 2400. This gives the robber incentives.

Zuckerberg's Photo Leak and Other Attacks on Facebook

In December, 2011, Facebook founder Mark Zuckerberg's photos were leaked by a relatively simple, presumably accidentally discovered, vulnerability in Facebook protocols [18]. The vulnerability was exploited when a user reported a victim's display photo as nude. Facebook responded by sending more photos of the victim, even from his private albums asking the reporting party, whether they were also nude. It took the leak of the personal photographs of its founder for Facebook, to fix

---

[7]Photo modified for didactic purposes.

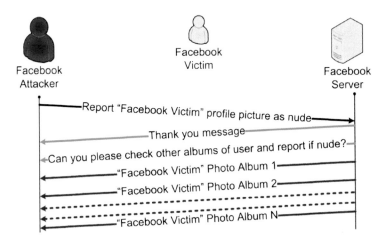

Fig. 2.7 Helping Facebook identify information leakage in their protocols [29]

the bug, which would have been far easier to identify and rectify if the protocol had been visualized as shown in Fig. 2.7. The vulnerability was fixed after the leak.

Similarly, Felt [20] presented a cross-site scripting vulnerability in the Facebook Markup Language which allowed arbitrary JavaScript to be added to the profiles of users of an application, which led to session hijacking. Facebook by default uses HTTP which helps the attacker with traffic analysis. The attacker does not have to decrypt the packet sniffed in transit. Dhingra and Bonneau independently provided limited hacks into Facebook photos [6, 15].

2.3.3   Implicit Flows of Information

Leakage of information on social networks does not always need to be explicit. There may be many more implicit flows of information, where one piece of information may leak information about another. For example, Dey et al. [14], in a study of 1.47 million user accounts on Facebook, found that only 1.5 % of users revealed their age. Using the high school graduation year and friend connections of these users, they were able to estimate the ages of 84 % of these users with a mean error of plus or minus 4 years.

Similarly, the "likes" of a user on Facebook or the videos that a user might watch on Facebook enable an attacker to deduce lots of implicit information about that user [10]. On Facebook we found 233 people liked Breivik's Wikipedia page,[8] and

---

[8]http://www.facebook.com/pages/Anders-B-Breivik/265349460157301 Accessed: July 13, 2012.

according to Facebook 13 people were talking about him when we checked that page. We also found 954 people liked Al-Awlaki's Wikipedia page[9] and according to Facebook 30 people were talking about him at that time [25]. These people may be much more likely to commit a violent act of terror than other Facebook users, on average. Moreover, "likes" of a particular brand of chocolate or medicine can also reveal a lot. For example, a user writing reviews about sweet chocolates is unlikely to be diabetic while someone who "likes" "Accu-Check," a brand of diabetes testing kit, is more likely to be either diabetic or closely involved with diabetic patients. Furthermore, a user who mostly Tweets in the evenings, and never in the morning, is likely to be either busy in the morning or sleeping. These are just a few examples. There is likely to be an enormous number of similar correlational flows of information. Unfortunately, this type of privacy leak is the hardest to defend against, because it adds an enormous number of not-clearly-known information flows to the threat model, which might otherwise be comprised of explicit flows of information to an adversary.

2.3.4   Clash of Interests

Most social networks, like other web services, are supported by the revenue generated from advertisements. This may create a conflict of interest on the part of the service provider when it comes to the rights of its users and the rights of advertisers. Without the advertiser's money, the service provider would not be able to function. Similarly, without a proper user base, it won't be able to generate lots of advertising money.

The users' interests and those of the advertisers do not always match. Users want their data to be inaccessible to anyone except the parties they have explicitly permitted under a well-documented, human-readable, and informed type of consent. Moreover, users want data to be used only for the exact purpose they have consented to. From a user's point of view, any deviation from these terms by the service provider is considered less than wholly honest at best, and actively dishonest at worst.

Advertisers, on the other hand, want to mine the maximum amount of data about a user over a long duration of time. They desire to keep the data for an indefinite time. More data will provide them with a competitive advantage and help advertisers in crafting better targeted advertisements. These well-crafted advertisements will increasingly influence users' choice of products resulting in making the advertisers happier. Thus, the advertisement networks will be able to

---

[9] http://www.facebook.com/pages/Anwar-al-Awlaki/102248169830078?rf=134722733227304 Accessed: July 13, 2012.

charge their customers higher fees in the future. Zuckerberg confirmed the threat from advertising companies during an interview by saying [31]:

I think that these companies with those big ad networks are basically getting away with collecting huge amounts of information …

Another issue of interest concerning advertisers is the use of behaviorally targeted advertising [47]. One may argue that personalization is useful for a person who is too busy to search through normal banner or textual ads, but it can have some seriously concerning implications. Suppose a neutral user stumbles across a right-wing article and opens it. Then, the next day the search engine recommends to her two such articles, and being curious the user opens them. After a few days all her search results may be those meant for extremely radicalized people, with advertisements pointing her to products that may potentially help her act on the radical views introduced to her by behavioral targeting. We wonder how many people have already been radicalized as a side effect of behavioral targeting, as the process is still ignored due to the monetary gains it makes for large corporations.

Considering the discussion above, it is hard to trust service providers when there is a conflict of interests. Computer scientist Lanier sums this up by saying:

Facebook says, 'Privacy is theft,' because they're selling your lack of privacy to the advertisers who might show up one day.

Social networks like Facebook charge 30 % of a third party application's revenue, for using its ecosystem [17]. These applications have their own privacy practices, which are not necessarily in line with those of Facebook. Facebook in their "terms of use" agree that they have no control over the data that is being stored by third party applications. Users have not necessarily given informed consent to such practices. Facebook will not risk losing some of the famous gaming applications, which provide a large chunk of Facebook's revenue, by asking them for better privacy practices. Such application providers may then further sell the data to potentially malicious parties without any control by the user, Facebook, or the application provider.

In a very similar way to advertisers, governments want more private information about people in order to gauge their sentiments and opinions. Facebook has recently announced that it will help advertisers by letting them target their adverts to users on the basis of their phone numbers. Advertisers using this mechanism can direct advertisements to customers who have previously used a service and provided that service with their phone numbers. Service providers, in several countries, are bound to provide governments with users' personal information, and in some cases are legally prohibited from informing users about such sharing of information.

This creates a situation where service providers either are forced by law to monitor users' actions on social networks or are economically incentivized to collect and store more information about users for advertisement and a government's profiling purposes.

2.4   Defense Mechanism

Anonymity, pseudonymity, and unlinkability [11,12] have been proposed as some of the well-researched solutions for privacy protection; yet, there are dozens of attacks proposed against some of the best systems being researched and used. Another approach proposed and adopted in some quarters is to ensure users' privacy through regulations. Unfortunately the time taken for regulations to be drafted and become effective may exceed the valuable life of information [39]. There is limited research on how to quantitatively analyze one's vulnerability to different attacks. Moreover, there is the problem of what happens if a particular set of information about a user is leaked. A user cannot realistically be expected to be careful about their privacy protection at all times, considering a user's inherent limitations, including bounded rationality and limited working memory. If we cannot realistically expect a user to protect their privacy at all times, is there another mechanism that might provide users with more rational options in a limited time? And is there a model which could help a user identify whether it is possible for her to annul the effect of at least some of the leakage of information, so as to substantially minimize any adverse effects?

In this section we discuss two such proposed tools for users' privacy protection.

2.4.1   Usable Privacy Through Visual and Interactive Flow of Information

The world of computer technology has evolved from stroking keys to point and click, and recently to touch and tap. Users need a more graphical and interactive view of developments in respect to privacy, so that they may better understand the currently obscure world of communications, up and down the stack.

Facebook, in January 2011, offered users an opt-in to secure browsing. Prior research has shown that users are reluctant to make changes to the default settings [23]. Normally it is too cumbersome for the user to search into all the options and visualize the impact of switching between them. It will be much easier for the user if they are provided with a view as shown in Fig. 2.8. With this user-friendly view, users will be encouraged to change from their current settings to another without making too much extra effort.[10]

Facebook, by default, emails notifications about almost any activity relevant to the user including any friend requests received, friends commenting on the users' wall, any photos in which the user has been tagged, any new message for the user on Facebook, etc. These email notifications contain links to the activity in question. Phishers also send users emails with links seemingly coming from legitimate

---

[10]This section is based on our paper [29].

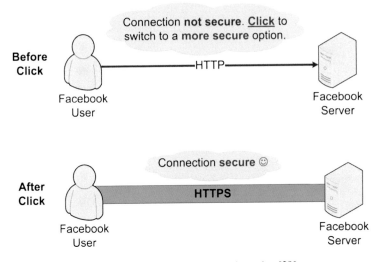

Fig. 2.8 Easier, graphical and interactive switch to secure browsing [29]

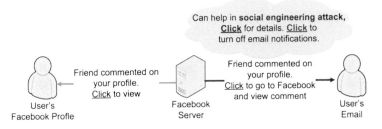

Fig. 2.9 Identify possible flows of information that may aid in social engineering attacks [29]

sources, but actually directing the user to the attackers' pages. These attacker pages steal their victims' credentials or launch a man-in-the-middle attack. When users regularly receive similar emails from Facebook it is easier for them to fall victim to phishers claiming to be Facebook. It is possible to turn off these email notification and thus reduce the risk of phishing attacks, but currently the user needs to search into several levels of options by first clicking the downward pointing arrow in the top right corner of the Facebook page, then clicking on "Account Settings" from the pull down menu, followed by clicking the third option in the left side labelled "Notifications," then clicking on "Facebook" under the "All Notifications" in the center of the page and, finally unchecking the desired notifications. It would be much more user-friendly if the flow were represented as shown in Fig. 2.9, and if deactivation were possible with a click on the graphic.

When Google+ was launched, one aspect that was widely appreciated, and marketed as a privacy feature, was the concept of circles. A user could divide his social network contacts into different circles and share content on a circle

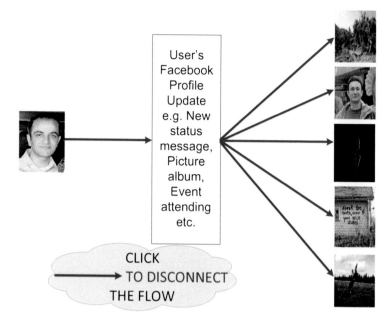

Fig. 2.10 Visualizing Facebook updates and providing users with interactive control [29]

by circle basis. Though Facebook termed all the user's social network contacts as "Friends," there was a possibility of classifying users into lists [26]. Google+ circles were appreciated more than Facebook lists (which were widely unknown to users) due to the graphical and interactive un-usability of the latter. In the real world, relationships with people change with time, i.e., some people get closer and others get emotionally distant. It is hard for users to keep on reflecting such relationship changes in our social network presence. A better way would be to visualize information flow on the basis of each update, as shown in Fig. 2.10. The picture on the left is that of the user who is updating his Facebook profile. The information is flowing, shown by the directed arrows, to the list of those friends of the user shown on the right, with whom the information is currently being shared, each represented by a thumbnail picture. Since some of the pictures may have been uploaded recently and not be readily recognizable by the user, the user can move the pointer over any of the arrows or pictures to get additional information, such as their name. Clicking on any of the arrows will disconnect the flow of that particular information to the specific person and save the user from regretting the sharing of private and embarrassing information.

These are only a few examples where the graphical and interactive view of information flow can be considered as a more usable means of ensuring users' privacy. In practice, the visual and interactive view may identify many currently obscure sources of leaks of information.

Advantages of Graphical and Interactive View of Information Flow

The use of graphical and interactive view of information has several advantages including the following:

Countering Limitations of the Working Memory. When a user logs into a social network or provides information on any other website, he is psychologically distracted in many ways. These distractions make users less conscious about privacy. With a graphical flow of information available, a user can be reminded of what information is being shared and with whom. This brings a concern about privacy back into the working memory.

Countering Bounded Rationality Problem. Human rationality in decision making is limited by the amount of time there is to make a decision, the amount of data available, the arrangement of this data, and the cognitive limitations of the mind. Decision makers should not be expected to make optimal decisions in complex scenarios, in limited time. When a user visits a web site, he should not have to spend hours reading the privacy policy, understanding the technical jargon, getting aware of all the parties with whom the information will be shared, etc. A more graphical representation of his information flow can put him in a better position for quick decisions.

As a Protocol Verification Technique. The graphical and interactive view of information flow can help privacy experts, allowing service providers testing a unit to identify leaks of information.

Transparency and User's Increased Trust in the System. Using graphical representation of information flow, users will feel more in command of their private data. This will build their trust and confidence in the system giving them better control when sharing information.

2.4.2 Rational Privacy Vulnerability Scanner

Users can be provided with a rational privacy vulnerability scanner, which is based on stochastic almost combinatorial games. Stochastic almost combinatorial games are stochastic games in a combinatorial setup. Here a user is shown the threat level to his privacy by assigning cost and probability to the possibility of an attacker accessing a user's information that they may share with a limited set of users on a social network. As a motivating example, let us suppose an attacker wants to make a hotel reservation with the credit card details of a victim. There are normally three units of information required for this attack to be successful, namely, the name of the cardholder (1), the address to which it is registered (2), and the card number (3) (occasionally the card expiry date and the three digit security code may be required, but these are not considered in this example). The attacker may acquire this information in various orders and "units".

One possible method for attacker is to buy all three units of information from an underground market [41] for a certain cost, $c_{\{\}->\{123\}}$ (cost of finding all three units of information from the initial state). This case is shown in Fig. 2.11j. The details bought may not always be valid, and thus there is a probability of success attached, represented by $p_{\{\}->\{123\}}$. The costs are incurred even if the details are invalid, as shown by the loop back arrow. The probability of the information being incorrect would be $1 - p_{\{\}->\{123\}}$. In this case, this is because of the illegal nature of underground markets, but even information trading companies may not guarantee accurate and up-to-date information.

Another possibility is that the attacker finds a card, with name and number, at an ATM machine. The cost $c_{\{\}->\{13\}}$ here is zero and the probability that the card is not yet cancelled is $p_{\{\}->\{13\}}$. This case of getting all 3 units of information in one go is shown in Fig. 2.11h. The attacker now needs to find the correct address, using the information she has. The address from a name could be found using social networks like Facebook, people-search services like www.123people.com or by other means. This information again will have a cost (time is money) and have a certain probability of being valid. There are several other possible combinations, as shown by the rest of the cases in Fig. 2.11. The attacker needs to find the most feasible path, where the determination of feasibility depends on the criteria of the attacker.

It is also possible that the victim comes to know about the compromise of her information. Suppose the victim knows that all three units of her information are compromised. To render an attack useless, the victim will need to annul only one bit of information, although changing name or address might cost more than ordering a new card with a new number. So the victim can push the attacker from state $s_{\{1,2,3\}}$ to a state $s_{\{1,2\}}$, thus diminishing the probability of an attack.

In this example, there were only three relevant units of information, and we were able to find the most practical transitions without any external aid. But when the number of relevant information units increases, the number of possibilities of transitions between the states also increases, creating the need for a framework to aid the user in such decisions.

The graph presented from the viewpoint of the attacker is called the *attack graph* while the graph presented from the viewpoint of the defender is called the *defense graph*. For three units of information Fig. 2.12 shows an attack graph and Fig. 2.13 shows a defense graph, with edges labelled with cost and probability of success. Both the views combined will create provide us with a stochastic almost combinatorial game scenario.

We envision a tool, which provides users with an estimated cost that an attacker will have to incur in order to acquire his information and launch an attack if he uploads that information on social networks. The tool can use the economic models in [28].

The aims of the scanner can be to mine all possible units of relevant information, gather it in one place, create links between the different units of information and provide a threat report. Using the provided data it can show the cost and success probabilities of any possible remedies, in case of a privacy breach. The tool can be

Fig. 2.11 All possible ways
to find 3 units of data

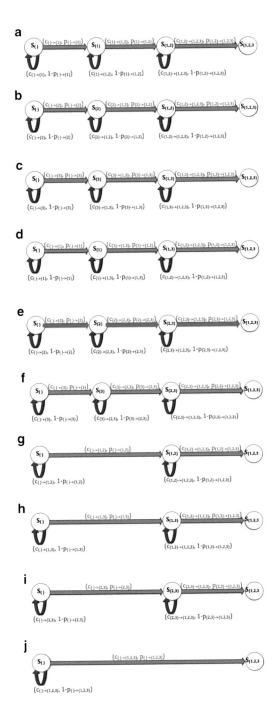

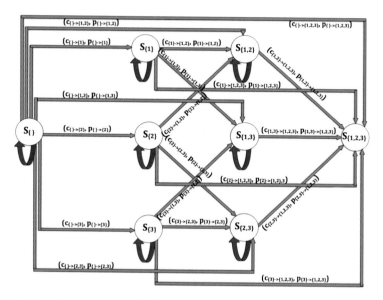

Fig. 2.12 Attack Graph for 3 units of information

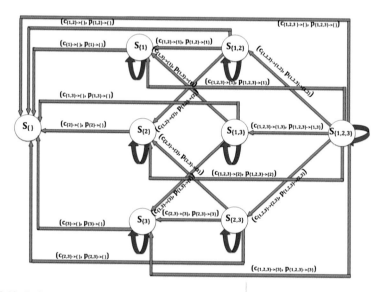

Fig. 2.13 Defense Graph for 3 units of information

used as a guide on what information to avoid giving in cases where provision of some information is optional. This will afford a ready answer to the question: How much do we lose by providing such non-compulsory units of information?

The dichotomy between privacy valuation and experimental observation is widely noted in literature. This dichotomy is observed under constraints of limited working memory and the bounded rationality of users. An interesting direction to take will be to use the framework outlined here and repeat some of the tests performed by Acquisti and Grossklags in [1]. Our hypothesis is that with the increased working memory and extended bounds of rationality due to the use of these tools, a user's dichotomous behavior may be reduced.

2.5 Summary

In a nutshell, in this chapter we discussed four different causes of privacy leaks in online social networks. These four causes include: users' limitations; design flaws and limitations; implicit flow of information; and clash of interests. Then we discussed two defense mechanisms involving the visual and interactive control of flow information, and economic modeling of the privacy threat, thus providing users with the means to make a more rational choice. Interested readers may follow the given references for a detailed review of the provided threats and defenses.

Acknowledgements The author would like to thank Professor Yvo Desmedt with whom he co-authored some of the work cited in this chapter. The author would also like to thank University College London for providing him financial support through the University College London PhD Studentship Program. This chapter is based on our work in [24–30].

References

1. Acquisti, A., Grossklags, J.: Uncertainty, ambiguity and privacy. In: WEIS, 2005
2. Baddeley, A.: Working memory. Science **255**(31), 556–559 (1992)
3. Barret, D., Saul, M.H.: "weiner now says he sent photos". Wall St. J. (2011)
4. Berger, P.L.: Models of Bounded Rationality, Vol. I–III. MIT Press, Cambridge, MA (1982)
5. Bilge, L., Strufe, T., Balzarotti, D., Kirda, E.: All your contacts are belong to us: automated identity theft attacks on social networks. In: WWW, pp. 551–560, 2009
6. Bonneau, J.: New facebook photo hacks. http://www.lightbluetouchpaper.org/2009/02/11/new-facebook-photo-hacks/, (2009). Accessed 15 July 2011
7. Bonneau, J., Anderson, J., Danezis, G.: Prying data out of a social network. In: ASONAM, pp. 249–254, 2009
8. Bonneau, J., Anderson, J., Stajano, F., Anderson, R.: Eight friends are enough: Social graph approximation via public listings. In: SNS, 2009
9. Boshmaf, Y., Muslukhov, I., Beznosov, K., Ripeanu, M.: The socialbot network: When bots socialize for fame and money. ACSAC, Sept 2011
10. Chaabane, A., Acs, G., Kaafar, M.: You are what you like! information leakage through users' interests. In: Proc. Annual Network and Distributed System Security Symposium, 2012
11. Chaum, D.: Untraceable electronic mail, return addresses, and digital pseudonyms. Comm. ACM **24**(2), 84–88 (1981)
12. Chaum, D.: Blind signatures for untraceable payments. In: CRYPTO, pp. 199–203, 1982
13. Cooper, B.: Italian drugs fugitive jailed after posting pictures of himself with Barack Obama waxwork in London on Facebook. Mail Online February 14, 2012
14. Dey, R., Tang, C., Ross, K.W., Saxena, N.: Estimating age privacy leakage in online social networks. In: INFOCOM, pp. 2836–2840, 2012

15. Dhingra, A.: Where you did sleep last night? ... thank you, i already know! iSChannel **3**(1) (2008)
16. Donald, A.M., Cranor, L.F.: How technology drives vehicular privacy. J. Law Pol. Inform. Soc. **2**, (2006)
17. Ebersman, D.A.: Facebook Inc., Form S-1 registration statement. United States Securites and Exchange Commission, February 1, 2012
18. Facebook bug sees Zuckerberg pictures posted online. BBC, December 7, 2011
19. Facebook Timeline: http://www.facebook.com/about/timeline. Accessed 16 May 2012
20. Felt, A.: Defacing Facebook: A security case study. 2007
21. Jagatic, T.N., Johnson, N.A., Jakobsson, M., Menczer, F.: Social phishing. Comm. ACM **50**(10), 94–100 (2007)
22. Lindamood, J., Heatherly, R., Kantarcioglu, M., Thuraisingham, B.M.: Inferring private information using social network data. In: WWW, pp. 1145–1146, 2009
23. Mackay, W.E.: Triggers and barriers to customizing software. In: CHI, pp. 153–160, 1991
24. Mahmood, S.: New privacy threats for Facebook and Twitter users. In: IEEE 3PGCIC, 2012
25. Mahmood, S.: Online social networks: The overt and covert communication channels for terrorists and beyond. In: IEEE HST, 2012
26. Mahmood, S., Desmedt, Y.: Poster: preliminary analysis of Google+'s privacy. In: ACM Conference on Computer and Communications Security, pp. 809–812, 2011
27. Mahmood, S., Desmedt, Y.: Online social networks, a criminals multipurpose toolbox (poster abstract). In: Balzarotti, D., Stolfo, S.J., Cova, M. (eds.) Research in Attacks, Intrusions, and Defenses, vol. 7462 of Lecture Notes in Computer Science, pp. 374–375. Springer, New York (2012)
28. Mahmood, S., Desmedt, Y.: Two new economic models for privacy. In: ACM SIGMETRICS/Performance Workshops, PER, 2012
29. Mahmood, S., Desmedt, Y.: Usable privacy by visual and interactive control of information flow. In: Twentieth International Security Protocols Workshop, 2012
30. Mahmood, S., Desmedt, Y.: Your Facebook deactivated friend or a cloaked spy. In: IEEE PerCom Workshops, pp. 367–373, 2012
31. MailOnline: Zuckerberg defends Facebook... by saying Microsoft, Google and Yahoo! are even worse at ignoring user privacy. Daily Mail, November 8, 2011
32. Henderson, M., de Zwart, M., Lindsay, D., Phillips, M.: Will u friend me? Legal risks of social networking sites. Monash University, 2011
33. Monkovic, T.: Eagles employee fired for Facebook post. New York Times, March 10, 2009
34. Obama advises caution in use of Facebook. Associated Press, September 8, 2009
35. Parent, W.: Privacy, morality and the law. Philos. Publ. Aff. **12**, 269–288 (1983)
36. Polakis, I., Kontaxis, G., Antonatos, S., Gessiou, E., Petsas, T., Markatos, E.P.: Using social networks to harvest email addresses. In: WPES, pp. 11–20, 2010
37. Privacy: Stanford Encyclopedia of Philosophy, 2002
38. Samaha, J.: Criminal Justice. Thomson Wadsworth, Belmont, CA (2006)
39. Spiekermann, S., Grossklags, J., Berendt, B.: E-privacy in 2nd generation e-commerce: privacy preferences versus actual behavior. In: ACM Conference on Electronic Commerce, pp. 38–47, 2001
40. Stelzner, M.: Social media marketing industry report. http://www.socialmediaexaminer.com/SocialMediaMarketingReport2011.pdf, 2011
41. The underground credit card blackmarket. http://www.stopthehacker.com/2010/03/03/the-underground-credit-card-blackmarket/, 2010
42. Warren, S.D., Brandeis, L.D.: The right to privacy. Harv. Law Rev. **4**(5), 193–220 (1890)
43. Wasserman, S., Faust, K.: Social Network Analysis. Cambridge University Press, Cambridge (1994)
44. Weeks, N.: Greek police detain 24 in athens immigrant clash after murder. http://www.bloomberg.com/news/2011-05-11/greek-police-detain-24-in-athens-immigrant-clash-after-murder.html, 2011
45. Westin, A., Blom-Cooper, L.: Privacy and Freedom. Bodley Head, London (1970)

46. Xu, W., Zhou, X., Li, L.: Inferring privacy information via social relations. In: International Conference on Data Engineering, 2008
47. Yan, J., Liu, N., Wang, G., Zhang, W., Jiang, Y., Chen, Z.: How much can behavioral targeting help online advertising? In: WWW, pp. 261–270, 2009
48. Yardi, S., Romero, D.M., Schoenebeck, G., Boyd, D.: Detecting spam in a Twitter network. First Monday 15(1) (2010)

Chapter 3
Online Social Networks and Terrorism: Threats and Defenses

Shah Mahmood

Abstract Over 1 billion active users of online social networks are evidence of the enormous growth of these technologies. Although the majority of online social network users use such services for ordinary social interactions, a very small number may possibly be misusing them for terrorism. In this chapter, we first provide the background, definition, and classification of terrorism. Second, we discuss how some terrorists may be using online social networks to: (1) recruit new members to a terrorist organization and maintain the loyalty of their existing sympathizers; (2) plan attacks and share information about them; (3) gather intelligence; (4) train recruits for specific attacks; (5) raise funds for their causes; (6) propagate fear amongst the enemy population; and (7) engage in counterintelligence to uncover undercover agents. Third, we discuss several mechanisms to detect terrorists using online social networks, including: (1) keyword-based flagging; (2) sentiment analysis; (3) honeypots; (4) social network analysis; (5) facial recognition; and (6) view escalation. We show that the keyword-only flagging mechanism used by US Department of Homeland Security to detect terrorists is potentially effective, but certainly produces a large number of false positives, making it possibly less efficient in practice. Finally, we propose the use of targeted advertisements to rehabilitate possible radicals using the online social networks.

Legal Notice: The content of this chapter are meant only to provide researchers and security personnel with information about threats of the use of online social networks for terrorism and how to counter those threats. It is in no way meant to help anyone do anything unethical or illegal.

S. Mahmood (✉)
Department of Computer Science, University College London, London, UK
e-mail: shah.mahmood@cs.ucl.ac.uk

R. Chbeir and B. Al Bouna (eds.), *Security and Privacy Preserving in Social Networks,*
Lecture Notes in Social Networks, DOI 10.1007/978-3-7091-0894-9_3,
© Springer-Verlag Wien 2013

3.1 Introduction

The popularity of online social networks has resulted in an unprecedented human connectivity where Facebook alone claims to have over 900 million active users per month. These users have more than 125 billion friend connections and on average upload 300 million photos per day. Every day around 3.2 billion comments and "likes" are also generated by these users [7]. Similarly, over 140 million users are exposing some of their thoughts as 340 million tweets every day on Twitter [29], 280,000 meetings of like minded people are arranged by 9 million users of Meetup [1], 4 billion videos are watched on YouTube on a monthly average [33], 80 million users are flicking through pictures uploaded by 51 million registered users of Flickr [8], around 15 million users have shared their 1.5 billion locations using Foursquare [10], over 90 million users can hang out on Google+ [4], and almost 175 million users are sharing their resumes and being connected to their professional contacts on LinkedIn [15]. The majority of this enormous amount of information is used for social interaction between ordinary citizens, but we cannot disregard the fact that a relatively small number of these information flows are used by those who plan to use violent means to promote their political ideology. Unfortunately, these small number of terrorists are capable of having an impact on billions of people in the world, as has been obvious from the 9/11 attacks where 19 hijackers shook the pillars of democracy, starting a new era of fear and unpredictability. Similarly, a lone attacker in Norway was enough to reveal the alarmingly "real" danger of militant neo-Nazi attacks. It is evident, from past incidents, that some of these terrorists have been active on online social networks. Anders Breivik, the Norwegian mass murderer, had a Facebook and Twitter account. Abu Jundal, the Indian terrorist, suspected to be linked to Pakistan's Lashker-e-Taiba was found to be using Facebook to recruit new members for his organization [27]. Although it is relatively easy to analyze the use of an online social network by a terrorist after the attack or after he is arrested, it is much more of a challenge to detect a terrorist from his activity on an online social network alone. The task of detecting terrorists from their activity on online social networks is challenging and complex because of the enormous number of implicit and explicit communication channels. On Facebook, terrorists can pass messages amongst them even if they are not connected through the 125 billion friend links. Similarly, they do not need to be fans of the same political pages or be subscribed to the same YouTube channel. They have got trillions of equally effective, implicit ways of communication through the online social networks.

In this chapter, our contributions are multifold. We first provide the definitions and classification of terrorism. Second, we discuss how some terrorists can take advantage of online social networks for: (1) recruitment and loyalty maintenance; (2) planning and information sharing; (3) intelligence gathering; (4) psychological warfare; (5) training; and (6) fund raising. We further discuss how online social networks can be used to identify undercover agents. As nearly 1 billion users are registered with online social networks, if an otherwise social undercover agent avoids using them, it could arouse the suspicion of an observing adversary.

Furthermore, due to improvement in face recognition technologies, it is easier to link the past lives of an undercover agent to their present ones, possibly putting them in harm's way. Third, we show six possible ways in which the use of online social networks by terrorists can be detected, along with their strengths and weaknesses. These include: (1) keyword-based flagging; (2) opinion mining and sentiment analysis; (3) social network analysis; (4) honeypots; (5) face recognition; and (6) view escalation. We also look into the lists of keywords and phrases used by the Department of Homeland Security to monitor online social network traffic and show how some of those keywords or phrases generate an enormous amount of false positives. Finally, we show how possible terrorists can be rehabilitated using online social networks especially with the help of targeted advertisements.

3.2 Terrorism: Background, Definition, and Classification

The word terrorism is derived from the French word "terrorisme." It was originally used to the use of terror tactics to intimidate the population during the Reign of Terror in France. Systematic use of terror as a policy was first recorded in English in 1798 [20]. In recent history, terrorism gained extensive media attention after the infamous 9/11 attacks on the USA in 2001.

There is no universally accepted definition of terrorism. The definition given in title 22, Unites States code, section 2656f(d) [30] is:

> The term 'terrorism' means premeditated politically motivated violence perpetrated against non-combatant targets by sub-national groups or clandestine agents, usually intended to influence an audience.

The definition according to the International Convention for the Suppression of the Financing of Terrorism is [2]:

> Any other act intended to cause death or serious bodily injury to a civilian, or to any other person not taking an active part in the hostilities in a situation of armed conflict, when the purpose of such act, by its nature or context, is to intimidate a population, or to compel a government or an international organization to do or to abstain from doing any act.

Although both these definitions are quite broad, the first definition exempts criminal drug gangs who may use terror to achieve their goals. Similarly, it exempts an act of terror performed by a state, by limiting the definition to sub-national groups or clandestine agents. The second definition is broader in some aspects as a state or criminal drug gangs can also be classified as terrorists, but they must cause serious bodily injury or death. So, an act of terror with no casualties may not qualify as terrorism according to this definition.

A much broader definition according to the UN Security Council Resolution 1566 is:

> criminal acts, including against civilians, committed with the intent to cause death or serious bodily injury, or taking of hostages, with the purpose to provoke a state of terror in the general public or in a group of persons or particular persons, intimidate a population or

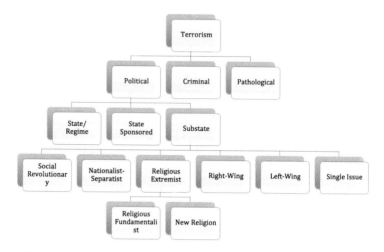

Fig. 3.1 Classification of terrorism (according to [12, 23, 24])

compel a government or an international organization to do or to abstain from doing any act, which constitute offenses within the scope of and as defined in the international conventions and protocols relating to terrorism, are under no circumstances justifiable by considerations of a political, philosophical, ideological, racial, ethnic, religious or other similar nature, and calls upon all States to prevent such acts and, if not prevented, to ensure that such acts are punished by penalties consistent with their grave nature.

Many countries do not have any special definitions for terrorism mainly because most crimes which can be classified as terrorism are already covered by their existing criminal laws.

Figure 3.1 provides a classification of terrorism, where it is first classified into three main types: political, criminal, and pathological.

Political Terrorism. A type of terrorist activity with a political motivation. Political terrorism is sub-classified into state or regime based terrorism, state-sponsored terrorism, and sub-state terrorism.

The examples of political terrorism includes the attacks by Al-Qaeda and Anders Breivik.

Criminal Terrorism. An act of terrorism with a criminal motivation. The motivations behind most criminal terrorist activities are monetary, promotion of illegal business, and revenge.

The examples of criminal terrorism include the killing of law enforcement officers by Mexican drug cartels.

Pathological Terrorism. It is an act of terrorism with no political or criminal motivation. A pathological terrorist acts only for the sheer joy of terrorizing others.

The examples of pathological terrorism include the 2011 shooting of the US Congresswoman Gabrielle Giffords and the 2012 Colorado movie theater shooting.

State/Regime Terrorism. State/Regime terrorism is an act of political terrorism performed by the state. The killing of Kurdish citizens by the government of Iraq under President Saddam Hussain is an example of such a terrorist act.

State Sponsored Terrorism. An act of political terrorism performed by a client agent under the directions of a state. These client agents include paramilitary and clandestine organizations. Pakistan and India have accused each other of sponsoring terrorists to harm the other.

Sub-State Terrorism. An act of political terrorism performed by a non-state organization. 9/11 attacks were example of sub-state terrorism. Sub-state terrorism could be further classified as: social revolutionary terrorism; nationalist–separatist terrorism; religious extremist terrorism; right-wing terrorism; left-wing terrorism; and single issue terrorism.

Social Revolutionary Terrorism. An act of sub-state terrorism performed to overthrow a social order adopted by the regime, for example, capitalist economy. Groups accused of social revolutionary terrorism include the Red Army Faction in Germany and Red Brigades in Italy.

Nationalist–Separatist Terrorism. An act of sub-state terrorism to achieve a nationalist or separatist goal, for example, the demand to establish an independent country. The examples of nationalist–separatist terrorist groups include the Tamil Tigers and the IRA.

Religious Extremist Terrorism. An act of sub-state terrorism motivated by a religious ideology or cause. Examples include the London bombings in 2005 and the Malegaon (India) bombing in 2008. Religious extremist terrorism can be further classified as religious fundamentalist terrorism and new religion terrorism. The latter is an act of terror to promote a new religion while the former attempts to reinstate the fundamental beliefs of an established religion.

Right-Wing Terrorism. An act of sub-state terrorism inspired by ideologies including neo-Fascism, neo-Nazism, racism, and opposition to immigration. Examples include the 2004 Cologne bombing and the murder of policewoman Michele Kiesewetter in 2007 by a German neo-Nazi group.

Left-Wing Terrorism. An act of sub-state terrorism inspired by the Marxist-Leninist ideologies. Revolutionary Armed Forces of Columbia are an organization involved in acts of left-wing terrorism.

Single Issue Terrorism. An act of sub-state terrorism where the members/terrorists want to resolve only a single issue instead of a widespread political change. Examples include terrorism by animal rights or anti-nuclear groups.

The classification of terrorism is not absolute, i.e., there could be acts of terrorism motivated by more than one categories of terrorism, for example, the attacks by Anders Breivik were both religious and right-wing.

Fig. 3.2 Use of online social
networks for terrorism

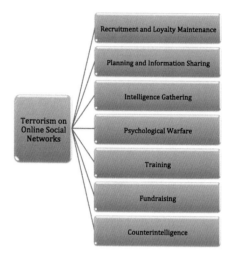

3.3   Online Social Networks and Terrorism

Terrorism is a multivariate problem where terrorists have to deal simultaneously
with different types of audiences. On the one hand, they have to manage the recruitment of new members and maintain the support of their existing sympathizers,
possibly by propagating their position as the victims of an oppressor enemy, while
on the other hand they have to continue the psychological warfare of keeping
the enemy publics in constant fear of an attack. Unfortunately, the online social
networks have made the job of a terrorist easier, as is discussed in this section.
Figure 3.2 shows the possible ways in which terrorists may be using the online
social networks.

3.3.1   Recruitment and Loyalty Maintenance

To recruit new members and maintain the loyalty of their existing supporters and
sympathizers, terrorists have to engage in marketing strategies. One common tactic
effectively used by ideological, nationalist or separatist terrorists is to propagate
their position as the oppressed victims of the enemy they are fighting. For example,
Al-Qaeda portrays the West as an oppressor, who has taken over their resources.
In a very similar manner neo-Nazi forces (e.g., Anders Breivik) violently protest
that Muslims (or Jews in yesteryear) have taken over their resources. Tamil Tigers,
Basque fighters, and the IRA complain that Sri Lanka, Spain, and the United
Kingdom, respectively, have taken over their land. Using their position as a victim,
these terrorist organizations propagate and justify their violent means. Martha
Crenshaw, in 1981, rightly said [5],

The most basic idea of terrorism is to gain recognition or attention.

Online social networks help these terrorists for recruitment and loyalty maintenance in a number of ways, for example, terrorists use YouTube to share videos about their causes. Anders Breivik uploaded his manifesto on YouTube a few hours before he killed scores of innocent civilians in Norway. The manifesto has been widely disseminated by his fans and followers on social media and has been viewed thousands of times. He first argues that multiculturalism is harmful for the Europe, then propagates the fear of Europe being colonized by Muslims, and then uses historical references to the Crusades to present himself as the savior of Europe. Some comments on his YouTube videos clearly shows its radicalization effect:

> Breivik, i love you...
> I wish you were French...

Another comment liked by 5 people is:

> This is 100 % right, left side b****** ! Wake the f*** up or in 20 years there will be no Germany nor France, Muslims are going to cut our throats and rape our women, this is not [sic] f****** joke. Act now, cause it can be too late... [sic]

The second commenter also uses the strategy of propagating fear by saying that lack of immediate action, presumably in a violent manner given that he calls Breivik's work 100 % right, will be responsible for a bigger loss in the future.

In a very similar way, Al-Qaeda and the Taliban have exploited several incidents where civilians have been tortured. For example, the killing of women and children in night raids in Afghanistan by the USA and its allies. Images from these incidents are widely disseminated through Facebook and Twitter and could potentially help Al-Qaeda and Taliban recruit new members. Moreover, messages of Al-Qaeda leaders including Osama bin Laden, Al Zawahiri, Anwar Al-Awlaki, etc., are also available and easy accessible through YouTube.

It is not always the charisma of radical leaders that influences new members to join terrorist organization or launch solo attacks. Sometimes, it is due to the "mistreatment" of others that radicalizes. For example, the images from the infamous incident at Abu Ghraib prison in Iraq where prisoners were tortured by the US military. Similarly, US Army Staff Sergeant Robert Bales brutally killed 16 innocent civilians, in Kandahar, Afghanistan, including several children and women at 3:00 am and then burnt some of the bodies. The Afghan government's investigation confirmed the claims by local people that the attack was carried out by a group of soldiers and not only Bales. US Government, on the other hand, said that the attack was carried out by Bales alone and refused to accept the demands of the Afghan government and people—to let them try Bales under the Afghan law for his crime. This created a sense of powerlessness amongst Afghans, providing a perfect reason for them to get radicalized. Few days after the incident, the Afghan government's representatives, including the President's brothers, who were attending a memorial at the village of the massacre were shot at by local militants, revealing their switched loyalties and showing the visible radicalization impact. It is very likely that the circulation of these images and videos due to the wide coverage by main stream media and extensive sharing on social media, could have helped in radicalizing other sympathizers globally or at least in Afghanistan.

Furthermore, the very nature of social networks make it easier for information to flow amongst users. This creates the possibility of polarization of the population. This polarization slowly and gradually leads to radicalization. To start the polarization, a minimum of one user has to make a radical comment. For example, Scot McHugh, a member of the British Armed Forces, wrote as his Facebook status message,

Go to afghan in a month and half! Carnt wait to shoot some towel heads ;) [sic][1]

McHugh's statement was liked by at least 30 of his friends. Following the statement, McHugh was dismissed by the British Army and was not sent to Afghanistan. Similarly, a 19-year-old British citizen, Azhar Ahmed wrote on his Facebook page,

People gassin about the deaths of Soldiers! What about the innocent familys who have been brutally killed.. The women have been raped.. The children who have been sliced up..! Your enemy's were the Taliban not innocent harmful [sic] families [sic]. All soldiers should DIE & got to HELL! ...! gotta problem go cry at your soldiers grave & wish him hell because thats where he is going.. [sic][2]

Azhar Ahmed was charged with treason over his comments [25]. His comment was liked by 8 of his friends. Those opposing Ahmed's views made a Facebook page called "Azhar Ahmed S**mbag!!!".[3] When we last accessed this page, it had over 3,090 fans. The page first called to kill Ahmed but then changed their demand to put him away, never to be let out. The page is a social media battle ground, between those who support Ahmed's views and those who oppose him, resulting in a fertile polarization ground with a potential of acting as a cradle for radicalization.

Let us now look into another tool, less explicit, that could be used by terrorists on the online social networks. It is the multiplayer online social games. These games are mostly constrained by bandwidth of a user, and to increase their user base, the application developers mostly make very light weight programs. They sacrifice features of monitoring their users' potentially malicious activities. It is understandable because the security and monitoring features do not provide them with economic incentives. Thus, this lack of control within these applications has provided terrorists with a new virtual world to recruit their new members. They can observe the behavior of other game players, interact with them, influence them by sharing propaganda contents and potentially radicalize them. Lone radicals can use these games to group with others to form groups and later launch sophisticated attacks. Online social games could create strong bonds between players as most of these games portray real life situation enabling a recruiter to test the reliance of other players in difficult situations. It also provides the recruiter with the option of vetting potential candidates without revealing their real identity.

---

[1] The spelling and grammar mistakes are left intact to show the exact message.

[2] The spelling and grammar mistakes are left intact to show the exact message.

[3] http://www.facebook.com/AzharAhmedScum, Accessed September 19, 2012.

For a sophisticated attack requiring specialists in a particular area, e.g., finding a biological weapons expert, terrorists may be using LinkedIn to find candidates with such expertise. With LinkedIn terrorists can check the professional experience of their target recruit and then using Facebook they may find the likes and interests of that person. With Meetup they may join an event where their potential recruit is planning to go and create a real life bond, slowly and gradually influencing them. If the potential recruit is not being influenced, terrorists can easily find their family members using Facebook, their address using social search engines like www.123people.com, or their work colleagues using LinkedIn. Using their other social contacts, a person may be coerced to help the terrorists. For example, the Mexican drug cartels have attacked (kidnapped or killed) the family members of law enforcement agencies as an act of revenge. Although there is no recorded incident of these attackers using online social networks, but such incidents are not far from being materialized.

Conversely, those who desire to be recruited can also use social networks to advertise themselves to potential organizations by displaying their activism for the terrorist group's cause.

3.3.2 Planning and Information Sharing

One of the most fundamental pillars of terrorism is fear. In 9/11 the hijackers created this fear by selecting some of the symbolically most important places in the USA e.g., the World Trade Center buildings were symbols of the US's free market belief and economy while the Pentagon is the symbol of national defense. Similarly, Anders Breivik decided to target the ruling Labour Party's summer camp and the capital's town center. In 1991, the Tamil Tigers expressed their power by assassinating the then Indian Prime Minister. All these and many other incidents give evidence that these terrorists wanted to convey a message that they can attack even their enemy's supposedly most secure places.

Today, terrorists can use online social networks for selecting their targets, communicating their plans, data mining about their targets, and spreading propaganda after an attack.

For target selection, terrorists can monitor events on Facebook. With the current settings on Facebook it is not possible for a user to hide that they are attending an event. Moreover, users can be traced back to their profiles from the event pages. Using the employers field on such users' Facebook page, an attacker can easily work out the number of servicemen belonging to important national security agencies attending an event. Terrorists can also impersonate some other users attending the event by faking their credentials or forcefully acquiring them.

The US Department of State on their Facebook page provide links to the pages of all the US foreign missions' Facebook pages. Following those links we were able to able to find that there[4] are 11,825 fans of the US Embassy in Afghanistan, 58,134 fans of the US Consulate General in Peshawar, Pakistan, and 480 fans of the US Embassy in Mali. These fans comment and like various messages by the respective US missions. A terrorist could visit these pages and start targeting the well wishers of the USA, which is very much possible for some militants in their strongholds in these three places.

Once the target is selected, terrorists may track their targets using applications like Foursquare where users willingly share their locations. More privacy conscious users may limit this sharing but they still cannot prevent leaks due to a system's design flaws as was obvious from the LinkedIn password compromise [22]. Moreover as a user's privacy can also be leaked by their friends, a terrorist can track them using their friends' Facebook account or other social network accounts.

After 9/11, the US government started monitoring the emails, calls and all other explicit means of communications between potential suspects of terrorism. Terrorists have now found means to circumvent such monitoring methods. They can communicate implicitly using online social network games. Using their actions in games they can share their plans. A terrorist can also see the reaction to his acts in a multiplayer social game by committing an act of terror and then monitor the reactions of other users. This way, they can also select targets with more impact and also get a feeling of the aftermath. Terrorists do not need to be part of the same fan page or be friends on Facebook and other social networks to share their plans. They may leave messages in code words on very busy fan pages and only visit them once in a while. Criminal organizations like the Mexican drug cartels and gangs in the USA have used such code words for their internal communications. It normally takes law enforcement agencies several years to decode and understand them. Furthermore, detecting steganographic messages, within the enormous 300 million photos uploaded everyday on Facebook alone, is in no way a trivial task for the national security agencies.

3.3.3 Intelligence Gathering

Online social networks are a goldmine of interesting information about individuals. This information is exploited by business and marketers to craft better targeted advertisements. In a very similar way, terrorists maybe using social networks to get a better insight into the psychology of their opponents, to unfold their plans, or to locate them. According to Gabriel Weinmann, Professor at Haifa University, Hezbollah and Hamas observe the Facebook status updates and other activities of Israeli soldiers. As a countermeasure, the Israeli Army has directed its soldiers to avoid sharing sensitive and extensive information on any social network. It is very hard to cover some flows of information, e.g., friend links, likes, and family, which does become sensitive.

---

[4]We accessed the pages on July 13, 2012.

Similarly, the Taliban have been suspected of befriending NATO soldiers on Facebook by pretending to be attractive women to gather intelligence [6].

3.3.4   Psychological Warfare

Terrorism is more psychological than physical warfare. Terrorists need to constantly propagate fear in their enemy publics. Social networks have been widely adopted for this cause. For example, the Taliban regularly update their Twitter page "@alemarahweb" with propaganda about their latest operations. They also regularly post pictures and videos to convince their readers about their claims. These pictures and videos are also shared on Facebook and YouTube. Taliban and the ISAF Press Office got engaged in a war of words on Twitter in late 2011 accusing each other of lying and killing civilians [28].

Similarly, the Somalia-based Al-Shabab are also present on Twitter "@HSM-Press." They have sent a total of 879 Tweets and have 15,774 followers.[5] A large number of the followers of these pages are journalists and officials from security agencies. Interestingly, the Al-Shabab Twitter account is not following any Twitter users, avoiding leaks about their interest in other organizations with similar interests. Two examples of their Tweets as psychological warfare are,

> To the apostate forces, Martyrdom operations remain a deeply traumatic memory; to the Western Kafir, a painfully unfathomable phenomenon - June 18, 2012

> History is a testament to the gruesome fate of invaders in Somalia & these African mercenaries are headed straight to that horrible graveyard - July 15, 2012

Al-Shabab is also spreading their fear through YouTube. "Blow by blow" their hip-hop and rap song is becoming famous on YouTube. The song opening lines are as follows,

> Month by month, year by year
> keeping those kaafirs in fear . . .

The adaptation of songs and poetry increases the bounds and reach of terrorist organization. Thus, their message gets conveyed to a larger population.

Anders Breivik's manifesto, shared on YouTube, his actions, and his choice of targets, were meant to be a warning to liberals in Europe to face the wrath of right wing extremism if they continue to support multiculturalism.

Mexican drug cartels are also spreading their terror by sharing their executions on social networks, according to a Live Leak report [19].

Lots of content shared on the social media is originally reported by the mainstream media. Unfortunately, the main stream media has played a very pivotal role in helping terrorists achieve the psychological power over the minds of billions

---

[5]We accessed the page on September 19, 2012.

of people. For example, it would not have been possible to spread the messages of Osama bin Laden without their widespread coverage by Al-Jazeera and then re-telecast by other news channels. Ray Surette in 1998 interestingly envisioned the fact, as obvious from his words,

> Terrorism has become a form of mass entertainment and public theatre and thus highly valuable to media organizations.

It will be an interesting research to see the increase in viewership and revenues of media organizations in the days following major terrorist attacks. The increase revenues and increased viewership creates incentives for these media organizations to further dramatize such events and exaggerate the fear, selling their programs while also helping the terrorists. The biased use of the word "Muslim," by the mainstream media, in every violent news, where a person belonging to the religion is involved, irrespective of his mental state or motivation for the crime is an evidence of the claim that "Islamophobia" is used as a psychological tool to exaggerate and sell fear.

3.3.5 Training

Terrorists can use online social networks to train for the skills they require for a relatively sophisticated attack. YouTube videos can train attackers in a wide range of skills, from how to clone RFID cards [17] to the creation of bombs in a Taliban bomb factory. The comments on these videos can further clarify any missing or confusing parts of the training. YouTube videos can also help attackers in personal fitness training, which may be required for one to one combat or for surviving under harsh conditions. Similarly, Meetup and Facebook can be used to organize physical training sessions under the guise of other events which may not arouse suspicion.

The US Army approved US $ 50 Million to develop games for the army for training purposes [26]. In a similar way, the online multiplayer games can be used to train for actual combat. The Sun newspaper in the UK claimed that terrorists are using "Call on Duty" to plan and train for actual combat [31]. The nature of combat games and the game developers' competition to provide users with a more realistic experience makes them more attractive for the terrorists. Also, it is harder for the security officials to differentiate between terrorist recruits and game lovers due to the nature of such games.

3.3.6 Fundraising

Terrorist organizations like other political organizations cannot sustain their day-to-day operations without the required funding. They have for long used the Internet, and since recently online social networks, to raise money for their causes either directly or indirectly. In the direct method they announce their bank accounts for the sympathizers to pay money into while in the indirect method they only spread their propaganda and the sympathizers through traditional means of going into a physical office or through a representative, contribute their share of fundings.

Neo-Nazis have used the YouTube revenue-sharing system to get advertisement revenue from Google. The ads placed on their videos include those by BT, Virgin media, and O2 [9]. Though these videos approved for advertisement revenue sharing were removed by YouTube, there is unfortunately no current mechanism for YouTube to completely stop such misuse of their services and the flow of money to these extremist organizations. Moreover, the current YouTube or AdSense terms of use do not mention any strict actions being taken against future extremist misusers.

3.3.7 CounterIntelligence: Easier to Blow Covers of Undercover Agents

Online social networks are making the life of undercover agents much harder by making it easier for the terrorist organizations' counterintelligence wings to blow an agent's cover. If an undercover agent does not use an online social network while being relatively social otherwise, it could arise suspicion from the terrorist organizations, as a significantly large population of users, especially belonging to the West, are actively present on one or more social networks. On the other hand if undercover agents use social networks they face a multitude of other problems. With the ever improving facial recognition software, it is becoming easier to link the past life of an agent to their current one. For example, let us say the agent has set up a Facebook account for his undercover assignment in the Middle East. Now, Facebook with its face recognition software can easily recognize the agents' photos uploaded by his real life friends in the USA. Such suggestions of tagging a user may be given to his friends on the social network. In our scenario, the agent may have infiltrated a suspected terror cell, thus leading to his cover being blown.[6] This is only one example, there could be many more implicit and explicit flows of information leaking a lot. Even with the best privacy measures, a flaw in the system, which is not very unlikely, could reveal all information about a user.

3.4 Detection Mechanisms: Strengths and Weaknesses

In the previous section we described several ways in which online social networks may be helping terrorists. In this section we discuss some mechanisms (shown in Fig. 3.3) to detect the use of online social networks by terrorists. Each of these mechanisms have their own strengths and weaknesses.

---

[6]Agents in the police and other law enforcement agencies may take up to 5 years undercover to reach the leaders of a group.

Fig. 3.3 Mechanisms to
detect terrorists in online
social networks

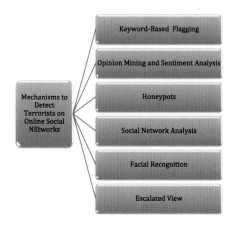

3.4.1 Keyword-Based Flagging

The first mechanism we consider is the keyword-based flagging. This mechanism
is used by the US Department of Homeland Security to monitor online social
networks. Some of the keywords they use for monitoring any terrorism related
activities are given in Table 3.1. We used Google's Search to check the number of
times, ten of the keywords appeared publicly on Facebook, Twitter, and YouTube.
The results are surprising, as most of these keywords are used millions of times on
each of these social networks. The word "terror" has been used 124 million times
on Facebook, 117 million times on Twitter, and 55.7 million times on YouTube.
It is very obvious that all these instances of using the word were not to plan a
terror attack. Similarly, the less likely words "dirty bomb" and "biological weapons"
have been used 2.69 and 0.497 million times on Facebook, 1.48 and 1.69 million
times on Twitter, and 0.241 and 0.919 million times on YouTube, respectively.
Such monitoring of terror attacks on the basis of keywords is bound to generate
an enormous number of false alarms. Moreover, an attacker can easily circumvent
the monitoring system by using a variant of the spelling. Also, it has been widely
known that many criminal organizations use their own codewords, which the law
enforcement agencies are not always totally familiar with. Such codewords can
circumvent the monitoring system based on keywords.

3.4.2 Opinion Mining and Sentiment Analysis

Terrorism is a war of opinions and sentiments waged by attackers who believe
they can achieve their political goals through violent means. A better detection
mechanism could be to monitor the opinions and sentiments of the users of

Table 3.1 DHS monitoring keywords on social networks (July 13, 2012)

Keyword	Used publicly on Facebook	Used publicly on Twitter	Used publicly on YouTube
Terror	124,000,000	117,000,000	55,700,000
Pakistan	374,000,000	355,000,000	175,000,000
Afghanistan	298,000,000	386,000,000	127,000,000
Taliban	71,400,000	66,500,000	13,900,000
Nuclear	209,000,000	192,000,000	69,300,000
"Dirty bomb"	2,690,000	1,480,000	241,000
"Ammonium Nitrate"	1,910,000	2,090,000	1,140,000
"Chemical weapon"	831,000	835,000	738,000
"Biological weapon"	497,000	1,690,000	919,000
Jihad	9,998,000	7,570,000	16,000,000

online social networks. On Facebook we found 233 people liked Anders Breivik's Wikipedia page[7] and according to Facebook 13 people were talking about him when we checked that page. We also found 954 people liked Anwar Al-Awlaki's Wikipedia page[8] and according to Facebook 30 people were talking about him at that time. These people may be much more likely to commit a violent act of terror than anyone using a keyword for information purposes. Opinion mining can also help security officials detect which events result in converting a non-violent user into a violent terrorist. Such results can help design better counter-terrorism mechanisms and policies.

On YouTube and Facebook, sentiment analysis can also be used to assign positivity and negativity scores to profiles and comments [32]. Though sentiment analysis has been successfully used for analyzing blogs and movie reviews, it will require more adjustments to accommodate the more informal structure of comments on social media.

Another important use of opinion mining for security officials is to help them detect which events result in converting a non-violent user into a violent terrorist. Such results can help design better counter-terrorism mechanisms and policies. This change in behavior could either be detected in real time if the user is under observation or after the radicalization of a user, by comparing his pre-radicalization activities to the post radicalization and observing other parallel events, with possible impacts, in the duration before the threshold point.

The opinion mining and sentiment analysis detection mechanisms also have their weaknesses. Terrorists in the past have deceived security officials by providing false information. For example, eight British terrorists who planned to blow up seven planes mid air used porn magazines to deter security [3]. With online social networks like Twitter there is a constraint on the number of words per tweet which

---

[7]http://www.facebook.com/pages/Anders-B-Breivik/265349460157301 Accessed: July 13, 2012.

[8]http://www.facebook.com/pages/Anwar-al-Awalki/102248169830078?rf=134722733227304 Accessed: July 13, 2012.

could further complicate the effectiveness at such detection. Moreover, Anders Breivik is believed to have setup his Facebook account and uploaded his manifesto only a few hours before he executed the mass murder. This could give an indication that he was aware of the authorities monitoring such content. Strong sentiments in a short duration of time, e.g. liking neo-Nazi music, white supremacist leaders, and writing about violence in quick succession might be an indication of an attack taking place very soon.

3.4.3   Honeypots

Honeypots are traps, long used as a defense, detection, or distraction mechanism. They are setup to deceive potential attackers by giving them fake information (which the attacker considers factual), in order to interact with them to extract sensitive information from them. The idea was first published in 1991 in two texts "The Cuckoos Egg" and "An Evening with Breford" where the authors wrote about their attempts of using honeypots to catch hackers. This concept can easily be adapted to online social networks. Undercover agents may play multiplayer online social network games and pretend their violent desires and wait for any terrorist organizations or lone terrorists to approach them. Once they find potential supporters of terrorist organizations, the security officials can remain in the game for long enough till they have got substantial amount of sensitive intelligence information. In a very similar way, security agencies may upload terrorist sympathetic YouTube videos to see who likes them and then trigger polarizing comments to observe which of the users show radical inclination. Alternatively, the security officials may upload YouTube videos against a radical group on any social network and observe if anyone attempts to radically defend them. On Facebook, fan pages for well-known terrorists could act as honeypots and attract potential terrorists.

Honeypots are very easy to implement and only collect direct data but are more of a reactive defense mechanism where the security officials wait for the attacker to fall into the trap. It may be harder to trap well-trained terrorists or careful lone attackers into a honeypot. Moreover, honeypots may provide attackers with signatures, alerting them about being monitored. For example, Sabu from the hacker group Lulzsec turned into an FBI informant. He acted as a honeypot to identify other members of the group, but Virus, another member of the group, identified a honeypot signature as obvious from their IRC discussion below [21]:

Virus (10:30:18 PM)   don't start accusing me of [being an informant]—especially after you disappeared and came back offering to pay me for sh**—that's fed tactics

Virus (10:30:31 PM)   and then your buddy, topiary, who lives in the most random place

Virus (10:30:36 PM)   who's docs weren't even public

Virus (10:30:38 PM) gets owned

Sabu (10:32:29 PM) offering to pay you for sh**?

Virus (10:32:55 PM) yeah, you offered me money for "dox"

Virus (10:33:39 PM) only informants offer up cash for sh**—you gave yourself up with that one

3.4.4   Social Network Analysis

Social network analysis conceptualizes the structure of a network with the ties between the members and a focus on the characteristics of the ties instead of individuals. Social network theory emerged as a key technique in modern sociology, used to determine the influential people amongst a social graph by analyzing ties with respect to the rest of the graph. It has since been adapted for use in economics, anthropology, psychology, computer sciences, etc. Systematic use of the term "social network" was started in 1954 by John Barnes.

In sentiment analysis and opinion mining we considered the individual, here we consider the connectivity between a group of users. This connectivity is measured by the centrality and between-ness of a user. Centrality relates to the extent of the user being central actor in a network, e.g., the centrality of someone connected on only one other user will be lower while that of someone connected to all other users directly will be the highest. Between-ness means the extent to which a user falls between two other users. Between-ness shows how influential a user is in the graph.

Social network analysis has been very widely used to detect the bonds amongst various types of social circles including terrorist organizations. Vladis Krebs mapped the social network for the 9/11 terrorists and was able to deduct that Mohammed Atta was their ring leader [13].

In Facebook the ties that can be considered for social network analysis are friendship between users, the number of messages exchanged between them, the events they attended together, the number of photos they are both tagged in, the number of applications they both use, etc. The direction of communication is also important for social network analysis. For example, an order giving leader will send more messages to everyone in the group, while a data collecting user will receive more messages and send less.

Although social network analysis is a very effective method to identify explicit links between terrorist groups and beyond, it is of lesser use when the terrorists are only using implicit means of communication. Social network analysis becomes very complex when we consider all the implicit communication channels because: (a) not all implicit communication channels are known and (b) this adds to the already large number of explicit communication channels. Furthermore, social network analysis only works when there is more than one entity involved, thus, if an attacker operates alone, it is impossible to detect him using this technique. It is always easier to perform social network analysis after the attack took place and the attackers are known as was done post 9/11.

3.4.5   Facial Recognition

In the previous section we mentioned that facial recognition can make the life of an undercover agent much harder. In this section we present a positive use of facial recognition, where it can be used to detect terrorists and their sympathizers. For the glorification of their leaders and fallen comrades, terrorists normally disseminate their photos and videos. It is not unlikely for a non-violent user to upload an article about the US's war on Terror, but it is very unlikely for a non-terrorist to upload pictures of top-tier, second-tier, and lower-tier terrorists. Using facial recognition, these photos can be compared to the directory of known terrorists and the profile of suspects profile flagged for further detailed monitoring. We mentioned different tiers of terrorists because the lower-tier are most likely known only to their close friends and organizational comrades, while the top-tier will be widely known. Thus detection of the photo upload of a lower-tier terrorist may make the uploader more likely to be a terror sympathizer. Similarly, picture frames in YouTube videos can be monitored for contents about known terrorists.

3.4.6   View Escalation

Government acts, for example, the PATRIOT Act in the USA enables law enforcement agencies to request complete information about a user from their service providers. This complete information provides the law enforcement officials with a better view enabling them to perform better analysis. However, if the service provider is outside the jurisdiction of the security officials they are left with a limited view. In such cases, they may use other methods to escalate their view.

Security analysts may use privacy loopholes like the Zuckerberg photo leak or deactivated friend attack for an escalated view. In the Zuckerberg photo leak, when a user's Facebook photo was reported as nude, Facebook provided the reporter with all photos from that user's profile to verify if they were also nude. This method was exploited to view the private photos of Mark Zuckerberg, the CEO of Facebook amongst others. Similarly, in the deactivated friend attack, any user (in our case a security analyst) could add his target as a Facebook friend and deactivate their profile [18]. While deactivated, the security analyst's profile is not visible to the target and the target cannot delete the analyst or impose any stricter privacy measures.

FBI and Homeland security may be using fake user accounts to be friends with users on Facebook to observe their actions [14].

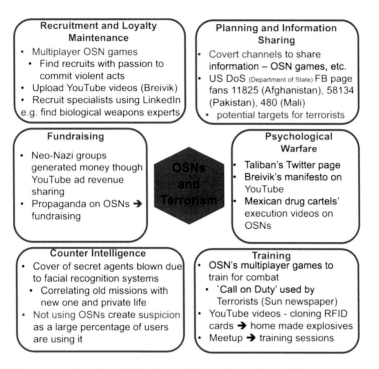

Fig. 3.4 Summary of ways in which online social networks can be used by terrorists

3.5 Terrorist Rehabilitation Using Online Social Networks

Unlike economically driven crimes, terrorism is based on beliefs and ideology, gained through an individual's exposure to terrorist propaganda and indoctrination. Incarceration, whether brief or long term is not expected to change the ideology and beliefs of a terrorist. To change the extremist ideology of a terrorist, we need to rehabilitate them. The rehabilitation could include: (1) religious rehabilitation; (2) psychological rehabilitation; (3) social rehabilitation; and (4) vocational rehabilitation [11]. Religious rehabilitation is most important for religious extremist terrorists. In such rehabilitation, the religious beliefs of a terrorist may be countered by providing him with examples, from the same religion, which opposes extremism. A *fatwa* (a ruling on a point of Islamic law) by a prominent religious scholar can be more convincing for the terrorist than a non-scholar providing an explanation. Ustad Sayyaf (Afghan religious scholar and member of the parliament) issued such a fatwa declaring suicide missions as "haram" (non-permissible in Islam). Psychological rehabilitation includes hearing out the genuine concerns of a terrorist about the sufferings of their people and then providing them with non-violent methods to counter the problems. A terrorist must be convinced that the non-violent method

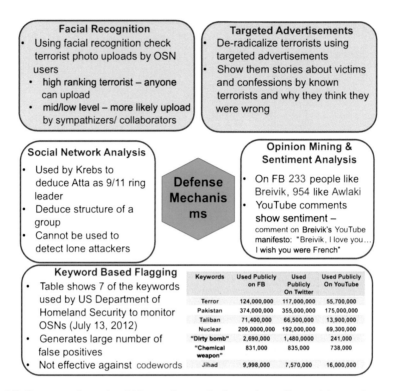

Fig. 3.5 Summary of ways in which terrorists can be detected on online social networks

is more beneficial for him and the group he is attempting to fight for. Social rehabilitation includes helping the family of the terrorist and integrating them to the society to avoid other extremists from reaching out and exploiting them. Vocational rehabilitation includes providing terrorists with skills required to get respectable jobs in the society and give them a chance to re-integrate to the society. Most insurgents in poor countries join the movements due to the lack of skills to earn a respectable living, thus enabling extremists to exploit them.

Online social networks can be used for all four types of rehabilitations. Any user detected to be a possibly radicalized could be sent with counter-propaganda, for example through targeted advertisements. They could be provided with non-violent religious decrees, countering the use of violence. They could be shown links to stories of the families of victims of previous terrorist attacks, or they could be shown videos of imprisoned terrorists who accept that they were fooled and brainwashed. Social networks and the Internet could be used to provide extremists with vocational training, enabling them to acquire the skills required for a respectable job.

Users attempted to be un-radicalized should be closely monitored for a period of time to avoid them being re-radicalized.

3.6 Summary

To summarize, in this chapter we discussed three things. First, in the light of the background and definitions of terrorism, we discussed how some terrorists may be using the online social networks for their activities, which is summarized in Fig. 3.4. Second, we discussed the pros and cons of several mechanisms to detect terrorists on online social networks, which is summarized in Fig. 3.5. Some of the terrorist profiling mechanisms may have a large number of false positives, possibly causing innocent users to be classified as terrorists. Finally, we should that the detected radicals could be rehabilitated using social networks.

Acknowledgments This chapter is an expanded version of our work in [16]. The author would like to thank the University College London for financial support through the UCL PhD Studentship Program.

References

1. About Meetup: http://www.meetup.com/about/. Accessed 20 February 2012
2. Article 2: International Convention for the Suppression of the Financing of Terrorism, 1999
3. Asher, J.: Accused terrorists used porn magazines to deter security. NineMSN, 5 April 2008
4. Barnett, E.: Google+ hits 90 million users. The Telegraph, 20 January 2012
5. Crenshaw, M.: The causes of terrorism. Comp. Polit. 379–399 (1981)
6. Deceglie, A., Robertson, K.: Taliban using Facebook to lure Aussie soldier. The Sunday Telegraph, 9 September 2012
7. Facebook: Facebook Statistics. http://newsroom.fb.com/content/default.aspx?NewsAreaId=22. Accessed 13 July 2012
8. Flickr: http://advertising.yahoo.com/article/flickr.html. Accessed 20 February 2012
9. Flveash, K.: Neo-Nazis scoop YouTube ad revenue from UK telcos. The Register, 22 June 2012
10. Foursquare: https://foursquare.com/about/. Accessed 20 February 2012
11. Gunaratna, R., Jerard, J., Rubin, L.: Terrorist Rehabilitation and Counter-Radicalisation New Approaches to Counter-Terrorism. Taylor & Francis, New York (2011)
12. Hudson, R.A.: Who Becomes a Terrorist and Why: The 1999 Government Report on Profiling Terrorists. Lyons Press, Guilford, Connecticut (2002)
13. Krebs, V.E.: Uncloaking terrorist networks. First Monday 7(4) (2002)
14. Lardner, R.: Your new Facebook 'friend' may be the FBI. MSNBC, 15 March 2010
15. Linkedin: http://press.linkedin.com/about. Accessed 19 September 2012
16. Mahmood, S.: Online social networks: The overt and covert communication channels for terrorists and beyond. In: IEEE HST, 2012
17. Mahmood, S., Desmedt, Y.: Online social networks, a criminals multipurpose toolbox (poster abstract). In: Balzarotti, D., Stolfo, S.J., Cova, M. (eds.) Research in Attacks, Intrusions, and Defenses, vol. 7462 of Lecture Notes in Computer Science, pp. 374–375. Springer, New York (2012)
18. Mahmood, S., Desmedt, Y.: Your Facebook deactivated friend or a cloaked spy. In: IEEE PerCom Workshops, pp. 367–373, 2012
19. Narco terror on social networks. Live Leak, 20 June 2012
20. Online ethymology dictionary. http://www.etymonline.com/index.php?term=terrorism. Accessed 25 September 2012

21. OPSEC for hackers: because jail is for wuftpd. http://www.slideshare.net/grugq/opsec-for-hackers. Accessed 25 September 2012
22. Parnell, B.-A.: Linkedin faces class action suit over password leak. The Register, 21 June 2012
23. Purpura, P.:   Terrorism and Homeland Security: An Introduction with Applications. Butterworth-Heinemann, Boston, MA (2007)
24. Schneider, B.R., Davis, J.A.: Avoiding The Abyss: Progress, Shortfalls, and the Way Ahead in Combating the WMD Threat. Greenwood Publishing Group, Westport, Connecticut (2006)
25. Seymour, R.: Azhar Ahmed – charged with treason over Facebook comments? 15 March 2012
26. Shachtman, N.: Army launches $50 million videogame push. Wired, 24 November 2008
27. Sharma, N.: Abu Jundal used Facebook to hunt for recruits for his terror mission. NDTV, 29 June 2012
28. Siddiqui, H.: Taliban and Nato-led forces engage in war of words on Twitter. Guardian, 14 September 2011
29. Twitter stats: https://business.twitter.com/basics/what-is-twitter/. Accessed 19 September 2012
30. U.S. Department of State, office of the coordinator for counterterrorism. Country Reports on Terrorism, 30 April 2007
31. Willetts, D., Wells, T.: Game fanatics. The Sun, 20 March 2012
32. Wilson, T., Wiebe, J., Hoffmann, P.: Recognizing contextual polarity in phrase-level sentiment analysis. In: HLT/EMNLP, 2005
33. YouTube: YouTube statistics. http://www.youtube.com/t/press_statistics. Accessed 16 May 2012

Part II
Access Control, Reputation and Semantic Policies in Social Networks

Chapter 4
User-Managed Access Control in Web Based Social Networks

Lorena González-Manzano, Ana I. González-Tablas, José M. de Fuentes, and Arturo Ribagorda

Abstract Recently, motivated by the expansion and the emergence of Web Based Social Networks (WBSNs), a high number of privacy problems and challenges have arisen. One of these problems that is currently attracting the attention of scientific community is the design and implementation of user-managed access control systems. In this regard, there exist a well-known set of requirements (relationship-based, fine-grained, interoperability, sticky-policies and data exposure minimization) that have been identified in order to provide a user-managed access control for WBSNs. These requirements, partially addressed by the works proposed in the literature, represent "building blocks" for a well defined user-managed access control model. In this chapter, we first provide a conceptualization of a WBSN to propose an access control model, called $SoNeUCON_{ABC}$, and a mechanism that implements it. A set of mechanisms among the recently proposed in the literature are selected such that, when deployed over $SoNeUCON_{ABC}$, the whole set of user-managed requirements can be fulfilled.

4.1 Introduction

Internet has become one of the main communication media, users as well as services have increased year after year.[1] Indeed, Web Based Social Networks (WBSNs) are one of the most recently ways of communication. According to [19], since the

---

[1] http://www.internetworldstats.com/emarketing.htm, last access November 2012.

L. González-Manzano (✉) · A.I. González-Tablas · J.M. de Fuentes · A. Ribagorda
Avda. de la Universidad, 30. Computer Science and Engineering Department, University Carlos III of Madrid, 28911 Leganés, Spain
e-mail: lgmanzan@inf.uc3m.es; aigonzal@inf.uc3m.es; jfuentes@inf.uc3m.es; arturo@inf.uc3m.es

R. Chbeir and B. Al Bouna (eds.), *Security and Privacy Preserving in Social Networks*, Lecture Notes in Social Networks, DOI 10.1007/978-3-7091-0894-9_4, © Springer-Verlag Wien 2013

beginning of social networks in 1997 and their establishment with Friendster in 2002, many of them have appeared and their population has intensively increased. For example, 3 years after the emergence of Facebook there were more than 52 millions of registered users, currently achieving about 850 million.[2] Likewise, Internet users in 2002 reached 587 millions and nowadays there is a population larger than 2,095 millions.[3] As a result, simplifying and only considering Facebook, notice that 35,79 % of Internet users make use of this WBSN. This matter leads us to wonder about the huge importance of analyzing and improving all kinds of features of this type of systems.

By contrast, given the huge quantity of users and data managed, a key issue comes into place, privacy. This feature is defined as "the condition of not having undocumented personal knowledge about one possessed by other" [60]. WBSNs are systems that store a great amount of personal information that must be carefully protected, even been an issue not deeply recognized and taken into account by users. For instance, J. Becker et al. studied that the total of Facebook users have never used any of the privacy mechanisms provided [12]. Likewise, Acquisti et al. analyzed that even Facebook users who are aware of privacy problems continue using it [2]. Namely, this matter may be related to the enormous appearance of perceived benefits, as well as, to the fact that people may be conscious about Internet security but not aware of threats [58]. By contrast, a much more recent study identifies that WBSN users, in regard to New York city Facebook users, have become much more private [36].

Despite interests and motivations of people, privacy is extremely relevant in everybody's life and recently it has been extensively studied and defined. For instance, in the Universal Declaration of Human Rights it is set the right to not have interferences with our privacy.[4] In fact, although users are not specially focused and worried about privacy issues, the chief point is that laws and rights highlight the necessity of protecting everybody's privacy. Furthermore, as C. Dwyer et al. pointed out, WBSNs need to provide enough data protection mechanisms to achieve the same level of privacy found offline [38].

Considering the importance of privacy, previously mentioned, together with the increase of WBSN users, a crucial question arises: Do WBSNs provide enough mechanisms to preserve privacy? This question has not got a simple answer, as even a previous question is still not clear: Which requirements must be fulfilled to appropriately protect users privacy in WBSNs? In 2007, Gates identified a set of requirements in order to provide user-managed access control in WBSNs [29] which remains that a proper access control management contributes to privacy preservation. These requirements establish that access control systems developed for WBSN must be *relationship-based*, *fine-grained*, *interoperable* and follow the

---

[2]http://www.internetworldstats.com/facebook.htm, last access November 2012.

[3]http://www.internetworldstats.com/stats.htm, last access November 2012.

[4]http://www.un.org/en/index.shtml, last access November 2012.

sticky-policy paradigm. In this work, Gates' requirement list is taken as a starting point and a new requirement, *data exposure minimization*, is identified in the light of recent trends in currently active WBSNs [41]. Since then, several research works have proposed access control systems that address some of the requirements identified by Gates, but not all of them or partially. Moreover, there is not a clear path that guides researchers and industry in the task of developing access control systems for WBSNs that address the whole set of requirements.

There are some contributions that try to analyze security in WBSNs. R. Ajami et al. [5] study security challenges of WBSNs, focusing on identifying privacy, anonymity and security risks, as well as, describing some security WBSN proposals, such as VIS or FlyByNight (also analyzed in this work). Likewise, in [41] breaches and security mechanisms for WBSNs are identified. Moreover, a set of possible attacks like information leakage, de-anonymization, and phising are noticed. From a similar point of view but based on WBSNs privacy, E. Zheleva and L. Getoor present a description of privacy breaches and attacks [82]. Nonetheless, regarding purposes of this work, the most similar approach is [22]. It focuses on access control in WBSNs, specifying requirements that, contrary to the user centric perspective described herein, are based on access control policy languages and access control mechanisms.

Therefore, identified problems are twofold, (1) the lack of a model that particularly addresses the whole set of requirements and (2) the lack of specific mechanisms to manage them. Due to this fact, this chapter presents three main contributions. The first one is the proposal of *SoNeUCON*$_{ABC}$, an access control model that allows the fulfillment (from a theoretical point of view) of the whole set of requirements as an extension of the *UCON*$_{ABC}$ usage control model [68], and a basic mechanism that implements it. The second contribution of this work is to select a set of mechanisms recently proposed in the literature that allow fulfilling the remaining three requirements on top of *SoNeUCON*$_{ABC}$. The third contribution is to analyze in detail to which degree recent academic proposals for WBSNs and currently deployed WBSNs satisfy the identified requirements or adopt any of the mechanisms that would facilitate the satisfaction of any of them. In this sense, 25 academic proposals and 9 WBSNs in use are analyzed.

The rest of the chapter is structured as follows. Section 4.2 contains a brief background on access control models and mechanisms. Section 4.3 provides a conceptualization of a WBSN, necessary to develop the contributions of this work. Then, Sect. 4.4 details the requirements identified by Gates in [29] and the new one added herein. In Sect. 4.5 the proposed access control model, *SoNeUCON*$_{ABC}$, and the specific mechanism selected to fulfill the whole set of requirements is presented. Next, in Sect. 4.6, the results of the analysis of WBSNs regarding the satisfaction of the set of user-managed access control requirements are summarized. Finally, Sect. 4.7 contains conclusions and open research issues.

4.2   Background

In this section, an overview of traditional and recent access control models is provided, as well as a brief depiction of the mechanisms that implement these models.

4.2.1   Access Control Models

Currently three traditional access control models can be identified: Mandatory Access Control (MAC), in which objects and subjects are classified according to security levels and access is granted in regard to them; Discretionary Access Control (DAC), in which access to information is carried out in respect to the user's identity and a set of authorizations or rules; and Role Based Access Control (RBAC), based on the definition of different roles, assigning permissions to roles, and, then, assigning roles to subjects [67]. More specifically, RBAC has been extended in other models: TrustBAC [65], based on trust instead of roles; TBAC [65], focused on dynamically managing permissions once tasks completeness; TRBAC [13], based on activating or deactivating a role in regard to time specifications; RelBAC [39, 42, 65] in which permissions are modeled as relationships between users and data while access policies are instances of them; LRBAC [64] that adds location constraints in the traditional approach; or GRBAC [34] that focuses on including environment and object roles to the basic approach in which only subject roles are tackled.

However, another new type of access control model has recently appeared, Attribute Based Access Control (ABAC) [21, 70] that is focused on the definition of policies considering attributes of subjects, resources, and the environment. Indeed, under this perspective, the Usage Control Model ($UCON_{ABC}$) [51, 68], extensively developed by J. Park and R. Sandhu, has been carried out. It focuses on presenting an unified framework for access control in which MAC, DAC, RBAC, and DRM (Digital Rights Management) systems are likely to belong to.

More specifically, the $UCON_{ABC}$ model considers eight components: *subjects (S)* that are entities that exercise rights on objects; *objects (O)* that are entities which subjects hold the right on; *subject attributes (ATT(S))* and *object attributes (ATT(O))* that refer to features associated with subjects and objects, respectively; *rights (R)*, which are recognized as privileges exercised on objects such as read or write; *Authorizations (A)* that correspond to predicates on subject and object attributes that are evaluated in order to decide whether the requested right on a specific object made by a certain subject should be allowed or denied; *oBligations (B)* that represent predicates that must be satisfied before, during or after the right is granted; and *Conditions (C)* that correspond to environmental or system factors not controlled by subjects and which are taken into account during the access decision process.

In $UCON_{ABC}$, *usage decision functions* are pointed out in order to permit or deny access. They are based on authorizations, obligations, and conditions, but two other factors are also considered: the *mutability of attributes* and the *continuity of decision*. On the one hand, mutability refers to the possibility of updating subject and object attributes at different times with respect to the moment at which the right is exercised, namely, before (pre), during (ongoing), or after (post), besides considering inmutable attributes. On the other hand, continuity of decision refers to the possibility of persistently verifying policies while the right is exercised, which is referred to as on-going decision; the other possibility is checking policies only before the right is exercised, which is referred to as pre-decision.

In the original $UCON_{ABC}$ model, it is assumed that an access control policy is defined by the system's administrator and this policy is applied to all users in the system. A recent work by Salim et al. proposes an administrative model, orthogonal to the $UCON_{ABC}$ model, where the attributes and rights of subjects and objects are established through assertions made by authorized subjects [66]. Specifically, this administrative model allows the specification of attributes and rights by the entity that has the authority to establish them.

4.2.2   Access Control Mechanisms

Access control models are implemented through access control mechanisms. The most traditional ones are access control lists (ACLs) and capabilities. ACLs inform about the authorized permissions in regard to a single resource and multiple users. On the contrary, capabilities attest the permissions granted to a single user for multiple resources.

Every access control system has an associated architecture. Usually, the core component is a reference monitor that consists of two elements, namely the Policy Decision Point (PDP) and the Policy Enforcement Point (PEP).[5] The former provides affirmative or negative responses in regard to the requested rights over objects according to the defined policies. The latter enforces decisions taken by the PDP. Both elements are considered trusted entities.

Several architectures may be adopted depending on the number and location of the PDP and PEP, and the nature of the interactions among the parties involved in an access request. Depending on the location of the PDP and PEP, architectures can be classified among server, client or client-and-server side. In the first case, the reference monitor is deployed in the server, in the second case, in the client, and in the third, the components of the reference monitor are deployed in both locations. Regarding the number of components involved in the reference monitor, different n-tier architectures can be identified. For example, a two-tier architecture

---

[5]http://tools.ietf.org/html/rfc2904, last access November 2012.

may consider a server-side reference monitor combined with a client-side reference monitor.

The $UCON_{ABC}$ model allows its implementation with several architectures depending of the concrete features selected from the model. If ongoing decision is required, the PDP and PEP need to be permanently interacting with each other and state-full. Thus, this continuous connection links both to attribute mutability and privilege revocation. In case any security policy is violated during a resource usage, the PDP communicates with the PEP in order to enforce the end of the usage process.

Apart from ACLs and capabilities, a third access control mechanism is referred to as "lock and key." In its cryptographic implementation, data is stored encrypted (locked) and it can only be disclosed if the appropriate description key is known [17]. This type of mechanism is recently attracting a lot of attention from data protection researchers. For instance, Personal Data Servers (PDS) are storages of encrypted personal data that intend to minimize the consequences of having compromised the storage device [7, 18, 56, 75]. In order to access to data, keys must be previously exchanged between contestants. If access control is cryptographically enforced, no reference monitor is required [37].

Additionally, there are other recent approaches which encrypt data with respect to attributes, known as Attribute Based Encryption (ABE) schemes. These proposals focus on creating a pair of keys, to encrypt and decrypt, in regard to an established group of attributes. Indeed, there are many ways of performing it, either using a single Central Authority (CA) to generate and distribute keys [9, 16, 31, 43], or using some CAs to decentralize key management [9, 31] or even developing a protocol to remove the necessity of a CA [30, 52]. ABE schemes can be divided into two groups, Ciphertext-Policy ABE (CP-ABE) [9, 16] and Key-Policy ABE (KP-ABE) [30, 31, 43, 52]. The former corresponds to the association of policies with ciphertexts and attributes to user keys and it is a remarkable technique in applications in which data is managed by multiple profiles, such as in hospitals or in the army. By contrast, the latter corresponds to the attachment of policies to user keys and attributes to ciphertexts, being useful in applications like auditing logs. The main difference between both approaches is that in CP-ABE attributes of key users are known, while in KP-ABE they are hidden. Besides, it is noticeable that both techniques have some limitation [74]: it is essential the existence of an authority to provide keys and flexibility of access control policy definition is currently restricted because disjunctions and conjunctions are the only operators used. As a result, much more work is required in spite of having appeared proposals to handle some of these problems [30, 52, 59].

Finally, revocation is a challenging feature that has to be managed in any access control mechanism. Specifically, it refers to the modification of access control policies. Some techniques are relatively simple such as using a revocation list [79, 83] or expiration times [71] while others are more sophisticated, for example the use of a PKI [20] or a particular scheme [10]. Besides, revocation is related to attribute modification. Once an attribute is modified the most appropriate procedure is to reevaluate policies and act accordingly.

4.3   Conceptualization of WBSNs

WBSNs allow their users to establish relationships each other and share their data. In this section, a conceptualization of WBSNs is presented to lay the bases for understanding the proposal and its later analysis. Commonly, WBSNs are modeled as graphs, being Harary who, in 1953, applied graph theory regarding group behavior, social pressure, cooperation, power, and leadership [46]. Indeed, Harary is considered the pioneering of the application of graph theory to network analysis.

More specifically, a graph is characterized by a huge quantity of entities, called nodes, and a vast quantity of connections between the nodes, called edges. In general terms, when modeling a WBSN as a graph, users correspond to the nodes and users relationships to the edges. This type of representation has been used by many authors in recent literature [23–26], being Carminati one of the most representative.

4.3.1   Data

In a WBSN, the set of considered data types may include photos, videos, wall messages, and personal messages that are private and directly written to a certain person or a group of people. Furthermore, according to Bruce Schneier, data in WBSNs can be further classified as: *service data*, data you give to the WBSN for using it; *disclosed data*, data posted in your own pages; *entrusted data*, data you post on other user's pages; *incidental data*, data other users post about you; *behavioral data*, data about your habits that sites collect; and *derived data*, data about yourself derived from what other users say [69]. In the conceptualization presented herein, only *service*, *disclosed* and *entrusted data* is considered since the rest of data types require a great deal of management complexity and such management is, indeed, far from being satisfactorily solved yet. The set containing all data will be referred to as D.

Additionally, data has a set of attributes associated with it, such that $dAT = \{dat_1, dat_2, \ldots, dat_{n_{dAT}}\}$ where n_{vAT} is the total number of data attributes. Data attributes can be classified into two groups. First group involves own features of data such as type of data, creation time, and size. Second group refers to any kind of characteristic that can be assigned to data, for example the fact of being private, confidential or public, or the topic of the data among others.

4.3.2   Actions

In a WBSN several actions can be performed on data. The set of defined actions is denoted as AC. Main four actions that can be performed over data are: *read*, equivalent to visualize any kind of content; *update*, equivalent to write down tags in videos or photos, or changing any commentary previously written; *insert* an element, equivalent to upload a photo or a video to the WBSN; and *delete* an element. Nonetheless, if needed, more actions can be considered.

4.3.3 Users

The set of users of a WBSN is identified herein as V. In [76] users are classified into three groups: *owners*, *originators* and *viewers*. Originators are those users that originally create and upload a specific data to the WBSN. Owners, who can be a single user or a group of users, administer access to the data they own and share ownership privileges with the data originator. Finally, viewers are those users that request access to a certain data. However, it is noticed that conflicts between owners and originators are a matter of vital concern regarding privacy in WBSNs. For example, if a user A takes a picture of a user B and uploads it to the data space of user C in the WBSN, who has ownership rights over the picture? In fact, the right question is not who has ownership rights, but who has administrator rights over the picture [77]. This issue is simplified in current WBSNs, such as the management of wall data in Facebook where it is established that if a user A writes in the wall of user B, B becomes the owner of the message. Another example where this problem is illustrated is that of a parent supervising access control and privacy preferences of his/her child in a WBSN. The parent is not the owner nor the originator of the data uploaded by the child to the WBSN, but he should be allowed to co-administer the access control policies of his child. This matter must be carefully studied and analyzed in future work; furthermore, it will be discussed in Sect. 4.7. Therefore, herein, for simplicity and without losing generality and considering a single piece of data, d_i, it is distinguished the data *administrator*, i.e., *administrator*(d_i), that is the user who administers the data access controls, and the data *requesters*, i.e., the remaining users of the WBSN that request access to that piece of data. Therefore, all the data administered by a user v_i is denoted as D_i, where $D_i \subseteq D$. If a single administrator is considered for each piece of data, it follows that $D = \bigcup_i D_i$ and $\forall\, i, j\ D_i \cap D_j = \emptyset$.

As it happened with data, users may have also a set of associated attributes. This set of attributes is referenced as $vAT = \{vat_1, vat_2, \dots, vat_{n_{vAT}}\}$ where n_{vAT} is the total number of users attributes. Some of these attributes can be reflected as data in the WBSN. On the one hand, a user profile links each user with his/her nationality, age, music preferences, and so on. On the other hand, there is other group of attributes, called user contextual attributes [33], that describe a user's personal mood, like happy, nervous, and so on, or his/her current activity like eating, running, etc. A user's location is a particularly relevant user contextual attribute in WBSNs like Google Latitude, in which this information is the main data managed, or in WBSNs such as Facebook Places, in which user's location can be associated with data.

4.3.4   Relationships

In a WBSN, a user can usually accept and withdraw the establishment of relationships with other users of the WBSN. As previously identified, WBSN relationships corresponds to edges of the social graph which connect directly or indirectly pairs of users.

Direct relationships between users of a WBSN are identified herein as E. A set of attributes can be associated to a direct relationship. This set is denoted as $eAT_i = \{eat_1, eat_2, \ldots, eat_{n_{eAT}}\}$. The most important of these attributes are the relationship direction and the relationship type. The former refers to the unidirectional or bidirectional nature of the relationships and the latter represents the relationship semantic meaning, which is also called the relationship role by some authors.

On the one hand, regarding relationship direction, relationships can be unidirectional or bidirectional [24]. The former corresponds to relationships in which a relationship request is only established in one direction. For example, given users A and B, if A makes a friend request to B and B accepts the request, then, A is said to be friend of B but it cannot be said the same in the other way round (i.e., B cannot be said to be friend of A). On the contrary, the latter, a bidirectional relationship, implies that both nodes associated with a relationship request, once accepted, have the same type of relationship in both directions. Specifically, bidirectional relationships can be identified as a pair of unidirectional relationships in which both nodes are issuers and recipients of the relationships at the same time. Indeed, this is the type used in current social networks.

On the other hand, semantic meaning or role of relationships imply having different relationships such as "friend," "professional," and "family." Some social networks focus on a specific role, for example LinkedIn is based on professional relationships. On the contrary, there are social networks in which different roles can be considered.

For example, in Facebook there exists the possibility of managing groups of people according to established preferences, e.g., a group of "family," a group of "bestFriends" and so on. Furthermore, multiple direct relationships may be considered between two nodes, each one with a different role.

Relationships may have other attributes such as creation time or history. Additionally, users may attach other attributes to their relationships such as a level of trust [24] and certain duration. Level of trust corresponds to a quantitative measure to determine the strength or weakness of a relationship. Concerning duration, it is the time during which the relationship remains valid.

Two users of a WBSN may not be directly connected but indirectly, that is, a direct relationship does not exist between them but a path, P, connecting both nodes can be found considering other users and their relationships. The path consists of an ordered set of edges or direct relationships such that $P = \{e_1, e_2, \ldots, e_n\}$. In particular, a relationship between two users v_i and v_j is denoted as P_{v_i, v_j}. The number of edges in the path is referred to the length of the relationship. Note that if length equals one, it is the case of a direct relationship, and indirect if length is greater than one [57].

Therefore, the concept of user relationships and their attributes can be generalized to consider both direct and indirect types. The set of such relationships is then denoted as P and their attributes $pAT = \left\{ pAT_1, pAT_2, \ldots, pAT_{n_{pAT}} \right\}$, where $pAT_i = \{eAT_1, eAT_2, \ldots, eAT_n\}$. These attributes are similar to those associated with direct user relationships and can be derived from the attributes of the set of edges that compose the path. That is, given attributes $pAT_j \in pAT$ and a path $p_k \in P$, each $eat_n \in eAT_n \in pAT_n$ is calculated as $eat_i(p_k) = f_{\{i = 1, \, length(p_k)\}}(e_i)$, where e_i are the edges that compose p_k. For example, given three users, v_1, v_2, and v_3, connected in pairs by $e_1 = \{v_1, v_2\}$ and $e_2 = \{v_2, v_3\}$, the path between v_1 and v_3 is $p_1 = \{e_1, e_2\}$. As proposed by Carminati et al. in [25], the level of trust of such relationship p_1 can be calculated as $trust(p_1) = \sum_{i = 1, \, length(p_1)} trust(e_i)$. The set of functions used to derive the value of the attributes $pat \in pAT$ will be denoted by F and its definition is far from being a minor issue.

4.3.5   Context

Context information is other aspect that may be considered when modeling a WBSN. The set of these features is denoted as CX. Dynamic, assorted, and external features such as communication network status, service availability or other quality parameters can be involved here.

4.3.6   Summary of the Conceptualization

The conceptualization provided above is summarized in this section and depicted in Fig. 4.1.

Thus, a WBSN can be conceptualized as: WBSN = $\{V, vAT, P, pAT, F, D, dAT, AC, CX\}$. V corresponds to the graph nodes, which represent the WBSN users. P are the graph edges which represent the relationships which connect two users. D represents data and AC corresponds to actions that can be performed on data. vAT, dAT, and pAT represent the set of user, data and relationship attributes, respectively, and F the set of functions that allow the derivation of attributes values associated with indirect relationships. Finally, CX represents the system's context.

In particular, eAT, given that $eAT \in pAT$, considers, summing up, direction, role, level of trust, duration and own features such as history or creation time; vAT mainly corresponds to profile data and location; and dAT are features assigned such as confidentiality or privacy labels, related topic, and other own features like creation time.

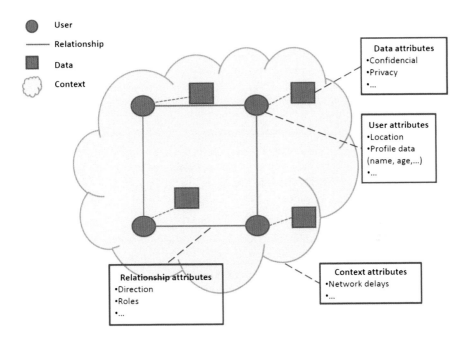

Fig. 4.1 Web based social network conceptualization

Note that, in the provided conceptualization, it is assumed that the set of user, data and relationship attributes is defined by the system, and that their value can be assigned by one of the involved parties. For example, the WBSN can assign to a piece of data its creation time, type, etc.; a user can assign the type, level of trust and duration to one of his/her relationship; external entities can be in charge of attesting real user attributes such as name, age, or driving license.

Finally, notice that as the conceptualization presented above is performed from a logical point of view, the way in which the storage is carried out, centralized like Facebook, decentralized such as Diaspora, encrypted or in plaintext, as well as, the way of specifying and managing relationships are not particularly considered.

4.4   Requirements for User-Managed Access Control in WBSNs

In [29] a set of requirements are identified as key in order to successfully develop user-managed access control in Web 2.0, which includes WBSNs. These requirements are described next.

1. *relationship-based*: data administrators (e.g., the data owner) control the release of data based on the established relationships with data requesters, instead of delivering information depending on the requester role or any other feature.
2. *fine-grained*: users must control their information, choosing who is able to access it and under which circumstances in a fine-grained way. It should be possible to define fine-grained policies for both data (a specific portion of data within a data structure) and requesters.
3. *interoperability*: users access multiple WBSNs and want that their data is used in a similar way in many of them. Access control systems should be interoperable between different WBSNs, so, it would be possible that user preferences follow the user whatever WBSN is used to access the user's data.
4. *sticky policies*: policies should follow the data to which they apply, preventing from uncontrolled data disclosures after being released.

In this work, a fifth desirable requirement is identified, *data exposure minimization* against honest-but-curious storage services. Servers may protect information stored against possible threats but users are unaware of the servers procedures and techniques to carry out this issue. Indeed, personal information is completely susceptible of being used by companies, specially advertising ones, to improve and develop their products according to users perceptions [20, 79]. Therefore, it is desirable that users have control over their personal information without letting servers access the user's data unknowingly.

4.5   Proposal for User-Managed Access Control in WBSNs

One of the purposes of this work is to explore access control solutions that satisfy the five requirements for WBSN described in Sect. 4.4. Alluding to the identified problems, which refer to the necessity of a model that satisfies the whole set of requirements and the definition of the related mechanisms, the main pair of contributions are described next:

- First, a model that, from a theoretical perspective, satisfies the whole set of requirements and a mechanism that, in practice, satisfies the requirements of *relationship-based* and *fine-grained* are presented. The model is an extension of the $UCON_{ABC}$ usage control model proposed by Park et al. [62] and of the $UCON_{ABC}$ administrative model proposed by Salim et al. [66]. The model is recognized as $SoNeUCON_{ABC}$, from Social Network $UCON_{ABC}$. The $SoNeUCON_{ABC}$ model is presented in Sect. 4.5.1 and the basic mechanism that implements it, as well as the architecture attached to it, are described in Sect. 4.5.2.
- Second, mechanisms that can be used on the previous basic system to satisfy the other three requirements (i.e., *interoperability*, *sticky policies*, and *data exposure minimization*) are discussed and the possibilities of integrating these mechanisms on the basic access control system are also noticed. Moreover, the particular

architecture attached to them are described. Sections 4.5.3, 4.5.4, and 4.5.5 address each of the mentioned requirements, respectively.

4.5.1 Extending the UCON$_{ABC}$ Model to Consider Relationships

At first sight, in order to address the *relationship-based* requirement, the RelBAC approach [39, 42, 65] (mentioned in Sect. 4.2) can be pointed out as the ideal proposal for social networks. However, WBSNs, though one of its main features is being relationship-based, need access control systems that are also capable of considering fine-grained policies in regard to attributes of data and requesters. For example, geo-social networks consider location, which is a feature directly related to users and difficult to handle through relationships.

On the other hand, access control systems that follow the ABAC approach consider attributes of users, data and (not always) context as the main policy elements to take the decision about whether a requested action is authorized or not. Indeed, the ABAC approach allows fine-grained access control but on the negative side, it sets aside relationship management. Therefore, in this work, the $UCON_{ABC}$ model is selected as it is the most representative and mature ABAC-based model in comparison with other approaches like [80] or [73].

On the one hand, in [80], subjects, resources, and contextual attributes are considered but other relevant elements such as rights or authorizations managed in $UCON_{ABC}$ are left aside. On the other hand, in [73], attributes are linked to the related entities without specifying particular types of them.

Under the assumption that the $UCON_{ABC}$ model is used, it is extended to include an exhaustive relationships management. Relationships can be included in the authorization decision process in different ways. For instance, relationships can be defined as an attribute of the origin entity of the relationship, i.e., a user Alice has a list of "friends," a list of "relatives," a list of "co-workers," etc., and these lists are attributes of the user Alice. This is the approach taken in some recent proposals for WBSN that build on $UCON_{ABC}$ [61, 63]. Even the decentralized administrative model of Salim et al., which has a direct application in WBSNs, takes this approach [66].

Nonetheless, WBSNs require the management of attributes as well as direct and indirect relationships. In this regard, even being an ongoing work, the study of WBSN features with respect to relationships management, at a starting point $UCON_{ABC}$ lacks indirect relationship management. To verify this issue Example 1 presents an access control policy that is tried to be defined by $UCON_{ABC}$.

Example 1. A user grants read access to users who are over 18 and who are friends of a friend of a friend.

According to $UCON_{ABC}$ the following subjects attributes $(ATT(S))$, objects attributes $(ATT(O))$ and predicates are required to define Example 1 access control policy:

- $ATT(S) = \{Age, Friends\}$ where Friends is the list of friends that the user has. Moreover, each list is composed of a set of attributes:

  - Friends $= \{User1_{Id}, relationshipTrust, \ldots\}$

- $ATT(O) = \{Object1_{Id}, \ldots\}$

- permit$(s \in S, o \in O, r)$: it grants access permission (r) over an object (o) to a particular subject (s).

- in$(s \in S, Friends of v \in S)$: it returns the existence of not of a friendship relationship between a pair of users $(s$ and $v)$. Notice that s has to be within the list of friends of v.

Then, considering that the administrator of the requested object o is referred to as a and s corresponds to the requester, the access control policy proposed in Example 1 and established by a is defined following [81]:

- $in(a, s.Friends) \wedge s.Age > 18 \longrightarrow permit(s, a.o, r)$

At first, it can be thought that an assorted set of predicates can be devised to construct policies but their specification is restricted regarding elements involved in $UCON_{ABC}$. As pointed out in [61, 63], $UCON_{ABC}$ addresses relationships through subjects attributes and thus they facilitate the management of direct relationships but prevent from managing the indirect ones. Consequently, a novel model, called $SoNeUCON_{ABC}$, is developed to address this issue. Nevertheless, a comparison and a deep analysis of relationships management with respect to existing access control models is a matter of future work (Sect. 4.7).

Therefore, the $SoNeUCON_{ABC}$ access control model, an extension of the $UCON_{ABC}$ model is proposed in this work (Fig. 4.2). According to $UCON_{ABC}$ a new independent entity, *relationship* (RT), and set of attributes, *relationship attributes* $(ATT(RT))$, are included. Note that either direct or indirect relationships are managed through attributes and, for instance, the attribute *length*, the distance between two nodes, can be applied to deal with the indirect ones. The point is that access control is managed through the establishment of policies in which $ATT(S)$, $ATT(O)$, $ATT(RT)$, and R are involved. The set of original entities, sets of attributes and functions considered in the $UCON_{ABC}$ model are also considered in the $SoNeUCON_{ABC}$ model. The elements of the WBSN conceptualization can then be related to those in the $SoNeUCON_{ABC}$ model:

- *Subjects* (S) are the WBSN users (V), previously identified as administrators and requesters; additionally, $ATT(S) \subseteq vAT$.
- *Objects* (O) are WBSN data (D), identified as photos, videos, wall messages and personal messages; additionally, $ATT(O) \subseteq dAT$.

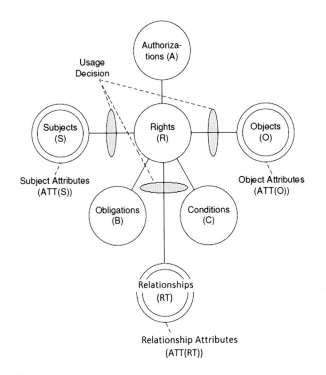

Fig. 4.2 *SoNeUCON*$_{ABC}$

- *Relationships* (RT) represent the set of relations that exist between a pair of users of the WBSN. Given a relationship between v_i and v_j, such set is noted as $P(v_i, v_j)$. This set is composed of two subsets, namely forwards paths set ($P''(v_i, v_j)$) and backward paths set ($P'(v_i, v_j)$), such that $P(v_i, v_j) = P''(v_i, v_j) \cup P'(v_i, v_j)$. Both subsets contain different paths $p_k(v_i, v_j)$, where a path is a collection of direct relationships that form together a way to connect v_i and v_j.

  As an example let us calculate $P(v_5, v_1)$ in the social network scenario depicted in Fig. 4.3, where v_1 is the requester and v_5 is the administrator. In such network, all direct relationships are bidirectional. Thus, given a direct relationship between v_1 (the requester) and v_2 (the administrator), the forward edge is referred to as $e_{2,1}$ whereas the backward one is noted as $e_{1,2}$. The forward paths set $P''(v_5, v_1)$ is given by: $P''(v_5, v_1) = \{\{e_{5,2}, e_{2,1}\}, \{e_{5,4}, e_{4,3}, e_{3,1}\}\}$.

  On the other hand, the backward paths set $P'(v_5, v_1)$ is: $P'(v_5, v_1) = \{\{e_{2,5}, e_{1,2}\}, \{e_{4,5}, e_{3,4}, e_{1,3}\}\}$.

  Taking into account both subsets, the set $P(v_5, v_1)$ is formed by their union, $P(v_5, v_1) = \{\{e_{5,2}, e_{2,1}\}, \{e_{5,4}, e_{4,3}, e_{3,1}\}\} \cup \{\{e_{2,5}, e_{1,2}\}, \{e_{4,5}, e_{3,4}, e_{1,3}\}\}$

  Finally, notice that $ATT(RT) \subseteq pAT$.

- *Rights* (R) refer to the actions (AC) that can be performed over WBSN data such as read, update, or delete;

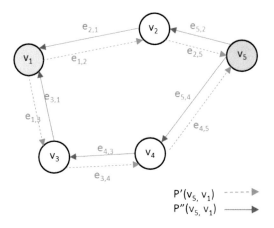

Fig. 4.3 Relationships example, $P(v_5, v_1)$

- *Authorizations* (A) are the rules defined as functional predicates that have to be satisfied in order to grant a subject a right on an object. Along this chapter these elements will be called policies;
- *Obligations* (B) refers to activities that have to be carried out by the user before or while the usage process.
- *Conditions* (C) correspond to the previously identified set of context features (CX), such as network availability.

Bearing in mind that the formalization of the model is not the goal of this chapter, but a future open research issue, the definition of policies is briefly detailed. In particular, in the construction of policies (ρ) it is specified that S can perform R on O with respect to $ATT(S)$, $ATT(O)$, and $ATT(RT)$.

Each policy ρ in *SoNeUCON*$_{ABC}$ model is generally depicted as follows:

- $\rho(\rho_s; \rho_o; \rho_{rt}; r)$

    - $\rho_s = \{subject\ predicates\}$
    - $\rho_o = \{object\ predicates\}$
    - $\rho_{rt} = \{length\ \{\{path_1\}, \{path_2\}, \ldots, \{path_n\}\}\}$ where it is defined as a set of paths composed of a set of direct relationships, together with the use of the mandatory attribute *length* that corresponds to the distance between a pair of nodes (the administrator and the requester). It takes value $k \in \aleph$ and $1 \le k \le |S|$ because the maximum length is the total number of WBSNs users and also, it can take value "*" to express that all kinds of relationships lengths are accepted. Moreover, paths can be described as unidirectional or bidirectional.

- $r = \{read\ |\ write\ |\ \ldots\}$

More specifically, the requested $r \in R$ over $o \in O$ is granted if $ATT(S)$ of the requester and $ATT(RT)$ between the requester and administrator satisfy the appropriate ρ. Notice that the request consists of the requester (s), the requested object (o) and the requested right (r) over the object, $Req = \{s, r, o\}$:

- $allowed(s, o, r) \Rightarrow \rho(\rho_s; \rho_o; \rho_{rt}; r)$

Finally, to have a better understanding of the creation of access control policies in $SoNeUCON_{ABC}$, the policy presented in Example 1 is defined:

- $\rho = (\text{age} > 18, \emptyset, \{\text{length} = 3 \{\text{role} = \text{friend}\},\{\text{role} = \text{friend}\},\{\text{role} = \text{friend}\}\}, r)$, where $\emptyset$ implies that any particular set of restrictions is specified in relation to requested objects.

As a result, though requiring a detail definition, this chapter presents $SoNeUCON_{ABC}$, a powerful access control model developed as an extension of the $UCON_{ABC}$ model to reach relationships management, involving both direct and indirect relationships.

4.5.2   Basic Mechanism

Relationship-based and *fine-grained* requirements are already satisfied if the $SoNeUCON_{ABC}$ model is followed. On the one hand, the management of relationships is performed through the establishment of paths between users as it is described in Sect. 4.5.1. On the other hand, fine-grained management is carried out by the appropriate use of subjects, objects and relationship attributes, reaching the creation of assorted and expressive policies. Thus, the basic approach focuses on both requirements. However, there exist several possibilities for implementing it. In this approach, a server-side architecture that considers a single domain is proposed. This approach is depicted in Fig. 4.4a. Both data and policies are located in a single domain corresponding to the WBSN. Thus, the WBSN takes the role of both the PEP and PDP elements of the reference monitor. Data administrators or system administrators establish usage or administrative policies, and according to them, the WBSN grants or denies requested access.

Nonetheless a significant point is the construction and management of access control policies. Regarding policies constructions, they could be defined according to all elements of the $SoNeUCON_{ABC}$ model, but the discussion in this work will be restricted to the authorizations element, as the issues related to the obligations and conditions elements do not present specific differences with respect to the original $UCON_{ABC}$ model and its implementations. Therefore, policies can be developed using any kind of user, relationship or data attribute, and actions. Note, however, that regarding the duration relationship attribute, it corresponds to an absolute time range during which the relationship exists. Analogous to a public key digital certificate, it can be described as "'valid-from ..., valid-until...'"; for example, "'valid-from 10th September 2011 12:00, valid-until 10th September 2011 14:00'".

On the other hand, according to policies management, this architecture requires that administrators establish policies in the storage service (the WBSN) with respect

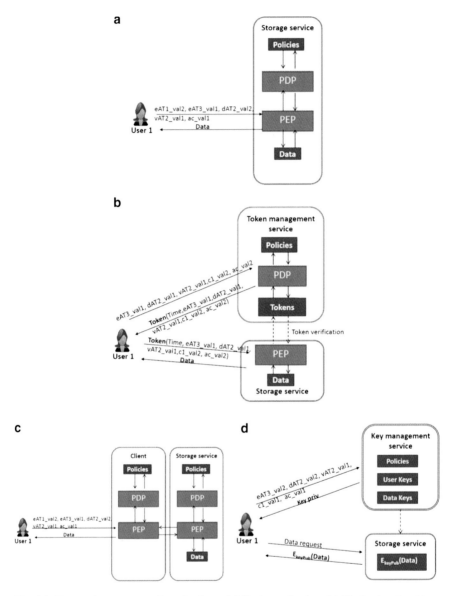

Fig. 4.4 Proposed access control mechanisms. (**a**) Basic mechanism (**b**) Mechanism to address interoperability (**c**) Mechanism to address sticky policies (**d**) Mechanism to address data minimization exposure

to their access preferences. Then, in case that a change in the policies is needed (e.g., caused by revocation), they simply update the policies and the changes will be taken into account in the next access request. Note that this issue is, whilst different, related to the continuity of decision functions that may be used if mutable attributes are considered.

Finally, a challenging issue is the simplicity of the mechanism either for administrators or users. Administrators only create, establish, and modify policies attached to data and requesters exclusively send requests to the PEP. However, there are a pair of relevant drawbacks. Firstly, as *data exposure minimization* requirement is not considered in this mechanism, data is not stored under user control and can be compromised. Then, trust in the storage device has to be assumed. Secondly, each time a user wishes to access to a specific data, policies must be satisfied. By doing this, policy enforcement is frequently performed and it can produce a bottleneck in the storage device (performance/workload).

4.5.3   Addressing Interoperability

WBSNs are independent services that have different (although potentially similar) access control models, with different privacy policy definition languages and specific types of data and actions particular for each WBSN. Currently, small interactions are allowed between different WBSNs, such as importing a list of contacts or propagating messages in other WBSNs after they have been published in one of them. Naturally, users belonging to different WBSNs would like to have the possibility of controlling their data and relationships in a global way, but, in the described situation, an interoperability problem exists. This problem is called in the literature the *Walled Garden Problem* [55].

The approach suggested in this work is to establish data usage policy man-agement and data storage in different domains. Administrators specify policies associated with data in a specific access control management service while data is stored in WBSNs. Therefore, the PDP is located in the first and the PEP is located in the second. The decoupling between both services should be so that access control management services do not know data details and data storage services do not know policy details; however, both types of services must count on a common data identification mechanism (note that it does not have to be for each independent data item, data sets are also appropriate) that allows the identification of the access control policies at stake given a certain access request. The decoupling should be also enough to allow a seamless application to use the same policies to different data sets independently of the WBSN in which data is stored in.

Access control tokens are a natural way to implement this approach. Particularly, the ITU-T X.812[6] recommendation describes tokens as elements possibly created by requesters and composed of multiple information. Besides, it is highlighted that they differ from certificates in the fact that certificates are delivered by a trusted authority. However, without losing generality and simplifying this issue, in this work, it is considered that tokens comprise tickets, certificates, or any other elements which may provide the sufficient access control information to the PEP. Therefore,

---

[6]http://www.itu.int/rec/T-REC-X.812/en, last access November 2012.

requesters are granted access only after they have acquired the appropriate token. Specifically, a requester first obtains the access token at the PDP and then he presents this token at the PEP (which corresponds to the WBSN) to access to the requested data. If the PEP can extract directly from the tokens the usage decision taken by the PDP, no more interactions are needed. However, depending on the specific design of tokens, an interaction between the PEP and the PDP may be needed. The architecture described here is depicted in Fig. 4.4b.

As in the previous basic mechanism (Sect. 4.5.2), features involved in policy construction are user, data and relationships attributes, and actions. However, there is a fundamental difference between the previous mechanism and this one, tokens can have a specific duration attached to them and once a token is acquired it can be used in several data requests until it expires. In this case the duration can consider an absolute time period or relative to the token creation.

Interestingly, in this mechanism it must be pointed out the way in which tokens are delivered. They can be delivered through out of band mechanisms such as emails and phone calls, or through a specific entity specially developed to achieve this goal. Besides, bearing in mind that tokens are usually valid for a particular period of time, the main way to distribute them is through an entity, for example called token management service.

In regard to administrative operations, administrators have to establish policies in the token service and relate them to the data stored in the chosen storage service. Indeed, this issue can be performed through multiple ways, for example using a table in the token service to associate policies, tokens, and data. Afterwards, when revocation is required different strategies can be applied. Tokens can be re-distributed or, due to the expiration time attached to them, requesters may, when necessary, acquire the corresponding new one.

This mechanism provides substantial benefits regarding the rest of mechanisms. It provides flexibility because policies are separated from data and they can be managed independently. Moreover, this decentralization leads to the reduction of overheads in the storage device. Tokens are valid during a period of time and policies do not need to be checked in each access request. Furthermore, this mechanism provides a simple technique to control revocation due to time stamps attached to each token. Nonetheless, there are some disadvantages. First, as in the basic approach, the problem of honest-but-curious storage services persists. Second, an entity to manage tokens is required. Thus, it is assumed an entirely trusted storage device and a trusted entity that acts as a token provider. In the end, more interactions are involved in this access control mechanism, making the process a little harder to perform. In particular, communications between the user, the token service and the storage device are expected.

A chief example of recent efforts to design this type of mechanisms is the one developed by the User-Managed Access (UMA) Working Group which presents an architecture, called UMA,[7] to give users control over their personal data. This

---

[7]http://kantarainitiative.org/, last access November 2012.

architecture focuses on keeping data separated from its access control management and requires a token to access data. Besides, another challenging example, directly related to WBSNs, is PrPl [71]. In this approach users store data in their personal computers or in chosen servers, called *butlers*, and the access is managed through the delivery of tickets, which are elements equivalent to tokens.

4.5.4 Addressing Sticky Policies

According to the *sticky policies* requirement, users must be able to control the usage of their data after access to it has been granted. A feasible way of facing this requirement is through the continuous performance of policy evaluation and enforcement, avoiding conflicts such as the execution of not granted activities (e.g., the printing of an object in which just a read permission is attached) [51]. However, this simple technique may produce high load of traffic or overload in WBSN servers, being indispensable the development of more efficient procedures. The fulfilment of this requirement is directly related to the specific architecture selected for implementing the access control mechanism.

Generally, to guarantee a full protection of data, some type of reference monitor must be deployed in the client (i.e., the location where data is going to be used). If combined with the basic approach described in Sect. 4.5.2, an architecture such as the one depicted in Fig. 4.4c can be used. The reference monitor in the client will be in charge of acting according to the indications of the reference monitor in the server and then, if a change in the policies (revocation rights) affects the data in use, the client reference monitor has to carry out the appropriate verifications by connecting to the server. In other words, if revocation of a requester (v_i) is performed in respect to a certain data (d_j) when v_i is using d_j, d_j must be immediately unavailable to v_i.

Indeed, any WBSN mechanism can (partially) fulfil this requirement but taking the following matter into account. As social networks are composed of web elements, an inherent property is that the web site content, once visualized, is available until reloading the page. Due to that fact, if v_i is using d_j and simultaneously access permission to d_j is denied to v_i, v_i will be able to manage d_j until the page is reloaded. A full compliance with the *sticky policy* paradigm would mean that after the change is made on the policies, their effects are immediately enforced.

4.5.5 Addressing Data Exposure Minimization

Servers store information using its own techniques and users are not directly involved in controlling data which may lead to unknown disclosures of information. As a result, the main applicable countermeasure is cryptography, providing therefore a cryptographic access control mechanism (called "lock and key" in Sect. 4.2).

Ideally, the storage service (the WBSN) has no access to clear data or policies associated with it. In fact, as mentioned in Sect. 4.2, under this approach a reference monitor is not specifically needed because data administrators encrypt data and it is stored encrypted in the storage service. Requesters may freely access to encrypted data, as only having the appropriate decryption key would grant them access to data in clear.

One of the approaches that can be taken is that administrators create user keys according to the keys used to encrypt the data whose access needs to be granted. The other approach, more complex, is the one considered by Attribute Based Encryption (ABE)d or any cryptographic encryption algorithm in which keys or ciphertexts involve a wide range of attributes. Anyhow, in both cases a user key is delivered to the user. Then, analogously to token-based access control (see Sect. 4.5.3), the way of delivering keys can take different forms. As in the previous case, they can be delivered through out of band mechanisms or through a specific entity. Without losing generality, it is assumed that an entity called key management service distributes available keys, and such architecture is depicted in Fig. 4.4d.

Besides, with respect to revocation, it can be managed on different ways. In general, administrators have to create new keys, perform the appropriate re-encryptions and distribute the new ones. However, depending on the approach this scheme can be modified, for example, a proxy can be used to perform the appropriate re-encryptions [48].

Lastly, pros and cons of this mechanism are pointed out. On the one hand, this mechanism has a relevant benefit that is users control their own data. On the contrary, there are several drawbacks. Revocation can be quite tedious due to re-distributions or re-encryptions. Also, as in token-based access control, it is indispensable trusting the entity which manages keys. Eventually, due to the use of cryptography, performance is reduced, though depending on the approach and even causing minimal negative effects.

Notice that apart from cryptography which is the most straightforward technique to devise, other ones could be applied (e.g., steganography), though they have not received significant research attention yet.

4.6   Analysis of Web Based Social Networks

Since the emergence of WBSNs, awareness of the importance of privacy has been steadily increasing. One of the goals of this work, which refers to the last contribution, is to analyze recent advances in user-managed access control systems for WBSN. Therefore, in this section, a set of 25 academic proposals that work on privacy regarding access control in WBSNs and 9 of the most currently used and active WBSNs are analyzed in regard to their adaptation to the identified WBSN's requirements, i.e., relationship-based, fine-grained, interoperability, sticky policies and data exposure minimization.

Tables 4.1 and 4.2 summarize the results of the analysis. In general terms, academic proposals largely differ from WBSNs in use. Academic contributions are specially focused on cryptographic access control enforcement and the development of novel techniques to provide user-managed data control, although leaving aside sticky policies. By contrast, the majority of active WBSNs tend to manage access control in a simpler way, neglecting important requirements such as *interoperability* or *data exposure minimization*. Moreover, it has to be noticed that neither academic proposals nor WBSNs in use, except for Badoo, handle conditions or obligations.

In order to structure this analysis, in each proposal and WBSN in use, the following ten features have been examined:

- *Elements in authorization predicates*: as mentioned in Sect. 4.5, authorization predicates should be composed of rights and user, data and relationship attributes, which is directly related to *relationship-based* and *fine-grained* requirements. In particular, the considered attributes of (direct) relationships are direction, roles, level of trust, duration and own features such as history or creation time. Regarding indirect relationships, as mentioned in Sect. 4.3, attributes are generally derived from those of the direct ones. This issue is noted with the symbol γ in Table 4.1. Mainly, to manage indirect relationships, academic proposals use a particular feature called *depth* [11, 27, 49, 50] that is included in policies (this is the feature referred to as *length* previously). Similarly, Facebook, Picasa, MySpace and Flickr also manage this issue but without giving details of the internal procedure. Data attributes considered in the analysis are data privacy (user assigned label) and own features like creation time. Moreover, considered user attributes are location and profile data. In some approaches, an identification attribute may be considered for data, users, and relationships. Although this attribute is not explicitly used in academic proposals. Likewise, it is also necessary to highlight that rights are only pointed out if they are considered in the policy itself. Indeed, rights are always involved in access control. Once giving access to a particular object, at least the read permission is granted. However, in these cases, in which rights are not directly involved in policies, the management of rights is not distinguished.
- *Interoperability*: indication of the management of tokens which, as previously studied, is a key step towards interoperability.
- *Sticky policies*: indication of the fact that policies follow the data to which they apply.
- *Data exposure minimization*: indication of a technique to prevent storage services from accessing data that users have not authorized. As pointed out in Sect. 4.5.5, this issue can be solved by different techniques but it is cryptography the only one applied. Therefore, the analysis of this feature refers to the identification of having or not used cryptography.
- *Policy definition*: specification of languages or tools used to develop policies, like XML.
- *Token element (if required)*: element used as a token. For instance, certificates and tickets.

Table 4.1 Analysis of academic proposals related to access control in WBSNs

Proposals	Features									
	Elements in authorization predicates	Interoperability	Sticky Policies	Data exposure minimization	Policy definition	Token element (If required)	Keys (If required)	Details on policy evaluation and enforcement mechanisms	Revocation techniques	Trust entity
D-FOAF (2006) [50]	eAT(trust, γ)	–	–	–	–	–	–	Resource granted if satisfying policies according to policy elements used an ACL	–	Group of users
A semantic web based framework (2009) [27]	eAT(role, trust, γ) + Rights	–	–	–	SWRL	–	–	Resource granted if satisfying policies according to policy elements	–	–
Granular framework (2009) [15]	eAT(role) + vAT	–	–	–	–	–	–	Resource granted if satisfying policies according to policy elements	–	–

SAC (2007) [6]	eAT(trust)	✓	–	–	Keys	Key portions	Resource is granted if having the appropriate token	–	Device (storage)
PrPI (2010) [71]	vAT + Rights	✓	–	–	Tickets	Key pair(private/public) per user	Resource is granted if having the appropriate token	Expiration time	Device (storage)
FaceVPSN (2011) [32]	–	✓	–	–	XML file	–	Resource is granted if having the appropriate token	–	–
Lockr (2009) [79]	eAT(role)	✓	–	–	Attestations	Key pair(private/public) per user	Resource is granted if having the appropriate token and verifying the token against special list	Re-sending tokens, through a external element (revocation list) or expiration time	–
Secure & Flexible framework (2010) [6]	eAT(role)	✓	–	–	Certificate	Key pair(private/public) per user	Resource is granted if having the appropriate token and verifying the regular expression attached to the token	Through a external element (certificate) attached a expiration time	–

(continued)

Table 4.1 (continued)

		Features								
Proposals	Elements in authorization predicates	Interoperability	Sticky Policies	Data exposure minimization	Policy definition	Token element (If required)	Keys (If required)	Details on policy evaluation and enforcement mechanisms	Revocation techniques	Trust entity
Vis-a-Vis (2010) [72]	vAT	✓	–	–	–	Group descriptor	Key pair(private/public) per user	Resource is granted if having the appropriate token and satisfying policies according to policy elements	–	Group of users + Device (elements provide by third parties)
Securing Social Networks (2011) [10]	eAT(role) + Rights	✓			Specific API	Signed pseudonym	Not implicitly specified	Resource is granted if having the appropriate token and verifying the token against a special list	Through a particular scheme (broadcast encryption)	X
NOYB (2008) [45]	–	–		✓	–	–	Depending on web services	Cryptographic algorithm based on dictionaries	Key updates + Re-encrypting	Group of users + Device (elements provided by third parties)

FlyByNight (2008) [53]	–	–	✓	–	–	Key pair(private/public) per user, other group pair and proxy keys per user per group	Cryptographic algorithm with proxy cryptography to handle "one-to-many" requests	Key updates + Re-encrypting	–
Safebook (2009) [35]	eAT(trust)	–	✓	–	–	Key pair(private/public) per user, attribute	Cryptographic public key algorithm and certificates	Key updates + Re-encrypting	Group of users
PeerSoN (2009) [20]	eAT(role)	–	✓	–	–	Not implicitly specified	Cryptographic algorithm	Through a external element (PKI)	
FaceCloak (2009) [54]	eAT(role)	–	✓	–	–	A master, an index and an access key per user and a master and index key per requester	Cryptographic algorithm	X	Group of users + Device (server)
Collaborative framework (2010) [83]	–	–	✓	–	–	Key pair(private/public) per user and other group pair	Cryptographic algorithm based on data blocks	Through a external element (revocation list)	Group of users

(continued)

Table 4.1 (continued)

Proposals	Elements in authorization predicates	Features				Token element (If required)	Keys (If required)	Details on policy evaluation and enforcement mechanisms	Revocation techniques	Trust entity
		Interoperability	Sticky Policies	Data exposure minimization	Policy definition					
Key-allocation scheme (2009) [40]	–	–	–	√	–	–	Key pair(private/public) per user and as many keys as the maximum path length	Cryptographic algorithm based on satisfying indirect paths	Key updates	Device(server)
LotusNet (2012) [3]	–	√	–	√	–	Certificate	Key per user	Certificate validation	Expiration time	Group of users
Persona (2009) [11]	eAT(role, γ) + vAT + Rights	–	√	√	Using attributes and rights	–	Key pair(private/public) per user and a key per group	Cryptographic algorithm based on CP-ABE	Key updates +Re-encrypting	Device(Key storage)
EASiER (2011) [48]	eAT(role)	–	√	√	Using attributes	–	Key pair(private/public)per user and proxy keys depending on attributes and users	Cryptographic algorithm based on CP-ABE	Through a external element (proxy)	Proxy

DECENT (2012) [47]	eAT(role)	–	✓	✓	Using attributes	Key pair(private/public) per user depending on attributes and users and a signature key	Cryptographic algorithm based on CP-ABE	Through a external element (proxy)	–
StegoWeb (2011) [14]	–	–	–	✓	–	–	cryptographic algorithm	Key updates	Group of users
Prometheus (2010) [49]	eAT(trust, role, γ)	–	–	✓	–	Key pair(private/public) per object or a key per object	Resource granted if satisfying policies according to policy elements and cryptographic public key algorithm	Through a external element (revocation list)	Group of users
LifeSocial.KOM (2011) [44]	–	–	–	✓	–	Key pair(private/public) per user and key per object	Cryptographic algorithm	–	–
Helloworld (2009) [1]	–	✓	–	✓*	Similar to OpenId URI	Key pair(private/public) per user	Message interchange protocol	–	Group of users

* Data partially encrypted

Table 4.2 Analysis of access control features in WBSNs in use

	Features									
Proposals	Elements in authorization predicates	Interoperability	Sticky Policies	Data exposure minimization	Policy definition	Token element (If required)	Keys (If required)	Details on policy evaluation and enforcement mechanisms	Revocation techniques	Trust entity
LinkedIn (2003)[a]	eAT(role, γ) + dAT(data privacy)	–	–	–	–	–	–	Resource granted if satisfying policies according to policy elements	–	–
Hi5 (2003)[b]	eAT(role) + dAT(data privacy) + Rights	–	–	–	–	–	–	Resource granted if satisfying policies according to policy elements	Through a external element (revocation list)	–
Facebook (2004)[c]	eAT(role) + Rights + vAT(location) + dAT(data privacy) + Rights	–	–	–	XACML	–	–	Resource granted if satisfying policies according to policy elements	Through a external element (revocation list)	–
Orkut (2004)[d]	eAT(role)+ vAT(email)	–	–	–	–	–	–	Resource granted if satisfying policies according to policy elements	Through a external element (revocation list)	–

Badoo (2006)[e]	dAT(data privacy)	–	–	–	–	–	Resource granted if satisfying policies according to policy elements	Through a external element (revocation list)	–
Twitter (2006)[f]	dAT(data privacy)	–	–	–	–	–	Resource granted if satisfying policies according to policy elements	Through a external element (revocation list)	–
Picasa (2002)[g]	dAT(data privacy) + eAT(γ)	✓	–	–	–	URL	Resource granted if satisfying the appropriate token	Through a external element (revocation list)	–
MySpace (2003)[h]	eAT(role, γ) + vAT(age) + dAT(data privacy)	✓	–	–	–	URL	Resource granted if satisfying policies according to policy elements or having the appropriate token	Through a external element (revocation list)	–

(continued)

Table 4.2 (continued)

Proposals	Elements in authorization predicates	Interoperability	Sticky Policies	Data exposure minimization	Policy definition	Token element (If required)	Keys (If required)	Details on policy evaluation and enforcement mechanisms	Revocation techniques	Trust entity
Flickr (2004)[i]	$eAT(role, \gamma)+$ dAT(data privacy,own features)+ Rights	√	–	–	XACML	URL	–	Resource granted if satisfying policies according to policy elements and/or having the appropriate token	Through a external element (revocation list)	–

[a] http://www.linkedin.com/, last access November 2012
[b] http://www.hi5.com/, last access November 2012
[c] http://www.facebook.com/, last access November 2012
[d] http://www.orkut.com/PreSignup, last access November 2012
[e] http://badoo.com/, last access November 2012
[f] https://twitter.com/, last access November 2012
[g] https://picasaweb.google.com/home, last access November 2012
[h] http://www.myspace.com/, last access November 2012
[i] http://www.flickr.com/, last access November 2012

- *Key element (if required)*: element(s) used as keys, specifying required types. For instance, a pair of public/private keys.
- *Details on policy evaluation and enforcement mechanisms*: description of ways to carry out policy evaluation and specific enforcement mechanisms.
- *Revocation techniques*: revocation considers the modification of policies. This issue can be commonly managed by any technique that makes use of policies by simply changing them. Nonetheless, in this analysis, revocation is exclusively studied in academic proposals and WBSNs in use that explicitly mention having tackled this issue.

  In general terms, techniques to perform revocation are classified as the following groups:

  - Changing policies: revocation can be simply carried out by changing policies attached to data, users, or relationships.
  - Expiration time: revocation can be performed by associating an expiration time with access control tokens. Consequently, once it expires, a new token is required.
  - Key updates: once a change in the policy is produced, keys are updated and appropriately distributed.
  - Re-encryption: using cryptography may imply several encryption and re-encryption operations. Then, once a change in the policy is produced, re-encryptions are indispensable.
  - Through an external element: revocation can be managed outside the system by making use of external elements. For example, it can be carried out through a revocation list, a proxy, a PKI, etc.
  - Through a particular scheme: revocation can be carried out through specific schemes such as particular protocols or cryptographic algorithms.

- *Trusted entity*: it can be a user, a group of users or a particular device, for instance, a storage device.

Furthermore, it is noticeable that all academic proposals and WBSNs in use manage the *relationship-based* requirement. However, even considering the management of relationships attributes, the establishment of relationships is generally managed out of band. For instance, the administrator directly communicates keys/tokens to his contacts regarding the level of trust put in them. A more practical example can be identified in the WBSN Picassa, where administrators send to his contacts the URL of his photo albums.

Finally, there are a pair of symbols to notice. Specifically, symbol "X" refers to the fact that a given approach has explicitly mentioned that this aspect is out of the scope of the proposal, and symbol "-" implies that such approach does not point out anything about a specific issue.

4.6.1   Academic Proposals for WBSNs

A total of 25 academic proposals focus on privacy in WBSNs. The summary of
the analysis is presented in Table 4.1. In general terms they can be divided into 3
groups, namely, (1) the approaches exclusively associated with the basic mechanism
for implementing $SoNeUCON_{ABC}$, (2) contributions which may satisfy the *interoperability* requirement, and (3) proposals particularly focused on cryptographic
access control enforcement and which are linked to the *data exposure minimization*
requirement.

Three of the studied proposals fall into the first group and they have been proposed from 2006 to 2009 [15, 27, 50]. All of them follow a server-side architecture.
In this case, the most remarkable feature is the fact that Access Control Lists
(ACLs) are used in 2 out of 3 approaches in order to manage policy decision and
enforcement. ACLs seem to be a common and feasible way to manage access control
in this basic approach. Besides, it is surprising that not a single proposal describes
some kind of revocation technique.

With respect to *interoperability*, 9 contributions meet this requirement and the
architecture is also server-side [1, 3, 4, 6, 10, 32, 71, 72, 79]. In general, requesters
obtain a token which can be represented in multiple ways, such as XML files [32]
or tickets [71], and can be composed of assorted features, like keys [32] or digital
signatures [79]. However, some of these contributions [4, 10, 72, 79] also manage
policies in the storage service, which may prevent them from being interoperable.
This is equivalent to the basic mechanism and, even using tokens, these approaches
have to be further studied to identify the possibility of adapting them to satisfy the
interoperability requirement. One relevant common point in some proposals is the
use of tokens to attest relationships and their attributes [3, 4, 6, 32, 79]. It perfectly
matches with the fact that relationships are chief WBSNs elements. Indeed, despite
requiring future work, relationship certificates can be a promising mechanism to
manage relationships in the way that relationships remain hidden from all users but
to the pair of them who are involved in it [25]. Lastly, notice that revocation is
managed by the majority of approaches and techniques applied are quite assorted.
For instance, in PrPl tokens have attached an expiration time and once expired new
tokens have to be created [71].

Only three proposals address the *sticky policy* requirement. Specifically, all of
them make use of ABE and, as mentioned in Sect. 4.5.4, ABE cryptography relies on
the involvement of policies in keys or ciphertexts, which naturally helps to control
access to data wherever it is used [11, 47, 48].

A total of 15 recent proposals make use of cryptography before storing data
in the server, therefore, they are in the position of fulfilling the *data exposure
minimization* requirement [1, 3, 8, 11, 14, 20, 35, 40, 44, 45, 47, 48, 53, 54, 83]. The
overall procedure is based on the use of a public–private key pairs in multiple
and assorted algorithms according to the defined policy decision and enforcement
functions. However, [44] proposes a hybrid scheme in which data is symmetrically
encrypted and asymmetric cryptography is applied in the data acquisition and

decryption process. Some proposals focus on requiring a particular policy in the key creation process [11,47,48], requiring proxies to forward data among different user groups [48] or requiring methods to store fake information [14,45,54]. Other identified issue is that the most common way of managing revocation is modifying keys and performing the appropriate re-encryptions.

A noticeable feature common to all proposals is the management of unidirectional relationships without considering a direction attribute as an authorization predicate element. Furthermore, this matter emphasizes that access control mechanisms are established from one user to another and the particular implementations are the ones which make relationships bidirectional.

One last point for this analysis is that approaches that do not handle cryptography require to trust multiple entities. However, there are some exceptions and several cryptographic proposals also trust particular devices but it is not due to lack of data confidentiality but to preserve data from unauthorized deletions [20], from dishonest revocations [47,48] or from unfairly key managements [40].

4.6.2   WBSNs in Use

Many WBSNs are currently used by thousands of people. Nevertheless, for the sake of simplicity, 9 of the most representative WBSNs in use have been studied according to features described above. The overall results are summarized in Table 4.2.

Firstly, it can be noticed that basic architectures similar to the one proposed in Sect. 4.5.2 are the most used, although the implemented access control models are much simpler than $SoNeUCON_{ABC}$. Moreover, it is remarkable the existence of three WBSNs that manage tokens and are candidates for satisfying interoperability, namely, Picasa, MySpace, and Flickr. Picasa, focused on photos management, is exclusively based on tokens which take the form of URLs. Likewise, Flickr, that also focused on photo management, uses URLs to provide access to photos. This WBSN is specially relevant concerning the management of public photos because everybody is able to access through the appropriate URL. Moreover, Flickr allows the management of rights in regard to notes, commentaries, and photos. Similarly, MySpace, a WBSN to share data such as photos, videos or music, uses URLs as tokens as well. However, MySpace only uses tokens for photos, applying simpler techniques for other types of data like wall messages. In sum, taking into consideration this analysis, tokens seem to be specially valuable regarding photo management.

In relation to attributes of authorization predicates that are directly related to the *fine-grained* requirement, there is a common pattern in which the role attribute (of relationships) and the data privacy attribute (of data elements) are key points. Besides, all analyzed WBSNs manage direction attribute in a bidirectional way, except for Twitter in which administrators request users to be followers of them but not in the other way round. Interestingly, this issue highlights again that mechanisms are intrinsically unidirectional and the bidirectional nature is provided by implementations. By contrast, other attributes such as the creation time of a relationship, size of data or nationality of users, are left aside. Also, similar to academic approaches, WBSNs in use base on relationships in which the administrator is who directly communicates, out of band, keys or tokens.

Concerning revocation, the most of proposals manage it through revocation lists consisting of blocking revoked users from accessing to data. This is the simplest technique among the identified ones but the most used.

Finally, regarding trusted entities, it must be noted that WBSNs do not detail whether some entity or object must be trusted or not. Nonetheless, as there is no evidence of the fact that WBSNs address *data exposure minimization*, trust must be specially put on storage services (i.e., the WBSN), as well as, on users to whom tokens are delivered.

4.7   Conclusions and Open Research Issues

One of the main problems that is currently attracting more attention in Web Based Social Networks (WBSNs) is the design and implementation of user-managed access control systems, as stated in [29]. For this purpose, four requirements have been previously identified. Moreover, in this work a fifth is added. This work contributes in this direction with several results. First, *SoNeUCON*$_{ABC}$, an access control model for WBSNs is proposed. This model allows to directly fulfil two of the requirements identified in [29], *relationship-based* and *fine-grained*. Based on this model, a set of mechanisms are selected so the three remaining requirements, *sticky policy*, *interoperability*, and *data exposure minimization*, can be fulfilled. As a result of this discussion, a pair of relevant conclusions are reached. On the one hand, it is quite feasible to implement in the short-term access control mechanisms for WBSNs that satisfy the requirements *relationship-based*, *fine-grained*, *interoperability* and *sticky policy*. On the other hand, it is also feasible, though computationally challenging, to implement in the medium term access control mechanisms for WBSNs that satisfy the requirements *relationship-based*, *fine-grained*, *sticky policy*, and *data exposure minimization* (some of them with limitations).

More to the point, a total of 25 academic approaches and 9 WBSNs in use have been analyzed against the mentioned requirements. Summing up, academic approaches leave aside *fine-grained* and *sticky policy* requirements while weaknesses of WBSNs in use are *interoperability* and *data exposure minimization*. Additionally, the analysis shows that conditions and obligations are not generally addressed.

Nonetheless, a set of open issues are identified. Firstly, in this chapter a general overview and applicability of *SoNeUCON*$_{ABC}$ is presented but its formalization is out of the scope of this work. For instance, regarding policy language and the corresponding policy construction, extended work is required. A first step towards an appropriate formalization is to follow approaches such as [51] or [62], related to $UCON_{ABC}$. Moreover, the formalization has to be supported by other authors specially focused on policy description [39]. Furthermore, indirect relationships must be particularly detailed by describing ways of calculating values of attributes involved [25]. For example, if the path between two nodes is n, the main point is to establish a method to enforce access control regarding the full relationship (path) that involves the n nodes. Therefore, who is the person that must establish these patterns, which factors must be considered or how particularly the calculus must be performed are some considerations to be determined.

Additionally, related to $SoNeUCON_{ABC}$ and the *fine-grained* requirement, WBSNs are systems which involve a huge quantity of users managing large sets of data by the establishment of access control policies. Then, a challenging issue is to identify demanding WBSN features, such as the possible specification of indirect relationships or cliques (e.g., a close group of users), and look for expressive access control policies to achieve a successful fine-grained access control management. Due to that fact, the expressiveness of access control models regarding policy languages is a future step [28].

As recently identified by J. Park et al. [61, 63] the distinction between users and sessions is an appealing matter. Policies may be defined in regard to the user session. For instance, a user opens different sessions from different computers, which means from different IPs, and access control mechanisms provide him with different permissions with respect to the session. Thus, future work runs towards the study of novel approaches associated with this matter [61, 63] and their integration of this issue into our model.

Other relevant point in WBSNs is the difficult definition of users and their administration rights over data. This is related to co-ownership. Although for simplicity this chapter recognizes as administrators any user who is the owner of an object or carries out administrative operations on it, differences between both must be identified. The following step then will be to define techniques to face co-ownership problem [78].

Furthermore, an interesting point refers to privacy according to the discovery of the WBSN structure. In other words, the identification of users and their relationships is a challenging problem. Some contributions and the whole of WBSNs in use do not consider the fact that relationships can be inferred from the social network structure.

Lastly, the main important issue to attain in future work is the satisfaction of the whole set of requirements. Indeed, bearing the persistent search of flexibility in mind and alluding to the *Walled Garden* problem, mechanisms based on tokens seem to be promising. Consequently, studies must be headed towards the analysis of the complete integration of tokens in the basic approach, as well as, the inclusion of cryptography to satisfy *data exposure minimization* requirement. As a final step, all developments must be appropriately designed and evaluated.

References

1. Ackermann, M., Ludwig, B., Hymon, K., Wilhelm, K.:   Helloworld: An open source, distributed and secure social network. In: W3C Wks. on the Future of Social Networking, 2009
2. Acquisti, A., Gross, R.: Imagined communities: awareness, information sharing, and privacy on the Facebook. In: Privacy Enhancing Technologies, vol. 4258 of Lecture Notes in Computer Science, pp. 36–58. Springer, Berlin/Heidelberg (2006)
3. Aiello, L.M., Ruffo, G.:   Lotusnet: Tunable privacy for distributed online social network services. Comput. Comm. **35**(1), 75–88 (2012)
4. Aiello, L.M., Ruffo, G.:   Secure and flexible framework for decentralized social network services. In: 2010 8th IEEE International Conference on Pervasive Computing and Communications Workshops (PERCOM Workshops), pp. 594–599, 2010
5. Ajami, R., Ramadan, N., Mohamed, N., Al-Jaroodi, J.: Security challenges and approaches in online social networks: A survey. Int. J. Comput. Sci. Netw. Secur. **11**, 1–12 (2011)
6. Ali, B., Villegas, W., Maheswaran, M.: A trust based approach for protecting user data in social networks, pp. 288–293, 2007
7. Allard, T., Anciaux, N., Bouganim, L., Guo, Y., Le Folgoc, L., Nguyen, B., Pucheral, P., Ray, I., Yin, S.: Secure personal data servers: A vision paper. Proc. VLDB Endow. **3**(1–2), 25–35 (2010)
8. Anderson, J., Diaz, C., Bonneau, J., Stajano, F.: Privacy-enabling social networking over untrusted networks. In: Proceedings of the 2nd ACM Workshop on Online Social Networks, pp. 1–6. ACM, New York (2009)
9. Attrapadung, N., Imai, H.: Conjunctive broadcast and attribute–based encryption. In: Proc. of the 3rd International Conference Palo Alto on Pairing–Based Cryptography, Pairing '09. Springer, New York (2009)
10. Backes, M., Maffei, M., Pecina, K.: A security API for distributed social networks. In: NDSS, vol. 11, pp. 35–51 (2011)
11. Baden, R., Bender, A., Spring, N., Bhattacharjee, B., Starin, D.: Persona: an online social network with user–defined privacy. SIGCOMM Comput. Comm. Rev. **39**, 135–146 (2009)
12. Becker, J., Chen, H.: Measuring privacy risk in online social networks. In: Proc. of W2SP 2009: Web 2.0 Security and Privacy, 2009
13. Bertino, E., Bonatti, P.A., Ferrari, E.: Trbac: a temporal role-based access control model. In: Symposium on Access Control Models and Technologies. Proc. of the Fifth ACM Wks. on Role-Based Access Control, pp. 21–30. ACM, New York (2000)
14. Besenyei, T., Földes, Á.M., Gulyás, G.G., Imre, S.: StegoWeb: towards the ideal private web content publishing tool. In: SECURWARE 2011, The Fifth International Conference on Emerging Security Information, Systems and Technologies, pp. 109–114, 2011
15. Besmer, A., Lipford, H.R., Shehab, M., Cheek, G.: Social applications: exploring a more secure framework. In: Proc. of the 5th Symposium on Usable Privacy and Security, SOUPS '09, pp. 2:1–2:10. ACM, New York (2009)
16. Bethencourt, J., Sahai, A., Waters, B.: Ciphertext-policy attribute-based encryption. In: Proc. of the 2007 IEEE Symposium on Security and Privacy, SP '07. IEEE Computer Society, Oakland, CA (2007)
17. Bishop, M.: Computer Security Art and Science. Addison-Wesley (2002)
18. Bouganim, L., Pucheral, P.: Chip-secured data access: confidential data on untrusted servers. In: Proc. of the 28th International Conference on Very Large Data Bases, VLDB '02, pp. 131–142. VLDB Endowment, 2002
19. Boyd, D.M., Ellison, N.B.: Social network sites: Definition, history, and scholarship. J. Comput. Mediat. Comm. **13**, 210–230 (2007)
20. Buchegger, S., Schiöberg, D., Vu, L.-H., Datta, A.: Peerson: P2p social networking: early experiences and insights, pp. 46–52, 2009

21. Capitani di Vimercati, S., Foresti, S., Samarati, P.: Authorization and access control. Security, Privacy, and Trust in Modern Data Management, pp. 39–53, 2007
22. Carminati, B., Ferrari, E.: Access control and privacy in web-based social networks. Int. J. Web Inform. Syst. **4**(4), 395–415 (2008)
23. Carminati, B., Ferrari, E.: Privacy-aware collaborative access control in web-based social networks. In: Proceeedings of the 22nd Annual IFIP WG 11.3 Working Conference on Data and Applications Security, pp. 81–96. Springer, New York (2008)
24. Carminati, B., Ferrari, E., Perego, A.: Rule-based access control for social networks. In: Proc. OTM 2006 Workshops (On the Move to Meaningful Internet Systems), vol. 4278 of LNCS, pp. 1734–1744. Springer, New York (2006)
25. Carminati, B., Ferrari, E., Perego, A.: Private relationships in social networks. In: Proc. of the 2007 IEEE 23rd International Conference on Data Engineering Wks., pp. 163–171. IEEE Computer Society, Oakland, CA (2007)
26. Carminati, B., Ferrari, E.: Access control and privacy in web-based social networks. Int. J. Web Inf. Syst. **4**(4), 395–415 (2008)
27. Carminati, B., Ferrari, E., Heatherly, R., Kantarcioglu, M., Thuraisingham, B.: A semantic web based framework for social network access control. In: Proc. of the 14th ACM symposium on Access control models and technologies, SACMAT '09, pp. 177–186. ACM, New York (2009)
28. Carreras, A., Rodriguez, L., Delgado, J., Maronas, X.: Access control issues in social networks, pp. 47–52, 2010
29. Carrie, Dr., Gates, E.: Access control requirements for web 2.0 security and privacy. In: Proc. of Wks. on Web 2.0 Security & Privacy (W2SP 2007, 2007
30. Chase, M., Chow, S.S.M.: Improving privacy and security in multi-authority attribute-based encryption. In: Proc. of the 16th ACM Conference on Computer and Communications Security, CCS '09, pp. 121–130. ACM, New York (2009)
31. Chase, M.: Multi-authority attribute based encryption. In: Proc. of the 4th Conference on Theory of Cryptography, TCC'07, pp. 515–534. Springer, New York (2007)
32. Conti, M., Hasani, A., Crispo, B.: Virtual private social networks. In: Proc. of the first ACM conference on Data and application security and privacy, CODASPY '11, pp. 39–50. ACM, New York (2011)
33. Covington, M.J., Sastry, M.R.: A contextual attribute-based access control model. In: Proc. of the 2006 International Conference on On the Move to Meaningful Internet Systems: AWeSOMe, CAMS, COMINF, IS, KSinBIT, MIOS-CIAO, MONET - Volume Part II, OTM'06, pp. 1996–2006, 2006
34. Covington, M.J., Moyer, M.J., Ahamad, M.: Generalized role—based access control for securing future applications. In: 23rd National Information Systems Security Conference, Citeseer, 2000
35. Cutillo, L.A., Molva, R., Strufe, T.: Safebook: Feasibility of transitive cooperation for privacy on a decentralized social network. In: 2009 IEEE International Symposium on a World of Wireless, Mobile and Multimedia Networks & Workshops, (217141):1–6, 2009
36. Dey, R., Jelveh, Z., Ross, K.W.: Facebook users have become much more private: A large-scale study. In: Proc. of SESOC 2012, 2012
37. Di Vimercati, S.D.C., Foresti, S., Jajodia, S., Paraboschi, S., Samarati, P.: A data outsourcing architecture combining cryptography and access control. In: Proceedings of the 2007 ACM Workshop on Computer Security Architecture, pp. 63–69. ACM, New York (2007)
38. Dwyer, C., Hiltz, S.R., Passerini, K.: Trust and privacy concern within social networking sites: A comparison of facebook and MySpace. In: AMCIS, p. 339 (2007)
39. Fong, P.W.L.: Relationship-based access control: protection model and policy language. In: Proc. of the first ACM conference on Data and application security and privacy, CODASPY '11, pp. 191–202. ACM, New York (2011)
40. Frikken, K.B., Srinivas, P.: Key-allocation schemes for private social networks. In: Proc. of the 8th ACM Wks. on Privacy in the Electronic Society, WPES '09, pp. 11–20. ACM, New York (2009)

41. Gao, H., Hu, J., Huang, T., Wang, J., Chen, Y.: Security issues in online social networks. IEEE Internet Comput. **15**, 56–63 (2011)

42. Giunchiglia, F., Zhang, R., Crispo, B.: Relbac: Relation based access control. In: Fourth International Conference on Semantics, Knowledge and Grid, 2008. SKG '08., pp. 3–11, 2008

43. Goyal, V., Pandey, O., Sahai, A., Waters, B.: Attribute-based encryption for fine-grained access control of encrypted data. In: Proc. of the 13th ACM Conference on Computer and Communications Security, CCS '06, pp. 89–98. ACM, New York (2006)

44. Graffi, K., Groß, C., Stingl, D., Hartung, D., Kovacevic, A., Steinmetz, R.: Lifesocial.kom: A secure and p2p-based solution for online social networks. In: Proc. of the IEEE Consumer Communications and Networking Conference. IEEE Computer Society, Oakland, CA (2011)

45. Guha, S., Tang, K., Francis, P.: Noyb: privacy in online social networks. In: Proc. of the First Wks. on Online Social Networks, WOSN '08, pp. 49–54. ACM, New York (2008)

46. Harary, F., Norman, R.Z.: Graph theory as a mathematical model in social science, 1953

47. Jahid, S., Nilizadeh, S., Mittal, P., Borisov, N., Kapadia, A.: Decent: A decentralized architecture for enforcing privacy in online social networks, 2012

48. Jahid, S., Mittal, P., Borisov, N.: Easier: encryption−based access control in social networks with efficient revocation. In: Proc. of the 6th ACM Symposium on Information, Computer and Communications Security, ASIACCS '11, pp. 411–415. ACM, New York (2011)

49. Kourtellis, N., Finnis, J., Anderson, P., Blackburn, J., Borcea, C., Iamnitchi, A.: Prometheus: user-controlled p2p social data management for socially-aware applications. In: Ifip International Federation For Information Processing, pp. 212–231, 2010

50. Kruk, S., Grzonkowski, S., Gzella, A., Woroniecki, T., Choi, H.-C.: D-foaf: Distributed identity management with access rights delegation. In: The Semantic Web? ASWC 2006, vol. 4185 of Lecture Notes in Computer Science, pp. 140–154. Springer, Berlin/Heidelberg, (2006)

51. Lazouski, A., Martinelli, F., Mori, P.: Usage control in computer security: a survey. Comput. Sci. Rev. **4**(2), 81–99 (2010)

52. Lin, H., Cao, Z., Liang, X., Shao, J.: Secure threshold multi authority attribute based encryption without a central authority. Inf. Sci. **180**, 2618–2632 (2010)

53. Lucas, M.M., Borisov, N.: Flybynight: mitigating the privacy risks of social networking. In: Proc. of the 7th ACM Wks. on Privacy in the Electronic Society, WPES '08, pp. 1–8. ACM, New York (2008)

54. Luo, W., Xie, Q., Hengartner, U.: FaceCloak: an architecture for user privacy on social networking sites. In: 2009 International Conference on Computational Science and Engineering, pp. 26–33, 2009

55. Au Yeung, C.M., Liccardi, I., Lu, K., Seneviratne, O., Berners-Lee, T.: Decentralization: The future of online social networking. In: W3C Wks. on the Future of Social Networking Position Papers, 2009

56. Mun, M., Hao, S., Mishra, N., Shilton, K., Burke, J., Estrin, D., Hansen, M., Govindan, R.: Personal data vaults: a locus of control for personal data streams. In: Proc. of the 6th International Conference, Co-NEXT '10, pp. 17:1–17:12. ACM, New York (2010)

57. Nin, J., Carminati, B., Ferrari, E., Torra, V.: Computing Reputation for Collaborative Private Networks, pp. 246–253. IEEE Computer Society, Oakland, CA (2009)

58. Oracle-Team: Online Security, A Human Perspective (2011)

59. Ostrovsky, R., Sahai, A., Waters, B.: Attribute-based encryption with non-monotonic access structures. In: Proc. of the 14th ACM Conference on Computer and Communications Security, CCS '07, pp. 195–203. ACM, New York (2007)

60. Parent, W.A.: Privacy, morality, and the law. Philos. Publ. Aff. **12**(4), 269–288 (1983)

61. Park, J., Sandhu, R.: A Position Paper: A Usage Control (UCON) Model for Social Networks Privacy, (2000)

62. Park, J., Sandhu, R.: The UCONabc usage control model. ACM Trans. Inf. Syst. Secur. **7**, 128–174 (2004)

63. Park, J., Sandhu, R., Cheng, Y.: A user-activity-centric framework for access control in online social networks. IEEE Internet Comput. **15**(5), 62–65 (2011)

64. Ray, I., Kumar, M., Yu, L.: LRBAC: a location-aware role-based access control model. In: Information Systems Security, vol. 4332 of Lecture Notes in Computer Science, pp. 147–161. Springer, Berlin/Heidelberg (2006)
65. Razavi, M.N., Iverson, L.: Towards usable privacy for social software. Technical report, University of British Columbia, 2007
66. Salim, F., Reid, J., Dawson, E.: An administrative model for UCONabc. In: Proc. of the Eighth Australasian Conference on Information Security, vol. 105 of AISC '10, pp. 32–38, 2010
67. Sandhu, R.S., Samarati, P.: Access control: principles and practice. Access 40–48 (1994)
68. Sastry, M., Krishnan, R., Sandhu, R.: A new modeling paradigm for dynamic authorization in multi-domain systems, pp. 153–158, 2007
69. Schneier, B.: A taxonomy of social networking data. IEEE Security Privacy **8**(4) (2010)
70. Scholl, M., Stine, K., Lin, K., Steinberg, D.: Security architecture design process for health information exchanges (HIEs). NISTIR 7497. National Institute of Standards and Technology
71. Seong, S.-W., Seo, J., Nasielski, M., Sengupta, D., Hangal, S., Teh, S.K., Chu, R., Dodson, B., Lam, M.S.: Prpl: a decentralized social networking infrastructure, pp. 8:1–8:8 (2010)
72. Shakimov, A., Lim, H., Li, K., Liu, D., Varshavsky, A.: Vis-a-Vis: privacy-preserving online social networking via virtual individual servers, (2010)
73. Shen, H., Hong, F.: An attribute-based access control model for web services. In: Seventh International Conference on Parallel and Distributed Computing, Applications and Technologies, 2006. PDCAT '06., pp. 74–79, 2006
74. Shi, W.: Attribute based encryption with pattern-awareness by attribute based encryption with pattern-awareness. Master's thesis, Inha University, 2010
75. Shilton, K., Burke, J.A., Estrin, D., Hansen, M.: Designing the personal data stream : enabling participatory privacy in mobile personal sensing. Work (September), 25–27 (2009)
76. Squicciarini, A.C., Shehab, M., Paci, F.: Collective privacy management in social networks. In: Proc. of the 18th International Conference on World Wide Web, WWW '09, pp. 521–530. ACM, New York (2009)
77. Squicciarini, A.C., Shehab, M., Wede, J.: Privacy policies for shared content in social network sites. VLDB J. 777–796 (2010)
78. Squicciarini, A.C., Shehab, M., Paci, F.: Collective privacy management in social networks. In: Proc. of the 18th International Conference on World Wide Web, WWW '09, pp. 521–530. ACM, New York (2009)
79. Tootoonchian, A., Saroiu, S., Ganjali, Y., Wolman, A.: Lockr: Better privacy for social networks. In: Proceedings of the 5th International Conference on Emerging Networking Experiments and Technologies, pp. 169–180. ACM, New York (2009)
80. Yuan, E., Tong, J.: Attributed based access control (ABAC) for web services. In: Proc. of the IEEE International Conference on Web Services, ICWS '05, pp. 561–569. IEEE Computer Society, Oakland, CA (2005)
81. Zhang, X., Park, J., Parisi-Presicce, F., Sandhu, R.: A logical specification for usage control. In: Proc. of the Ninth ACM Symposium on Access Control Models and Technologies, SACMAT '04, pp. 1–10. ACM, New York (2004)
82. Zheleva, E., Getoor, L.: Social Network Data Analytics, chapter Privacy in Social Networks: A Survey. Springer, New York (2011)
83. Zhu, Y., Hu, Z., Wang, H., Hu, H., Ahn, G.-J.: A collaborative framework for privacy protection in online social networks. Organization 1–15 (2010)

Chapter 5
UPP+: A Flexible User Privacy Policy for Social Networking Services

Ramzi A. Haraty and Sally Massalkhy

Abstract Social networking services are having a major impact on people's daily lives. Ordinary users have taken these social networking facilities as basis for their businesses and for keeping track of their families and friends. In doing so, they add personal information, videos, pictures, and other data that is fundamentally unprotected due to the user's unawareness and the rigidity of the privacy policies of these facilities. Since users usually sign the privacy policy, granting their ownership of data to the site's owners, privacy concerns surface. In this paper, we present a privacy policy model—UPP+—for enhancing privacy and security for ordinary users. We use the Alloy language to formalize the model and the Alloy Analyzer to check for any inconsistencies.

5.1 Introduction

In the past few years, social network services have become major admirations in people's lives. Almost everyone who has access to the Internet has become addicted to certain social networking service sites such as Facebook, Google+, MySpace, or Twitter. In doing so, they add personal information, videos, and posts that are profoundly vulnerable due to the user's obliviousness and the stringency of the privacy policies of these facilities. Since users usually sign the privacy policy, granting their ownership of data to the site's owners, privacy concerns surface. Additional concerns arise when naïve users encounter [1]:

- Privacy policies that are hard to understand and assign particular policy settings that might conflict with each other.

R.A. Haraty (✉) • S. Massalkhy
Department of Computer Science and Mathematics, Lebanese American University,
Beirut 1102 2801, Lebanon
e-mail: rharaty@lau.edu.lb

R. Chbeir and B. Al Bouna (eds.), *Security and Privacy Preserving in Social Networks*,   139
Lecture Notes in Social Networks, DOI 10.1007/978-3-7091-0894-9_5,
© Springer-Verlag Wien 2013

- Privacy policies that constantly change and keep on changing; thus, confusing the users.
- Privacy policies that are explained in an informal way and in an incomplete manner, which cannot provide consistent and complete account of the privacy.

The user's profile is usually the most important feature in a social network service. The owner of the profile is the one in control of the contents visible in the profile to others. Social network services often offer an access control panel that helps users control the privacy of their profiles by providing privacy policy levels whereby the user chooses a level and then categorizes her friends accordingly. This work presents an enhanced user privacy model, which we call UPP+. UPP+ is a flexible and easy to understand policy privacy for social networking services.

Before developing a policy, one needs to describe formally its components and the relationships between them by building a model. The model needs to be analyzed and checked to figure out possible bugs and problems. Thus, formalizing privacy security models helps designers building a consistent system that meets its requirements and respects the goals of discretion. This objective can be achieved through the Alloy language.

Alloy is a structural modeling language for software design. It is based on first order logic that makes use of variables, quantifiers, and predicates (Boolean functions) [2]. Alloy, developed by MIT (Daniel Jackson and his team), is mainly used to analyze object models, translates constraints to Boolean formulas (predicates), and then validates them using the Alloy Analyzer [3] by checking the code for conformance to a specification. Alloy is used in modeling policies, security models, and applications, including name servers, network configuration protocols, access control, telephony, scheduling, document structuring, and cryptography [4]. Alloy's approach demonstrates that it is possible to establish a framework for formally representing a program implementation and for formalizing the security rules defined by a security policy, enabling the verification of that program representation for adherence to the security policy [5, 6]. Additionally, it allows users to describe a system design and check that there is no misunderstanding before writing the code.

This remainder of this chapter is organized as follows: Sect. 5.2 provides related work. Section 5.3 presents the Alloy language, the Alloy Analyzer and their features. Section 5.4 presents the model descriptions and discusses the consistency proof and Sect. 5.5 concludes the work.

5.2 Related Work

There has been a plethora of work that deals with security and privacy policies. McLean [7] claimed that models are "used to describe any formal statement of a system's confidentiality, availability, or integrity requirements." Privacy models provide a detailed and precise means of formally describing privacy policies and proving their validity. Formalizing policy models provides system designers with evidence that they are constructing a consistent system that will meet its specifications when implemented.

Fong et al. [8] proposed a privacy preservation model for social network sites like Facebook. In their paper, they analyzed and formalized the mechanism of the access control for Facebook social network. They imitated the Facebook's access control mechanism by taking into consideration its most important features which are its predicates.

Danezis [9] introduced a machine learning approach that was used to automatically find the privacy settings of users and give a readymade privacy policies package to the users; then this mechanism was evaluated. This approach is aimed to aid the end users when they want to restrict access from certain contacts. Their purpose is to infer user contexts, context assignment, and privacy policy per context.

Dania [10] introduced a formal model for social network privacy and used Facebook as her test case. Secure-UML was used as the formal language to the model. In [11], the author discusses the architecture, security policy, and protection mechanisms of four National Security Agency—certified systems. The author formally compares their techniques used for protecting data against users. In [12], the authors present a temporal multilevel secure data model. The model combines the characteristics of temporal data models and multilevel secure data models. The main focus of the model is mandatory access control, polyinstantiation, and secure transaction processing, while at the same time providing time support to record historical, present, and future data.

Hassan and Logrippo [13] proposed a method to detect inconsistencies of multiple security policies mixed together in one system and to report the inconsistencies at the time when the system is designed. The mixed models are checked for inconsistencies before real implementation. Inconsistency in a mixed model is due to the fact that the used models are incompatible and cannot be mixed. They demonstrated their method by mixing Bell–LaPadula with role-based access control (RBAC) [14] in addition to separation of concerns.

Shaffer in [15] described a security domain model (DM), designed for conducting static analysis of programs to identify illicit information flows, such as control dependency flaws and covert channel vulnerabilities. The model includes a formal definition for trusted subjects, which are granted privileges to perform system operations that require mandatory access control policy mechanisms imposed on normal subjects but are trusted not to degrade system security. The DM defines the concepts of program state, information flow, and security policy rules and specifies the behavior of a target program.

Misic and Misic in [16] addressed the networking and security architecture of healthcare information system. This system includes patient sensor networks, wireless local area networks belonging to organizational units at different levels of hierarchy, and the central medical database that holds the results of patient examinations and other relevant medical records. In order to protect the integrity and privacy of medical data, they targeted the Clinical Information System Security Policy and proposed the feasible enforcement mechanisms over the wireless hop.

The Clinical Information System Policy was recently formalized by Haraty and Naous [17].

The authors of [18, 19] presented a method to validate access control policy. They were mainly interested in higher level languages where access control rules can be specified in terms that are directly related to the roles and purposes of users. They discussed a paradigm more general than the RBAC in the sense that the RBAC can be expressed in them.

5.3 Formal Privacy Policy Model in Alloy

In this section, we overview the Alloy language and demonstrate how a model can be checked for consistency using Alloy and apply our method to our proposed UPP+ model.

5.3.1 The Alloy Language

To formalize the security models we use the Alloy language and its analyzer. Alloy is a lightweight modeling formalism using a first order predicate logic over the domain of relations. These relations are similar to relational algebra and calculus. It is a textual language developed at MIT. Alloy originates from Z. It is used for analyzing object models by checking for consistency of multiplicities and generating instances of models or a counterexample. Alloy Analyzer translates constraints to Boolean formulas and then applies SAT solvers.

5.3.2 Alloy Language Features

The following features present a subset of the full Alloy language that we used in formalizing our security models.

An Alloy model consists of one or more files, each containing a single module. A module consists of a header identifying the module, some imports and some paragraphs:

module::=header import*paragraph*

A model can be contained entirely within one module. The paragraphs of module are signatures, facts, functions, predicates, assertions, run commands, and check commands.

Alloy uses the following multiplicity keywords: *lone*: zero or one; *one*: exactly one; *some*: one or more; *set*: zero or more. These keywords are used as quantifiers in quantified formulas, quantified expressions, in set declarations, in relation declarations, and in signature declarations.

A signature represents a set of atoms and is declared using the "sig" keyword—such as *sig A {}* to define a signature named *A*. The types of signatures are: *subset*, *top-level*, and *abstract*, and a signature with a *multiplicity* keyword:

- A top-level signature represents mutually disjoint sets that does not extend another signature: *sig A{}*
- A subset signature represents a set of elements that is a subset of the union of its parents: *one sig B extends A{}*
- An abstract signature represents only the elements that belong to one of the signatures that extend it: *abstract sig A{}*
- A signature with *multiplicity* keyword constrains the signature's set to have the number of elements specified with the keyword.

Facts, functions, and predicates are packages of constraints. A fact is a constraint that always holds. A predicate is a template for a constraint that can be instantiated in different contexts. A function is a template for an expression, and an assertion is a constraint that is intended to follow from the facts of a model. Examples of facts, predicates, and assertions are:

fact {no iden & parent}

pred access(state: State, next: state, u: User, r: Resource)
{next.accessed = state.accessed + u ->r}

assert example1 {
A.sens = SecretNT
B.sens = SecretT}

Run and *Check* commands are used to instruct the Alloy Analyzer to perform various analyses, a *run* command causes the analyzer to search for an instance that shows the consistency of a function or a predicate, whereas a check command causes it to search for a counterexample showing that an assertion does not hold:

check example
run UPP+Model

5.4   User Privacy Policy Plus (UPP+) Model

Aïmeur et al. [20], in their work, introduced the user privacy policy (UPP) model. UPP is a privacy model, which enables its users to control who can access their data in social networks. To understand UPP, the authors introduced a framework to the social networking sites that consists of user privacy concern, profile viewers, privacy levels, and tracking levels. In our work, we extend the UPP model to UPP+ and take into consideration part of UPP's privacy concern—the profile viewers and the privacy levels; however, the tracking levels will not be of importance and will not be implemented in our model.

The user privacy concern is split into three different categories. The first category is security, which is a major concern of social networking sites that deal with user's security risks such as identity theft and impersonation, hackers, phishing, and many others that may harm the user's data and information. The second category is reputation and credibility, which involves the reputation of the user—both online and in the real world, since bad reputation will lead to affecting the credibility of this user in the society or at work. The third category is profiling, which involves product companies building profiles on users from social network information found online without the user's knowledge in order to sell them products. In brief, tracking is following a user through a friend list or a name tag. There are three levels of tracking: strong tracking—a user is tracked on the social network; weak tracking—the user's name appears on the list of friends but not in tags; and no tracking—the user is not mentioned anywhere in his/her friends' profile.

Almost everyone now uses social networks; therefore, we have a variety of users. This variety of users leads to different concerns regarding the privacy. Aïmeur et al. [20] proposed four different types of privacy settings regarding the data of the user:

- *Healthy Data:* information, if shared, would return no harm to the user or even track him/her down such as nickname, music interest, and other similar data.
- *Harmless Data:* data, if shared, may contain data that helps in profiling for companies that market products. Such information are religion, gender, and interests.
- *Harmful Data:* information and pictures that belong to the user that, if shared, may lead to bad reputation and credibility.
- *Poisonous Data:* data that belongs to the user, if shared, may lead to security risks. Such information are user's home address, phone number, and other similar information.

After the process of data partitioning, friends partitioning is in order. Friends partitioning refers to who can access the different types of data. This is categorized into different groups of people depending on the relationship between the user and the friends s/he has on the social network site and according to the trust s/he has in his/her friends. In UPP, the people in the user's social network site are divided into four different groups:

- *Best Friends:* people who are considered the closest to the user such as parents or best friends.
- *Normal Friends:* people who are considered friends with the user but not necessarily close such as relatives and groups of friends.
- *Casual Friends:* people who are considered as somewhat strangers to the user yet known, like people the user met twice or friends of friends.
- *Visitors:* people who are strangers to the user. These people are not necessarily in the friends list.

UPP also introduces privacy that the user can choose from in order to have privacy setting to his/her social network page. The privacy levels are split into four types. Each level has its own rules on each group of users. The rules are made in order to read the data sets:

- *No Privacy Rules*

    – Best friends can view all types of data set.
    – Normal friends can view all types of data set.
    – Casual friends can view all types of data set.
    – Visitors can view all types of data set.

- *Soft Privacy Rules*

    – Best friends can view all types of data set.
    – Normal friends can view Healthy data, Harmless data, Harmful Data but cannot view Poisonous data.
    – Casual friends can view Healthy data, Harmless data, Harmful Data but cannot view Poisonous data.
    – Visitors can view Healthy data and Harmless data but cannot view Harmful Data and Poisonous data.

- *Hard Privacy Rules*

    – Best friends can view all types of data set.
    – Normal friends can view Healthy data, Harmless data, Harmful Data but cannot view Poisonous data.
    – Casual friends can view Healthy data and Harmless data but cannot view Harmful Data and Poisonous data.
    – Visitors can view Healthy data but cannot view Harmless data, Harmful Data, and Poisonous data.

- *Full Privacy Rules*

    – Best friends can view all types of data set.
    – Normal friends can view Healthy data and Harmless data but cannot view Harmful Data and Poisonous data.
    – Casual friends can view Healthy data and Harmless data but cannot view Harmful Data and Poisonous data.
    – Visitors cannot view any type of data set.

    Table 5.1 summarizes the privacy settings, privacy levels, and users.

5.4.1   UPP+ Model Implementation

In order to implement UPP model, we will need to list the privacy data set, the privacy levels, the different types of users, and the constraints or the rules used in the model. This section will explain the implementation of this model. Table 5.2

Table 5.1 User privacy policy model

Privacy levels	Privacy settings				
	Healthy data	Harmless data	Harmful data	Poisonous data	Users
No Privacy	Yes	Yes	Yes	Yes	Best Friends
	Yes	Yes	Yes	Yes	Normal Friends
	Yes	Yes	Yes	Yes	Casual Friends
	Yes	Yes	Yes	Yes	Visitor
Soft Privacy	Yes	Yes	Yes	Yes	Best Friends
	Yes	Yes	Yes	No	Normal Friends
	Yes	Yes	Yes	No	Casual Friends
	Yes	Yes	No	No	Visitor
Hard Privacy	Yes	Yes	Yes	Yes	Best Friends
	Yes	Yes	Yes	No	Normal Friends
	Yes	Yes	No	No	Casual Friends
	Yes	No	No	No	Visitor
Full Privacy	Yes	Yes	Yes	Yes	Best Friends
	Yes	Yes	No	No	Normal Friends
	Yes	Yes	No	No	Casual Friends
	No	No	No	No	Visitor

Table 5.2 Privacy data set levels

Privacy data set levels	Description
PrivacyDS	Privacy Data Set
NoP	No Privacy Data Set
SoftP	Soft Privacy Data Set
HardP	Hard Privacy Data Set
FullP	Full Privacy Data Set

lists the privacy data set, while Table 5.3 lists the privacy data sets according to each level. Table 5.4 lists the user groups' data set and Table 5.5 lists the user groups according to each level.

The Privacy policies are split into four levels. In the No Privacy Level (NoP), a user in any category of No Privacy Users (Nusers), which are Best Friend (NBF), Normal friend (NNF), Casual friend (NCF), or Visitor (NV), has the right to read all four types of data, which are Healthy Data (NoPHealthyD), Harmless Data (NoPHarmlessD), Harmful Data (NoPHarmfulD), and Poisonous Data (NoPPoisonousD).

In the Soft Privacy Level (SoftP), a user (Nusers) from category Best friend (SBF) has the right to read all types of data; as is the case for Normal friend (SNF) and Casual friend (SCF)—they have the right to read Healthy Data (SoftHealthyD), Harmless Data (SoftHarmlessD), and Harmful Data (SoftHarmfulD) but cannot read Poisonous Data (SoftPoisonousD). A Visitor (SV) cannot read SoftHarmfulD and SoftPoisonousD.

Table 5.3 Privacy data set according to each level

No privacy data set	Description	Soft privacy data set	Description
NoPHealthyD	No Privacy Healthy Data	SoftPHealthyD	Soft Privacy Healthy Data
NoPHarmlessD	No Privacy Harmless Data	SoftPHarmlessD	Soft Privacy Harmless Data
NoPHarmfulD	No Privacy Harmful Data	SoftPHarmfulD	Soft Privacy Harmful Data
NoPPoisonousD	No Privacy Poisonous data	SoftPPoisonousD	Soft Privacy Poisonous Data
Hard privacy data set	Description	Full privacy data set	Description
HardPHealthyD	Hard Privacy Healthy Data	FullPHealthyD	Full Privacy Healthy Data
HardPHarmlessD	Hard Privacy Harmless Data	FullPHarmlessD	Full Privacy Harmless Data
HardPHarmfulD	Hard Privacy Harmful Data	FullPHarmfulD	Full Privacy Harmful Data
HardPPoisonousD	Hard Privacy Poisonous Data	FullPPoisonousD	Full Privacy Poisonous Data

Table 5.4 User group data set

Users group set	Description
Nusers	No Privacy User group
Susers	Soft Privacy User group
Husers	Hard Privacy User group
Fusers	Full Privacy User group

Table 5.5 Users groups according to each level

No privacy users group	Description	Soft privacy users group	Description
NBF	No Privacy Best Friend	SBF	Soft Privacy Best Friend
NNF	No Privacy Normal Friend	SNF	Soft Privacy Normal Friend
NCF	No Privacy Casual Friend	SCF	Soft Privacy Casual Friend
NV	No Privacy Visitor	SV	Soft Privacy Visitor
Hard privacy users group	Description	Full privacy users group	Description
HBF	Hard Privacy Best Friend	FBF	Full Privacy Best Friend
HNF	Hard Privacy Normal Friend	FNF	Full Privacy Normal Friend
HCF	Hard Privacy Casual Friend	FCF	Full Privacy Casual Friend
HV	Hard Privacy Visitor	FV	Full Privacy Visitor

In the Hard Privacy Level (HardP), a user (Husers) from category Best friend (HBF) can view all types of data. A Normal friend (HBF) cannot view Poisonous Data (HardPoisonousD). A Casual friend (HCF) and Visitor (HV) can view Healthy Data (HardHealthyD), Harmless Data (HardHarmlessD) but cannot view Harmful Data (HardHarmlessD) and HardPoisonousD.

In the Full Privacy Level (FullP), Best friend (FBS) can view all types of data, while Normal friend (FNF) and Casual friend (FCF) can view Healthy Data (FullHealthyD) and Harmless Data (FullHarmlessD) but cannot view Harmful Data (FullHarmfulD) and Poisonous Data (FullPoisonousD). As for Visitor (FV), s/he cannot view any type of data.

Table 5.6 UPP+ ownership

Ownership	Description
NO	No Privacy Owner
SO	Soft Privacy Owner
HO	Hard Privacy Owner
FO	Full Privacy Owner

Table 5.7 User privacy policy plus (UPP+) model

Privacy levels	Privacy settings				
	Healthy data	Harmless data	Harmful data	Poisonous data	Users
No Privacy	Yes (r/w/nc)	Yes (r/w/nc)	Yes (r/w/nc)	Yes (r/nw/nc)	Best Friends
	Yes (r/w/nc)	Yes (r/w/nc)	Yes (r/w/nc)	Yes (r/nw/nc)	Normal Friends
	Yes(r/w/nc)	Yes (r/w/nc)	Yes (r/w/nc)	Yes (r/nw/nc)	Casual Friends
	Yes(r/w/nc)	Yes (r/w/nc)	Yes (r/w/nc)	Yes (r/nw/nc)	Visitor
Soft Privacy	Yes (r/w/nc)	Yes (r/w/nc)	Yes (r/w/nc)	Yes (r/nw/nc)	Best Friends
	Yes (r/w/nc)	Yes (r/w/nc)	Yes (r/w/nc)	No (nr/nw/nc)	Normal Friends
	Yes (r/w/nc)	Yes (r/w/nc)	Yes (r/nw/nc)	No (nr/nw/nc)	Casual Friends
	Yes (r/w/nc)	Yes (r/nw/nc)	No (nr/nw/nc)	No (nr/nw/nc)	Visitor
Hard Privacy	Yes (r/w/nc)	Yes (r/w/nc)	Yes (r/w/nc)	Yes (r/nw/nc)	Best Friends
	Yes (r/w/nc)	Yes (r/w/nc)	Yes (r/nw/nc)	No (nr/nw/nc)	Normal Friends
	Yes (r/w/nc)	Yes (r/nw/nc)	No (nr/nw/nc)	No (nr/nw/nc)	Casual Friends
	Yes (r/nw/nc)	No (nr/nw/nc)	No (nr/nw/nc)	No (nr/nw/nc)	Visitor
Full Privacy	Yes (r/w/nc)	Yes (r/w/nc)	Yes (r/w/nc)	Yes (r/nw/nc)	Best Friends
	Yes (r/w/nc)	Yes (r/w/nc)	No (nr/nw/nc)	No (nr/nw/nc)	Normal Friends
	Yes (r/nw/nc)	Yes (r/nw/nc)	No (nr/nw/nc)	No (nr/nw/nc)	Casual Friends
	No (nr/nw/nc)	No (nr/nw/nc)	No (nr/nw/nc)	No (nr/nw/nc)	Visitor

In our proposed UPP+ model, we added constraints to the UPP model to make it more plausible. The UPP model contains the Readby constraints, which shows the different levels of privacy and different types of users and suggests who can read the different types of data; however, nothing in the model suggested who can change these data or who can share them. Our contribution came by adding the changedby and the sharedby constraints. We maintained the readby constraints of the UPP model. Moreover, the type of users, the privacy levels, and the type of data being used in the UPP model were not changed.

Table 5.6 shows the ownership of the account in the UPP+ model, with NO standing for No Privacy Owner, SO standing for Soft Privacy Owner, HO standing for Hard Privacy Owner, and FO standing for Full Privacy Owner.

Our contribution is adding constraints to the model to see the consistencies when having the data changed or shared. "Sharedby" stands for the rules added to the model to grant or deny access to the users to share the data. "Changedby" stands for the rules added to the model to grant or deny access to the users to change and modify the data. Table 5.7 illustrates the constraints added. "r" indicates that a user

```
one sig NoPHealthyD,NoPHarmlessD,NoPHarmfulD,NoPPoisonousD extends NoP
                        {readby: some Nusers,sharedby: some Nusers, changedby: one Nusers}

one sig SoftPHealthyD,SoftPHarmlessD,SoftPHarmfulD,SoftPPoisonousD extends SoftP
                        {readby: some Susers,sharedby: some Susers, changedby: one Susers}

one sig HardPHealthyD,HardPHarmlessD,HardPHarmfulD,HardPPoisonousD extends HardP
                        {readby: some Husers,sharedby: some Husers, changedby: one Husers}

one sig FullPHealthyD,FullPHarmlessD,FullPHarmfulD,FullPPoisonousD extends FullP
                        {readby: some Fusers,sharedby: some Fusers, changedby: one Fusers}
```

Section 5.1 UPP+ declaration of privacy data sets in each level

```
//Declaration of Owner in Privacy Policy
one sig NO extends Nusers{}
one sig SO extends Susers{}
one sig HO extends Husers{}
one sig FO extends Fusers{}
```

Section 5.2 UPP+ owner's declaration set

```
//Declaration of owner's instances
one sig NO1 in NO{}
one sig SO1 in SO{}
one sig HO1 in HO{}
one sig FO1 in FO{}
```

Section 5.3 UPP+ owner's instance declaration set

can read the data, "nr" means a user cannot read the data, "w" indicates that a user can share the data, "nw" means a user cannot share the data, "c" indicates that the user can change the data, and "nc" means a user cannot change the data.

Section 5.1 shows the Privacy Data sets in each level as part of the Privacy levels (in the Alloy language). It displays the access rights given to the users.

Section 5.2 shows the declaration of the Ownership "Owner" in privacy policy which extends from the user's levels.

Section 5.3 shows the owner's instances that belong to each user level, which is similar to the users' instances.

Section 5.4 shows that at the No Privacy Level, NBF, NNF, NCF, and NV cannot share poisonous data, while all the users can share all other types of data. At the Soft Privacy Level, SBF, SNF, SCF, and SV cannot share poisonous data, SCF and SV cannot share harmful data, and SV cannot share harmless data, while the rest of the data is shared by the rest of the users.

Section 5.5 shows that at the Hard Privacy Level, HBF, HNF, HCF, and HV cannot share poisonous data; HNF, HCF, and HV cannot share harmful data; HCF and HV cannot share harmless data; and HV cannot share healthy data, while the rest of the users can share the rest of the data.

Section 5.4 UPP+ system
sharedby constraints (part 1)

```
//Sharedby constraints on No Privacy data set

NoPPoisonousD.sharedby!=NBF
NoPPoisonousD.sharedby!=NNF
NoPPoisonousD.sharedby!=NCF
NoPPoisonousD.sharedby!=NV

//Sharedby constraints on Soft Privacy data set
SoftPPoisonousD.sharedby!=SBF
SoftPPoisonousD.sharedby!=SNF
SoftPPoisonousD.sharedby!=SCF
SoftPPoisonousD.sharedby!=SV

SoftPHarmfulD.sharedby!=SCF
SoftPHarmfulD.sharedby!=SV

SoftPHarmlessD.sharedby!=SV
```

Section 5.5 UPP+ system
sharedby constraints (part 2)

```
//Sharedby constraints on Hard Privacy data set
HardPPoisonousD.sharedby!=HBF
HardPPoisonousD.sharedby!=HNF
HardPPoisonousD.sharedby!=HCF
HardPPoisonousD.sharedby!=HV

HardPHarmfulD.sharedby!=HNF
HardPHarmfulD.sharedby!=HCF
HardPHarmfulD.sharedby!=HV

HardPHarmlessD.sharedby!=HCF
HardPHarmlessD.sharedby!=HV

HardPHealthyD.sharedby!=HV
```

Section 5.6 UPP+ system
sharedby constraints (part 3)

```
//Sharedby constraints on Full Privacy data set
FullPPoisonousD.sharedby!=FBF
FullPPoisonousD.sharedby!=FNF
FullPPoisonousD.sharedby!=FCF
FullPPoisonousD.sharedby!=FV

FullPHarmfulD.sharedby!=FNF
FullPHarmfulD.sharedby!=FCF
FullPHarmfulD.sharedby!=FV

FullPHarmlessD.sharedby!=FCF
FullPHarmlessD.sharedby!=FV

FullPHealthyD.sharedby!=FCF
FullPHealthyD.sharedby!=FV
```

Section 5.6 shows that at the Full Privacy Level, FBF, FNF, FCF, and FV cannot share poisonous data; FNF, FCF, and FV cannot share harmful data; FCF and FV cannot share harmless data; FCF and FV cannot share healthy data, while the rest of the users can share the rest of the data.

Section 5.7 shows that at the No Privacy Level and in Soft Privacy Level, all users of the model cannot change data, while the owner can change all types of data.

```
//Changedby constraints on No Privacy set
NoPPoisonousD.changedby!=NBF
NoPPoisonousD.changedby!=NNF
NoPPoisonousD.changedby!=NCF
NoPPoisonousD.changedby!=NV

NoPHarmfulD.changedby!=NBF
NoPHarmfulD.changedby!=NNF
NoPHarmfulD.changedby!=NCF
NoPHarmfulD.changedby!=NV

NoPHarmlessD.changedby!=NBF
NoPHarmlessD.changedby!=NNF
NoPHarmlessD.changedby!=NCF
NoPHarmlessD.changedby!=NV

NoPHealthyD.changedby!=NBF
NoPHealthyD.changedby!=NNF
NoPHealthyD.changedby!=NCF
NoPHealthyD.changedby!=NV
```

```
//Changedby constraints on Soft Privacy set
SoftPPoisonousD.changedby!=SBF
SoftPPoisonousD.changedby!=SNF
SoftPPoisonousD.changedby!=SCF
SoftPPoisonousD.changedby!=SV

SoftPHarmfulD.changedby!=SBF
SoftPHarmfulD.changedby!=SNF
SoftPHarmfulD.changedby!=SCF
SoftPHarmfulD.changedby!=SV

SoftPHarmlessD.changedby!=SBF
SoftPHarmlessD.changedby!=SNF
SoftPHarmlessD.changedby!=SCF
SoftPHarmlessD.changedby!=SV

SoftPHealthyD.changedby!=SBF
SoftPHealthyD.changedby!=SNF
SoftPHealthyD.changedby!=SCF
SoftPHealthyD.changedby!=SV
```

Section 5.7   UPP+ system changedby constraints (part 1)

```
//Changedby constraints on Hard Privacy set
HardPPoisonousD.changedby!=HBF
HardPPoisonousD.changedby!=HNF
HardPPoisonousD.changedby!=HCF
HardPPoisonousD.changedby!=HV

HardPHarmfulD.changedby!=HBF
HardPHarmfulD.changedby!=HNF
HardPHarmfulD.changedby!=HCF
HardPHarmfulD.changedby!=HV

HardPHarmlessD.changedby!=HBF
HardPHarmlessD.changedby!=HNF
HardPHarmlessD.changedby!=HCF
HardPHarmlessD.changedby!=HV

HardPHealthyD.changedby!=HBF
HardPHealthyD.changedby!=HNF
HardPHealthyD.changedby!=HCF
HardPHealthyD.changedby!=HV
```

```
//Changedby constraints on Full Privacy set
FullPPoisonousD.changedby!=FBF
FullPPoisonousD.changedby!=FNF
FullPPoisonousD.changedby!=FCF
FullPPoisonousD.changedby!=FV

FullPHarmfulD.changedby!=FBF
FullPHarmfulD.changedby!=FNF
FullPHarmfulD.changedby!=FCF
FullPHarmfulD.changedby!=FV

FullPHarmlessD.changedby!=FBF
FullPHarmlessD.changedby!=FNF
FullPHarmlessD.changedby!=FCF
FullPHarmlessD.changedby!=FV

FullPHealthyD.changedby!=FBF
FullPHealthyD.changedby!=FNF
FullPHealthyD.changedby!=FCF
FullPHealthyD.changedby!=FV
```

Section 5.8   UPP+ system changedby constraints (part 2)

Section 5.8 shows that at the Hard Privacy Level and in Full Privacy Level, all users of the model cannot change data, while the owner can change all types of data.

At this stage, we are ready to implement of the UPP+ model, using the Alloy language and its analyzer, in order to show its consistency. A Meta Model and instances are generated for the UPP+ model. Figure 5.1 depicts the Meta Model of UPP+. The figure shows that as in UPP model, PrivacyDS contains the four subsets: FUllP, HardP, NoP, and SoftP. Each of these levels contains the four types of data which are healthy, harmless, harmful, and poisonous, each of which extends

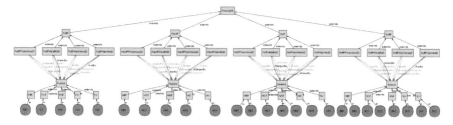

Fig. 5.1 The UPP+ meta model

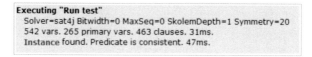

Fig. 5.2 UPP+ consistency output using Alloy Analyzer

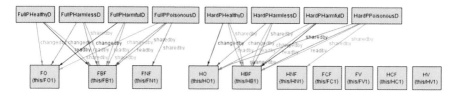

Fig. 5.3 UPP+ model instance 1 (part 1)

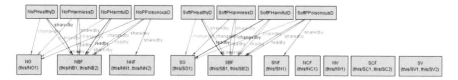

Fig. 5.4 UPP+ model instance 1 (part 2)

from the Privacy level. Since it is a Meta Model, it does not show the constraints of each user. The privacy data are read by, changed by, and shared by different types of users which are: Nusers, Susers, Husers, Fusers. Each type of the users extends to BF, NF, CF, V, and O.

After showing the Meta Model of UPP+, we need to test the model and show its constraints by running the predicate test. The result, depicted in Fig. 5.2, shows that instance is found and that the predicate is consistent. The time taken to check for consistency and to find an instance is 47 ms.

By clicking on Instance, the Alloy Analyzer will yield Figs. 5.3 and 5.4. More instances can be generated by clicking "next." Tables 5.8 and 5.9 show the instances using "changedby" and "sharedby."

Table 5.8 UPP+ model checking changedby consistencies

User type	Can change	Consistent
FO	FullPoisonousD	Yes
	FullHealthyD	Yes
	FullHarmlessD	Yes
	FullHarmfulD	Yes
HO	HardPoisonousD	Yes
	HardHealthyD	Yes
	HardHarmlessD	Yes
	HardHarmfulD	Yes
NO	NoPPoisonousD	Yes
	NoPHarmlessD	Yes
	NoPHarmfulD	Yes
	NoPHealthyD	Yes
SO	SoftPoisonousD	Yes
	SoftHealthyD	Yes
	SoftHarmlessD	Yes
	SoftHarmfulD	Yes

Table 5.9 UPP+ model checking sharedby consistencies

User type	Can share	Consistent
FBF	FullPoisonousD	Yes
	FullHealthyD	Yes
	FullHarmlessD	Yes
	FullHarmfulD	Yes
HBF	HardPoisonousD	Yes
	HardHealthyD	Yes
	HardHarmlessD	Yes
	HardHarmfulD	Yes
NBF	NoPPoisonousD	Yes
	NoPHarmlessD	Yes
	NoPHarmfulD	Yes
	NoPHealthyD	Yes
SBF	SoftPoisonousD	Yes
	SoftHealthyD	Yes
	SoftHarmlessD	Yes
	SoftHarmfulD	Yes

After showing that the system is consistent, we try different constraints that are wrong predicates, which should produce inconsistency. We ran a test example by stating that No Privacy visitor can change Harmless data as in Section 5.9 and that Full Privacy Best friend can share poisonous data, which proves the results of the predicate is inconsistent, as depicted in Fig. 5.5.

```
pred test()                          pred test()
{                                    {|
NoPHarmlessD.changedby=NV            FullPPoisonousD.sharedby=FBF
}                                    }
run test                             run test
```

Section 5.9 UPP+ inconsistent predicates

```
Executing "Run test"
Solver=sat4j Bitwidth=0 MaxSeq=0 SkolemDepth=1 Symmetry=20
0 vars. 0 primary vars. 0 clauses. 16ms.
No instance found. Predicate may be inconsistent. 0ms.
```

Fig. 5.5 UPP+ inconsistent output using Alloy Analyzer

5.5   Conclusion

In this work, we presented the UPP+ model, which carries significant enhancements over the UPP policy model. We used system examples based on the defined privacy model. We formalized the system according to the model and then checked its consistency and inconsistency. Since Alloy allows expressing systems as set of logical constraints in a logical language based on standard first order logic, we used it to define the system and its policy. When creating the model we specified the system users and data then Alloy compiles a Boolean matrix for the constraints, and we asked it to check if a model is valid, or if there are counterexamples. However, there exist many other privacy models for social networks that need to be formalized and analyzed to show their correctness. We plan to study these models in a more formal way to ascertain that they provide adequate privacy for users. We also believe that more work can be done to integrate multiple models in a mixed mode and formalize them to find potential interactions.

References

1. Danah, B.M., Ellison, N.: Social network sites: definition, history, and scholarship. J. Comput. Mediat. Commun. **13**(1), 210–230 (2007)
2. Jackson, D.: Alloy: a lightweight object modeling notation. Technical report 797. MIT Laboratory for Computer Science, Cambridge (2000)
3. Jackson, D., Schechter, I., Shlyakhter, I.: Alcoa: the alloy constraint analyzer. In: Proceedings of the International Conference on Software Engineering, Limerick, 2000
4. Wallace, C.: Using alloy in process modelling. Inf. Softw. Technol. J. **45**, 1031–1043 (2003). ISSN 0950-5849
5. Jackson, D.: Alloy 3.0 Reference Manual. Retrieved on April 18, 2013 from: http://alloy.mit.edu/reference-manual.pdf (2012)
6. Seater, R., Dennis, G.: Tutorial for Alloy Analyzer 4.0. Retrieved on April 18, 2013 from: http://alloy.mit.edu/tutorial4 (2012)
7. McLean, J.: In: Marciniak, J. (ed.) Encyclopedia of Software Engineering. Wiley, New York (1994)

8. Fong, P.W.L., Anwat, M., Zhao, Z.: A privacy preservation model for Facebook - style social network systems. In: Proceedings of the 14th European Symposium on Research in Computer Security (ESORICS'09), Saint Malo. Lecture Notes in Computer Science, vol. 5789, pp. 303–320 (2009)

9. Danezis, G.: Inferring Privacy Policies for Social Networking Services. CCS Computer and Communications Security, pp. 5–10. ACM, New York (2009)

10. Dania, C.: Modeling social networking privacy. In: Doctoral Symposium of the International Symposium on Engineering Secure Software and Systems (ESSoS), pp. 49–54. CEUR, The Netherlands (2012)

11. Haraty, R.A.: C2 secure database management systems – a comparative study. In: Proceedings of the ACM Symposium on Applied Computing, San Antonio, 1999

12. Haraty, R.A., Bekaii, N.: Towards a temporal multilevel secure database. J. Comput. Sci. **2**(1) (2006). ISSN 1549-3636

13. Hassan, W., Logrippo, L.: Detecting inconsistencies of mixed secrecy models and business policies. Technical report. University of Ottawa, Ottawa (2009)

14. Ferraiolo, D.F., Kuhn, D.R.: Role-based access control. In: Proceedings of the 15th National Computer Security Conference, Baltimore, pp. 554–563, 1992

15. Shaffer, A., Auguston, M., Irvine, C., Levin, T.: A security domain model to assess software for exploitable covert channels. In: Proceedings of the ACM SIGPLAN Third Workshop on Programming Languages and Analysis for Security, pp. 45–56. ACM, Tucson (2008)

16. Misic, J., Misic, V.: Implementation of security policy for clinical information systems over wireless sensor networks. Ad Hoc Netw. J. **5**, 134–144 (2007). ISSN 1570-8705

17. Haraty, R.A., Naous, M.: Modeling and validating the clinical information systems policy using alloy. In: Proceedings of the Second International Conference on Health Information Science. Lecture Notes in Computer Science, pp. 1–17. Springer, London (2013)

18. Hassan, W., Logrippo, L.: Detecting inconsistencies of mixed secrecy models and business policies. Technical report. University of Ottawa, Ottawa (2009)

19. Haraty, R.A., Naous, M.: Role-based access control modeling and validation. In: Proceedings of the Fifth IEEE International Workshop on Performance Evaluation of Communications in Distributed Systems and Web based Service Architectures (PEDISWESA'2013), Split, 2013

20. Aïmeur, E., Gambs, S., Ho, A.: UPP: user privacy policy for social networking sites. In: Proceedings of the Fourth International Conference on Internet and Web Applications and Services, pp. 267–272, 2009

Chapter 6
Social Semantic Network-Based Access Control

Serena Villata, Luca Costabello, Fabien Gandon, Catherine Faron-Zucker, and Michel Buffa

Abstract Social networks are the bases of the so-called Web 2.0, raising many new challenges to the research community. In particular, the ability of these networks to allow the users to share their own personal information with other people opens new issues concerning privacy and access control. Nowadays the Web has further evolved into the Social Semantic Web where social networks are integrated and enhanced by the use of semantic conceptual models, e.g., the ontologies, where the social information and links among the users become semantic information and links. In this chapter, we discuss which are the benefits of introducing semantics in social network-based access control. In particular, we analyze and detail two approaches to manage the access rights of the social network users relying on Semantic Web languages only, and we highlight, thanks to these two proposals, what are pros and cons of introducing semantics in social networks access control. Finally, we report on the other existing approaches coupling semantics and access control in the context of social networks.

6.1 Introduction

One of the key features of the Social Web is the ability to publish, and thus find a lot of personal and professional information about people. With the advent of the Social Semantic Web this is even more evident, as underlined by Breslin et al. [5]. The availability of personal and nonpersonal data of the users has both positive

S. Villata (✉) · L. Costabello · F. Gandon
INRIA, Sophia Antipolis, Nice, France
e-mail: serena.villata@inria.fr; luca.costabello@inria.fr; fabien.gandon@inria.fr

C. Faron-Zucker · M. Buffa
I3S, Université Nice Sophia Antipolis - CNRS, Nice, France
e-mail: faron@polytech.unice.fr; buffa@unice.fr

R. Chbeir and B. Al Bouna (eds.), *Security and Privacy Preserving in Social Networks*,
Lecture Notes in Social Networks, DOI 10.1007/978-3-7091-0894-9_6,
© Springer-Verlag Wien 2013

and negative sides. On the one hand, this allows people to share their data, e.g., photos, videos, posts, with their friends and the persons they know. On the other hand, semantic forms of the users' profiles like FOAF profiles and data can be reused elsewhere, e.g., what happened with FOAF search engines and aggregators as Plink, or FoaFSpace. This leads to the need for mechanisms where users can restrict the access to their data by specifying the attributes the accessors must satisfy to have the access granted.

In particular, security, protection, and access control represent a major challenge in content management systems. This issue is central also in collaborative social Web sites, where the collaborative editing and sharing of the documents raises the question of the definition of access rights. Moreover, access control is important to lead to a diffusion of Social Semantic Web platforms to make them able to guarantee the same kind of authorizations as in standard Social Web platforms like Facebook, or Google+. Managing the access to the resources is thus one of the major challenges facing the Social Semantic Web.

In this chapter, we address the following research question: *What are the benefits of adopting Semantic Web models and languages for social network-based access control?* Policies, norms, and the Semantic Web *Trust Layer*, as shown in Fig. 6.1, are usually presented as a set of rules and constraints that model the intended behaviors of the users. Within W3C, the Policy Languages Interest Group is the forum that coordinates the efforts of the community around the definition of policy languages, frameworks, and use cases. Apart from W3C activities, one of the most prominent standard for modelling policies is the eXtensible Access Control Markup Language (XACML). Policies can be defined at community level but they can also be defined at the individual level, e.g., my privacy policies in a social network and e-mail filtering policies. Policies on the Semantic Web build the foundation for privacy and access rights of personal or community data, whereas norms in general establish best practices, e.g., how to publish the data. The definition of both private and community policies is useful for various further applications such as checking compliance or conformance, policies alignment, or checking the internal consistency of policies.

We answer the research question by presenting two approaches for defining the access control policies using Semantic Web languages only. These two approaches are applied to social semantic networks and show pros and cons of introducing semantics in social network-based access control.

First, we consider content management systems based on Semantic Web servers, and we propose an approach for managing access rights to resources based on Semantic Web models and techniques [6]. We present an ontology dedicated to the representation of the access rights given on a document to some users or user classes. We call this ontology AMO, an acronym meaning *Access Management Ontology*. AMO is made of a set of classes and properties for annotating the resources and a base of inference rules modeling the access control policy. When applied to the annotations of resources, these rules enable to control access according to a given strategy. This declarative modeling as a rule base ensures an easy adaptation of the ontology to different access control policies and thus avoids modifying annotations

Fig. 6.1 The Semantic Web
stack (from http://www.w3.
org/sw)

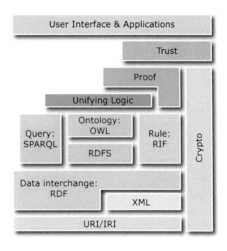

of documents in the case of a change of strategy. In the ISICIL[1] research project,
we use the AMO ontology to manage access to resources shared by a network of
technical watchers: documents produced by content management tools, wikis or
blogs, static HTML documents produced by web scraping (i.e., firefox extensions
similar to "Scrapbook"), bookmarks, etc. One of the issues of this project oriented
to Web 2.0 and Semantic Web techniques concerns the management of access to
the resources shared by the social network of watchers. Among the documents
produced by the watchers there are those of a collaborative web site run by the
semantic wiki SweetWiki that we developed [7] and that is used in this chapter
to illustrate the use of AMO. SweetWiki integrates Semantic Web technologies to
improve structure, search, and navigation. More specifically, it associates with wiki
pages RDF/S annotations that make the content of these pages processable by the
semantic engine CORESE [11].

Second, we describe the Social Semantic SPARQL Security for Access Control
vocabulary (S4AC[2]), a lightweight ontology which allows to specify fine-grained
access control policies for RDF data [15, 44, 45]. We adopt exclusively Semantic
Web languages and recycle, when possible, already existing vocabularies. In this
model, we avoid the usual access control lists (ACLs), often maintained by a sole
authority, because we cannot specify the access restrictions to any particular user,
in a context where the user information is so dynamic. We rely on social tags
assigned by the users to data and other users. Moreover, contextual information
is also considered in this model to grant access to users by considering not only
their personal information but also additional attributes, e.g., time and location
constraints. We adopt the PRISSMA[3] vocabulary [14] to model the user context

---

[1]http://isicil.inria.fr/

[2]http://ns.inria.fr/s4ac/

[3]http://ns.inria.fr/prissma/

in which the access request takes place. Following the widely adopted formalization of context provided by Dey [21], PRISSMA defines the information context as the sum of the following three dimensions: the *User*, modelling the target mobile user (stereotypes or specific instances), the *Device*, which represents the mobile device in use, and the *Environment*, the dimension dealing with the physical context where the consumption takes place. The overall access control framework allows data providers to specify lightweight access policies to protect their data, at *named graph* granularity [9]. The Access Control Manager (ACM) verifies which named graphs are accessible by the user, so that the user's query is run on those graphs only. The system evaluation shows that access control comes with a cost, and that performance loss is acceptable when dealing with sensitive data.

For the time being, our lightweight framework assumes the trustworthiness of the information sent by the data consumer. Moreover, our approach focuses only on SPARQL endpoints. Other access strategies are out of the scope of this work. Despite the amount of proposals of access control models [1,8,22–26,33,35,41,42], none of them presents a Social Semantic access control model based on Semantic Web languages only, a pluggable and easy-to-integrate filter for generic SPARQL endpoints without modifying the endpoint itself, providing access conditions from triple granularity level up to dataset granularity level, and taking into account the social tags assigned by the users to their data and other users and the contextual information. Moreover, we rely on W3C recommendations only, as we do not introduce any new language or technology.

The two proposals we describe do not deal with access control for the Social Web in general, but we present two frameworks suitable for the Social Semantic Web. Our aim is not to provide a privacy manager or a cryptography system, but we are interested in formalizing, developing, and evaluating access control frameworks which authorize or not the access of the users to the data of the other users, without considering personal information only.

The structure of the chapter is as follows. Section 6.2.1 presents the ontology AMO and the use of AMO in SweetWiki, highlighting the adaptability of AMO to different access control policies. Section 6.2.2 presents the S4AC ontology, defined for social access control using the SPARQL 1.1. language, coupled with the PRISSMA ontology for the user context definition, with the aim to propose a contextual access control model for social semantic networks. Section 6.3 is dedicated to the positioning with respect to the existing work.

6.2 Semantic Approaches to Social Network-Based Access Control

6.2.1 *Ontology Based Access Management*

We present in this section an access control model where access control is based on an ontology modeling access rights to resources and access control strategies.

The AMO Ontology

In a file system or in a content management system, roles (administrator, owner, etc.) are associated with users or user groups and different types of access to resources (writing, reading, etc.) are defined, access to resources varying from one user to another depending on its role. This analysis led us to define a set of classes and properties to describe the access rights to resources. This is what we describe in section "AMO Classes and Properties."

Content management systems share the same general principles for access control to resources; however, they adopt strategies that may vary from one system to another. To allow easy adaptation of the ontology supporting the management of access to resources according to the chosen strategy, this latter is declaratively modeled in AMO as a base of inference rules that can be modified at leisure without affecting the annotations of the resources to manage. We describe in section "AMO Inference Rules" a rule base that modelizes one strategy for the access control of documents in the semantic wiki SweetWiki.

AMO Classes and Properties

AMO is based on some basic principles shared by all content management systems:

- *Agents* of a content management system are the users, user groups, services that interact with the system.
- These agents have *roles*. In the case of collaborative editing systems such as wikis or CMS, these roles are those of guest (agent not registered in the system), contributor, administrator. Other roles can be modeled depending on the kind of system.
- Each role is associated with a list of authorized *actions*. In the case of collaborative editing systems, the possible actions on a resource are creation, reading, modification and destruction of content, modification of access rights, modification of the list of agents allowed on a resource, change of the access type defined for a resource. Other actions can be modeled for other kinds of systems.
- There are different *types* of access to resources. We choose to implement a strategy popular in some collaborative editing systems: a resource can be public (all users have reading and writing access), private (only authorized agents have reading and writing access), or semi-private (free reading access, writing access only to authorized agents). Again, other types of access can be added for other types of systems.
- Finally, the actions authorized to an agent on a resource depend on the role of the agent and/or the type of access defined for the resource.

The AMO ontology presented in Fig. 6.2 provides the concepts necessary to represent this knowledge. The three classes `Role`, `Action`, and `AccessType` are central to AMO. `Role` is the meta-class of classes `Administrator`,

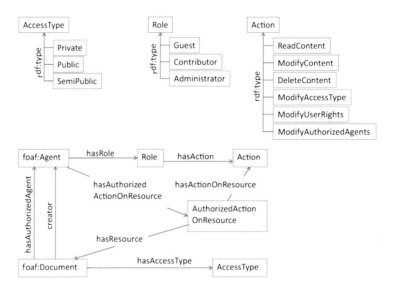

Fig. 6.2 AMO classes and properties

Contributor, and Guest. Action is the meta-class of classes ReadContent, ModifyContent, DeleteContents, ModifyUserRights, Modify AccessType, and ModifyAuthorizedAgents. Finally, AccessType is the meta-class of class Private, Public, and SemiPublic.

Three classes of the FOAF vocabulary—the standard for social web discussed in Sect. 6.3—are also central in AMO: Agent and its sub-class Group and Document. They are used as domain or range of properties of AMO and also in the rules of AMO.

Properties creator and hasAuthorizedAgent associate an agent with a document (they have for domain the class Document and for range the class Agent); hasRole associates a role with an agent and hasActionOnResource an action to a role; property hasAccessType associates an access type to a document.

In addition, to represent into a model of binary properties the ternary relation which states that an agent is authorized to perform an action on a resource, we have reified this relationship by introducing the subclass AuthorizedActionOnResource specializing the class Action, a property hasAuthorizedActionOnResource that associates an instance of AuthorizedActionOnResource with an agent, and the properties hasDocument and hasAction that associate with an instance of AuthorizedActionOnResource, respectively, a document and an action.

AMO is an RDFS vocabulary which can be used to annotate the RDF resources whose access we want to control.

	Public	Semi-Public	Private
Guest	ReadContent	ReadContent	
Contributor	ReadContent	ReadContent ModifyContent DeleteContent	
AuthorizedAgent	ReadContent ModifyContent DeleteContent ModifyAuthorizedAgents ModifyAccessType		
Administrator	ReadContent ModifyContent DeleteContent ModifyAuthorizedAgents ModifyAccessType ModifyUserRights		

Fig. 6.3 An access control policy modeled in AMO

AMO Inference Rules

Content management systems adopt access control strategies to resources that can vary from one system to another. Rather than varying the annotations of resources depending on the control strategies, we propose to model declaratively the control strategy in the AMO ontology, as a base of inference rules. Some rules may vary depending on the strategy modeled while the annotations remain unchanged. The rule base presented here is that of SweetWiki whose strategy of access control is similar to that of the widely used open source wiki Mindtouch Deki.[4]

By default, administrators have all rights on all resources. The contributors have all rights relative to the content of resources, those reported as agents of a resource by the author thereof also have some administrative rights on it. Guests are only allowed to read the content of resources. Figure 6.3 summarizes the access rights to a resource depending on the type of access and the role of the user who tries to access the resource (horizontally are the types of access resources, vertically user roles).

We model this strategy in AMO *declaratively* by six inference rules, each corresponding to a situation described in Fig. 6.3. For example, Rule 1 below specifies the rights granted to agents of a given resource. Other rules describe general laws such as *a member of a group inherits the roles assigned to her group* (Rule 2) or *creator of a resource is an agent of this resource* (Rule 3).

These rules are expressed in the SPARQL language, by using the query pattern CONSTRUCT/WHERE: such a query enables to *construct* RDF graphs by replacing the variables of its clause CONSTRUCT by the values that satisfy the clause WHERE (they are retrieved by searching for potential matches to clause

---

[4]http://www.mindtouch.com/

WHERE with the RDF data available in the content management system). A query
CONSTRUCT/WHERE can therefore be seen as a rule applied in forward chaining,
with clause WHERE the premise and clause CONSTRUCT the conclusion. These
rules, however, can also be used in backward chaining, as is the case in the semantic
engine Corese.

Rule 1:
```
CONSTRUCT {
    ?agent amo:hasAuthorizedActionOnResource ?a
    ?a amo:hasResource ?resource
    ?a amo:hasActionOnResource amo:ReadContent.
    ?a amo:hasActionOnResource amo:ModifyContent.
    ?a amo:hasActionOnResource amo:DeleteContent.
    ?a amo:hasActionOnResource amo:ModifyAccessType.
    ?a amo:hasActionOnResource amo:ModifyAuthorizedAgents }
WHERE {
    ?resource rdf:type foaf:Document.
    ?resource amo:hasAuthorizedAgent ?agent }
```

Rule 2:
```
CONSTRUCT {
    ?agent amo:hasRole ?role }
WHERE {
    ?group amo:hasRole ?role
    ?group foaf:member ?agent }
```

Rule 3:
```
CONSTRUCT ?resource amo:hasAuthorizedAgent ?agent
WHERE ?resource amo:creator ?agent
```

This *declarative* modeling of the strategy of access rights management ensures
easy maintenance. Changing rights of a class of users—and this for all resources
involved—will only require the addition or deletion of triples statements in the
conclusion of a rule. Similarly, the addition of new roles will only require the
addition of a class representing this role and the rules representing the access rights
associated with that role.

Access Rights Management in SweetWiki

The AMO ontology has been used in the ISICIL project to annotate resources
shared by a social network of business watchers. The management of access to
these resources in the engine SweetWiki was based on (1) the exploitation of these
semantic annotations, (2) inferences on these annotations based on AMO rules, and
(3) the formulation of SPARQL queries to retrieve knowledge about the authorized
access to a specific user on a given resource. In SweetWiki, annotations of resources
are based on FOAF, SIOC, and AMO ontologies, and SPARQL queries are used
in most of the features implemented: RDF annotations feed the semantic engine
CORESE embedded in SweetWiki. In particular, by using the approximate search
possibilities of Corese[10, 12] and a system of semantic tagging of documents,
SweetWiki offers an "intelligent" browsing mechanism enhanced by suggestions.

Annotation of Resources with AMO

When creating a wiki page, the identity of its creator is registered and also the type of access to the page that is decided by her and possibly one or more agents authorized on the page, also designated by the creator. In SweetWiki this knowledge is represented into RDF annotations associated with the created pages. For example, Annotation 1 below results from the creation of a private wiki page by the user AnnaKolomoiska who stated that agent MichelBuffa is authorized on this page. This annotation uses the AMO properties creator, hasAuthorizedAgent and hasAccessType (and the class WikiArticle of the SIOC vocabulary).

Annotation 1:

```
<rdf:RDF xmlns="http://seetwiki.i3s.unice.fr/AMO.rdfs#" ... >
    <sioc:WikiArticle rdf:about="#TestPage">
    ...
    <creator rdf:resource="#AnnaKolomoiska"/>
    <hasAuthorizedAgent rdf:resource="#MichelBuffa"/>
    <hasAccessType rdf:resource="#Private"/>
  </sioc:WikiArticle>
</rdf:RDF>
```

When registering a user in SweetWiki, this information is represented in an RDF annotation. For example, Annotation 2 below states that MichelBuffa is a contributor to the wiki. It uses the AMO class Contributor and AMO property hasRole (and the class Agent of the FOAF vocabulary discussed in Sect. 6.3).

Other annotations express knowledge relative to the user groups of the wiki. For example, Annotation 3 states that AnnaKolomoiska and CatherineFaron are members of the administrator group of the wiki. It uses for that the AMO property hasRole (and the FOAF classes Group and Agent and the FOAF property member).

Annotation 2:

```
<rdf:RDF xmlns="http://seetwiki.i3s.unice.fr/AMO.rdfs#" ... >
  <foaf:Agent rdf:about="#MichelBuffa">
    ...
    <hasRole rdf:resource="#Contributor"/>
  </foaf:Agent>
</rdf:RDF>
```

Annotation 3:

```
<rdf:RDF xmlns="http://seetwiki.i3s.unice.fr/AMO.rdfs#" ... >
 <foaf:Group rdf:about="#AdminGroup">
  <foaf:member>
   <foaf:Agent rdf:about="#AnnaKolomoiska"/>
  </foaf:member>
  <foaf:member>
   <foaf:Agent rdf:about="#CatherineFaron"/>
  </foaf:member>
```

```
  <hasRole rdf:resource="#Admin"/>
  </foaf:Group>
</rdf:RDF>
```

Inferences with the Rule Base of AMO

Applied to the annotations of resources, AMO rules enable to infer the rights of the wiki users on these resources. For example, consider again Rule 1. Its premise matches with Annotation 1 that illustrates section "Annotation of Resources with AMO": the resource TestPage is of type WikiArticle— a class of the SIOC vocabulary, subclass of the class Document of the FOAF vocabulary—and TestPage is related to the user MichelBuffa with the hasAuthorizedAgent property. Applied on Annotation 1, Rule 1 allows to conclude that MichelBuffa has the *read*, *modify*, and *delete* permissions on the content of the annotated resource TestPage and the *modify* permission on its type of access and its list of agents.

Similarly, Rule 2 applied on Annotation 3 allows to conclude that user CatherineFaron has the administrator role. Another rule of AMO (not provided here) describes general rights of an agent having the administrator role on any resource. It enables to conclude that CatherineFaron owns all the rights on the specific resource TestPage.

Finally, Rules 1 and 3 applied on Annotation 1 enable to conclude that user AnnaKolomoiska, creator of resource TestPage, has the rights of an agent of that resource: *read*, *modify*, and *delete* rights on its content and *modify* right on its type of access and its list of agents.

SPARQL Requests for Access Rights Management

Access to a particular resource by a given user depends, as all the actions in SweetWiki, on the answers to a SPARQL query provided by the Corese engine launched on the base of resource annotations. For this, Corese combines backward chaining on the AMO rule base and matching of queries with the annotation base. For example, the answer to the following SPARQL query will indicate whether the user CatherineFaron is allowed to modify the content of the resource TestPage:

Query 1:
```
  prefix amo: <http://sweetwiki.unice.fr/AMO.rdfs#>
  ASK {
   <http://sweetwiki.unice.fr#CatherineFaron>
                     amo:hasAuthorizedAccessOnResource ?x
   ?x amo:hasActionOnResource amo:ModifyContent
   ?x amo:hasResource <http://sweetwiki.unice.fr#TestPage> }
```

Other SPARQL queries are formulated to support all the functionalities of SweetWiki. For instance, the processing of the following query will provide the list of all the users having some rights on resource `TestPage` and for each of them it will state the list of her authorized actions on `TestPage`:

Query 2:
```
prefix amo: <http://sweetwiki.unice.fr/AMO.rdfs#>
SELECT ?agent ?action {
 ?agent amo:hasAuthorizedAccessOnResource ?x
 ?x amo:hasActionOnResource ?action
 ?x amo:hasResource <http://sweetwiki.unice.fr#TestPage> }
order by ?agent
```

6.2.2 Context-Aware Access Control for Semantic Social Networks

We present in this section a new access control model where access control is based on the features of the user accessing the protected data, and context has an important role in determining whether the user is granted access or not. This new model called *S4AC-PRISSMA* enhances the expressive power of the AMO model and allows the data provider to protect in a finer-grained way her resources.

The Context-Aware Access Control Model

In this section, we present our access control model. The access control model is built over the notion of Named Graph [9], thus supporting fine-grained access control policies, including the triple level (enforcing permission models is an envisioned use case for RDF named graphs[5]). We rely on named graphs to avoid depending on documents (one document can serialize several named graphs, one named graph can be split over several documents, and not all graphs come from documents[6]). At conceptual level, our policies can be considered as access control conditions over g-boxes[7] (according to W3C RDF graph terminology), with semantics mirrored in the SPARQL language.

The model is grounded on the Social Semantic SPARQL Security for Access Control Ontology (S4AC). An overview of S4AC lightweight vocabulary is provided in Fig. 6.4. Our access control model is integrated with the models adopted in

---

[5]http://bit.ly/w3rdfperm

[6]The discussion about the use of named graphs in RDF 1.1 can be found at http://www.w3.org/TR/rdf11-concepts

[7]http://bit.ly/graphterm

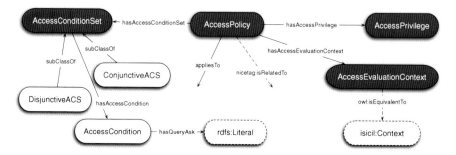

Fig. 6.4 An overview of the S4AC ontology. Core classes are in *grey*

the Social Semantic Web. In particular, S4AC reuses concepts from SIOC,[8] SKOS,[9] WAC,[10] NiceTag,[11] SPIN,[12] Dublin Core,[13] and the access control model as a whole is grounded on further existing ontologies, such as FOAF[14] and RELATIONSHIP.[15]

The main component of the S4AC model is the Access Policy, as presented in Definition 1. Roughly, an Access Policy defines the constraints that must be satisfied to access a given named graph or a set of named graphs. If the Access Policy is *satisfied*, then the data consumer is allowed to access the data. Otherwise, the access is not granted. The constraints specified by the Access Policies may concern the data consumer, i.e., the user or the environment in which the user is querying the SPARQL endpoint, or any given combination of these dimensions.

Definition 1 (Access Policy). An Access Policy (P) is a tuple of the form

$$P = \langle \text{ACS}, \text{AP}, T, R, \text{AEC} \rangle$$

where (i) ACS is a set of Access Conditions to satisfy, (ii) AP is an Access Privilege, (iii) T is the tag of the set of resources to be protected by P, (iv) R is the resource(s) to be protected by P, and (v) AEC is the Access Evaluation Context of P.

An Access Condition, as defined in Definition 2, expresses a constraint which needs to be verified in order to have the Access Policy satisfied. Notice that both T and R represent the data to protect. The difference is that T refers to the set of

---

[8]http://rdfs.org/sioc/spec/

[9]http://www.w3.org/TR/skos-reference/

[10]http://www.w3.org/wiki/WebAccessControl

[11]http://ns.inria.fr/nicetag/2010/09/09/voc.html

[12]http://spinrdf.org/

[13]http://dublincore.org/documents/dcmi-terms/

[14]http://xmlns.com/foaf/spec/

[15]http://vocab.org/relationship/

named graphs associated with a certain tag, and R refers to the URI(s) of the specific named graph(s).

Definition 2 (Access Condition). An Access Condition (AC) is a condition which tests whether or not a query pattern has a solution.

In the S4AC model, we express an Access Condition as a SPARQL 1.1 ASK query.[16]

Definition 3 (Access Condition Verification). If the query pattern has a solution (i.e., the ASK query returns *true*), then the Access Condition is said to be *verified*. If the query pattern has no solution (i.e., the ASK query returns *false*), then the Access Condition is said *not* to be *verified*.

Example 1. An example of Access Condition, which is verified only if the data consumer is a *collaborator* of the provider of the resource to protect, is the following:

```
PREFIX dcterms: <http://purl.org/dc/terms/>
PREFIX rel: <http://purl.org/vocab/relationship/>

ASK {?resource dcterms:creator ?provider.
     ?provider rel:collaboratesWith ?consumer.}
```

Each Access Policy P is composed by a set of Access Conditions, as defined in Definition 4.

Definition 4 (Access Condition Set). An Access Condition Set (ACS) is a set of access conditions of the form ACS $= \{AC_1, AC_2, \ldots, AC_n\}$.

Roughly, the verification of an Access Condition Set returns an answer of the kind *true/false*. We consider two standard ways to provide such an evaluation: conjunctively and disjunctively.

Definition 5 (Conjunctive Access Condition Set). A Conjunctive Access Condition Set (CACS) is a logical conjunction of Access Conditions of the form CACS $= AC_1 \wedge AC_2 \wedge \ldots \wedge AC_n$.

Definition 6 (Conjunctive ACS Evaluation). A CACS is verified if and only if every contained Access Condition is verified.

Definition 7 (Disjunctive Access Condition Set). A Disjunctive Access Condition Set (DACS) is a logical disjunction of Access Conditions of the form DACS $= AC_1 \vee AC_2 \vee \ldots \vee AC_n$.

---

[16]http://www.w3.org/TR/sparql11-query/#ask

Definition 8 (Disjunctive ACS Evaluation). A DACS is verified if and only if at least one of the contained Access Conditions is verified.

The second component of the Access Policy is the Access Privilege. The privilege specifies the kind of operation the data consumer is allowed to perform on the resource protected by the Access Policy.

Definition 9 (Access Privilege). An Access Privilege (AP) is a set of allowed operations on the protected resources of the form AP = {*Create, Read, Update, Delete*}.

We model the Access Privileges as four classes of operations in order to maintain a close relationship with CRUD-oriented access control systems. This relationship allows a finer-grained access control than simple read/write privileges as in WAC, and it suggests to the data providers how to specify the access privileges, following the example of CRUD-oriented systems, as we will discuss in relation to the user interface. The idea is that in the Social Semantic Web, there is a difference in allowing the users who ask to access my data to update my data or to delete my data. We distinguish the Update, Create, and Delete operation to let the user to specify with a deeper degree of detail what the consumers are allowed to perform on her data. Moreover, we relate the four privilege classes to the SPARQL 1.1 query and update language. This matching is realized with the `skos:related` property through the SPIN ontology. The latter models the primitives of the SPARQL query and update languages (e.g., `SELECT`, `INSERT DATA`) as SPIN classes. We show how this matching is actually used by our framework in section "The Access Control Manager".

As previously explained, policies protect data at named graph level. We offer two different ways of specifying the protected object: the provider may target one or more given named graphs, or it may target a set of named graphs with a common tag. The former is achieved by providing the URI(s) of the named graph(s) to protect using the `s4ac:appliesTo` property. The latter is accomplished by listing the tags of the named graphs to protect with the property `nicetag:isRelatedTo`. In this case, the assumption is that the named graphs have been annotated with such metadata.

The Access Policy is associated with an Access Evaluation Context. The latter provides an explicit link between the policy and the actual context data (in the case of the mobile context it is modeled with `PRISSMA`) that will be used to evaluate the Access Policy.

Definition 10 (Access Evaluation Context). An Access Evaluation Context (AEC) is a list of predetermined bound variables of the form AEC = $(\langle var_1, val_1 \rangle, \langle var_2, val_2 \rangle, \ldots, \langle var_n, val_n \rangle)$.

In this chapter, we focus on the mobile context, thus the Access Evaluation Context list is composed only by a couple AEC = $(\langle ctx, URI_{ctx} \rangle)$. We map therefore the variable *ctx*, used in the policy's Access Conditions, to the URI identifying the actual user context in which the SPARQL query has been performed. More specifically, the Access Evaluation Context is implemented as a SPARQL 1.1

BINDINGS Clause[17] to constrain the ASK evaluation, i.e. "BINDINGS ?ctx { (URI_{ctx}) }".

The choice and the design of a context model necessarily need a context definition first. We agree on the widely-accepted proposal by Dey [21]:

Definition 11 (Context). "*Context* is any information that can be used to characterize the situation of an entity. An entity is a person, place, or object that is considered relevant to the interaction between a user and an application, including the user and applications themselves" [21].

More specifically, we rely on the work by Fonseca and colleagues,[18] that we adopt as a foundation for our proposal. The mobile context is seen as an encompassing term, an information space defined as the sum of three different dimensions: the mobile *User* model, the *Device* features and the *Environment* in which the action is performed.

The Social Semantic Web scenario favors the adoption of an ontology-based model. As pointed out by Korpipää and Mäntyjärvi [28], an ontological approach leads to simple and extensible models. Linked Data on the Web heavily relies on lightweight vocabularies under the open world assumption (i.e. new ontologies can be added at anytime about anything) and model exchange and re-use are welcomed and promoted at Web scale. A large number of ontology-based context models relying on Dey's definition have been proposed in the later years, as summarized by Bolchini et al. [4] (e.g., CoBrA, CoDaMoS, SOCAM). These works are grounded on RDF and provide in-depth context expressivity, but for chronological reasons they are far from the best practices common on the Social Semantic Web (e.g., lightweight approach, heavy interlinking with other vocabularies), thus discouraging the adoption and reuse in the Web community. Our context-aware access control framework adopts PRISSMA, a lightweight vocabulary originally designed for context-aware adaptation of RDF data [14]. PRISSMA has been originally designed to express the contextual conditions under which activate a given representation for RDF [14]. In this chapter we propose context-based access policies, and we therefore need a vocabulary to model mobile context. We thus reuse classes and properties of the PRISSMA vocabulary for a different purpose, i.e. to represent contextual conditions for accessing RDF graphs. PRISSMA provides classes and properties to model core mobile context concepts but is not meant to deliver yet another mobile contextual model: instead, well-known lightweight vocabularies and recent W3C recommendations are reused (Fig. 6.5). Moreover, it does not provide a comprehensive, exhaustive context representation: the approach is to delegate refinements and extensions to domain specialists. The overall context is modeled by the class prissma:Context and is determined by the following dimensions:

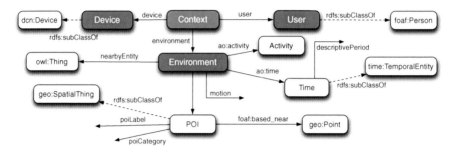

Fig. 6.5 The PRISSMA vocabulary at a glance

Definition 12 (User Dimension). The *User* represents the mobile requester associated with a *Context* and consists in a foaf:Person sub-class. It can model both stereotypes and specific users.

Definition 13 (Device Dimension). The *Device* consists in a structured representation of the mobile device used to access the RDF store.

The Device class inherits from W3C Delivery Context Ontology[19] dcn:Device, providing an extensible and fine-grained model for mobile device features and enabling device-specific access control.

Definition 14 (Environment Dimension). The *Environment* is the model of the physical context in which the resource consumption takes place.

Different dimensions are involved in modeling the surrounding environment. Location is modeled with the notion of Point of Interest (POI). The POI class consists in a simplified, RDFized version of the W3C Point of Interest Core specifications.[20] Time is modeled extending the time:TemporalEntity class.[21] Other dimensions are considered: the motion property associates any given high-level representation of motion to a Environment. The proximity of an object might determine access restrictions: nearby objects are associated with the Environment with the nearbyEntity property. The Activity class consists in a placemark aimed at connecting third-party solutions focused on inferring high-level representations of user actions (e.g., "running," "driving," "shopping,"). Further refinements and extensions are delegated to domain specialists (e.g., if dealing with indoor location, the room vocabulary[22] could be easily integrated).

---

[19]http://bit.ly/dc-ontology

[20]http://www.w3.org/TR/poi-core/

[21]http://www.w3.org/TR/owl-time

[22]http://vocab.deri.ie/rooms

Example 2. We now present an example of Access Policy with a conjunctive Access Condition Set associated with an `Update` privilege (Fig. 6.6a). The policy protects the named graph `:alice_data` and allows the access and modification of the named graph only if the consumer (i) knows Alice, and (ii) is not located near Alice's boss. Figure 6.6b visualizes a sample mobile context featuring all the dimensions described above. The user, Bob, knows Alice and is currently at work, near his and Alice's boss. Bob is using an Android tablet with touch display and he is not moving.

When dealing with mobile context, other issues need to be considered beyond context-model definition, such as context fetch, context trustworthiness, and privacy. The present chapter assumes that context data is fetched and pre-processed beforehand. `PRISSMA` supports both raw context data fetched directly from mobile sensors (e.g., GPS location, mobile features) and refined information processed on board or by third-party, server-side services (e.g., POI resolution or user activity detection). The trustworthiness of contextual information sent by mobile consumers should not be taken for granted. The `User`'s identity needs to be certified: this is an open research area in the Web, and initiatives such as WebID[23] specifically deal with this issue. Hulsebosch et al. [26] provide a survey of context verification techniques (e.g., heuristics relying on context history, collaborative authenticity checks). A promising approach is mentioned in Kulkarni and Tripathi [31], where context sensors are authenticated beforehand by a trusted party. We plan to tackle the issue of context-verification in future work. Privacy concerns arise while dealing with mobile user context. We are aware that sensible data such as current location must be handled with a privacy-preserving mechanism. In the present proposition, we do not address this issue, nor the problem of context integrity.

Further details of the `S4AC` model include, among others: the specification of the creator of the policy (`sioc:hasCreator`) to keep track of this information in the Social Semantic platform, the creation date (`dcterms:created`), the specification of the variables used in the access conditions of the policies and their description in natural language adopted in the user interface of the framework to help the provider reusing others' policies, and a `skos:prefLabel` property associated with the Access Conditions to provide a sort of "explanation" to the consumer in case she cannot access the data (following the example of AIR [27]). Moreover, we are able to manage the fact that only a maximum number of accesses is granted, as in Giunchiglia et al. [24], by means of an Access Condition, and we can grant random access to a resource (e.g. `ASK{FILTER(rand()>0.5)}`).

The semantics of our Access Control Policies is mirrored in the semantics of the SPARQL language, in particular concerning the `ASK` query and the `BINDINGS` clause. The result of the verification of each access condition is composed, in case of multiple conditions, conjunctively or disjunctively, and this combination is the overall result of the policy evaluation. The Access Privilege and the resource to

---

[23]http://www.w3.org/2005/Incubator/webid/spec/

a

```
:policy1 a s4ac:AccessPolicy;  ACCESS POLICY
           s4ac:appliesTo :alice_data;      RESOURCE TO PROTECT
           s4ac:hasAccessPrivilege [a s4ac:Update]; ACCESS PRIVILEGE
           s4ac:hasAccessConditionSet :acs1.

:acs1 a s4ac:AccessConditionSet;
        s4ac:ConjunctiveAccessConditionSet;
          s4ac:hasAccessCondition :ac1,:ac2.   ACCESS CONDITIONS
                                                   TO VERIFY
```

```
:ac1 a s4ac:AccessCondition;
       s4ac:hasQueryAsk
       """ASK {?context a prissma:Context.
               ?context prissma:user ?u.
               ?u foaf:knows ex:alice#me.}""".

:ac2 a s4ac:AccessCondition;
       s4ac:hasQueryAsk
       """ASK {?context a prissma:Context.
               ?context prissma:environment ?env.
               ?env prissma:based_near ?p.
               FILTER (!(?p=ex:ACME_boss#me))}""".
```

b

```
@prefix : <http://example/contextgraphs/bobCtx>
[other prefixes omitted]
<http://example/contextgraphs/bobCtx>{
```

```
:ctx a prissma:Context;
      prissma:user :usr;         THE CONSUMER'S
      prissma:device :dev;           CONTEXT
      prissma:environment :env.
```

```
:usr a prissma:User;
      foaf:name "Bob";           THE USER DIMENSION
      foaf:knows ex:alice#me.
```

```
:dev a prissma:Device;
      hard:deviceHardware :devhw;
      soft:deviceSoftware :devsw.
:devhw a hard:DeviceHardware;
       dcn:display hard:TactileDisplay. THE DEVICE DIMENSION
:devsw a soft:DeviceSoftware;
       soft:operatingSystem :devos.
:devos a soft:OperatingSystem;
       common:name "Android".
```

```
:env a prissma:Environment;
      prissma:motion "no";
      prissma:nearbyEntity :ACME_boss#me;
      prissma:currentPOI :ACMEoffice.      THE ENVIRONMENT
:ACMEoffice a prissma:POI;                     DIMENSION
            prissma:poiCategory example:Office;
            prissma:poiLabel example:ACMECorp.
}
```

Fig. 6.6 The Access Policy protecting :alice_data (**a**) and Bob's sample mobile context in TriG notation (**b**)

protect are components of the policy which do not concur to its verification. All the semantics of our Access Policies relies on the semantics of the ASK queries combined with the contextual BINDINGS.

Conflicts among policies might occur if data provider adds Access Conditions with contrasting FILTER clauses. For instance, it is possible to define positive and negative statements such as ASK{FILTER(?user=<http://example#bob>)} and ASK{FILTER(!(?user=<http://example#bob>))}. If these two Access Conditions are applied to the same data, a logical conflict arises. This issue is handled in our framework by evaluating policies applied to a resource in a disjunctive way. This means that in the example above, if the consumer satisfies one of the two access conditions, then the access is granted to her. This is not satisfactory in many situations, thus we expect to add a mechanism, following the example of [20], to avoid the insertion of conflicting policies as a future work.

Example 3. Consider the following scenario. Alice is attending a music festival and she uploads some content to the social platform. She prefers to share these contents to all the people knowing her but not to those who are friends of her boss. The policy (Fig. 6.6a) protects the named graph containing Alice's reviews of concerts (:alice_reviews visualized in Fig. 6.7).

cond1	`ASK { ?resource dcterms:creator ?provider .` `?provider rel:hasColleague ?user . }`
cond2	`ASK { ?resource dcterms:creator ?provider .` `?provider rel:hasFriend ?user . }`

cond3	`ASK { ?resource dcterms:creator ?provider .`
	`?provider dcterms:creator ?g .`
	`GRAPH ?g { ?user nicetag:hasCommunitySign ?tag }}`

cond4	`ASK { ?user a foaf:Person .` `FILTER(! (?user= <http://MyExample.net#bob>))}`
cond5	`ASK { FILTER(rand()>0.5) }`
cond6	`ASK { ?user a foaf:Person .` `FILTER(?user= <http://MyExample.net#bob>)}`

cond7	`ASK { ?resource dcterms:creator ?provider .`
	`?provider sioc:member_of ?g .`
	`?user sioc:member_of ?g . }`

The table above presents some examples of ASK queries which may be associated with the access conditions. *Cond1* grants the access to those users who have a relationship of kind "colleagues" with the provider. *Cond2* grants the access to the friends of the provider. *Cond3* is more complicated.[24] It grants the access to those users that are marked with a specified tag. To specify the tag, we use again the

---

[24]The GRAPH keyword is used to match patterns against named graphs.

```
ex:29900 a bibo:Article;
        dcterms:title "Great concert with Bob!";
        dcterms:date "2010";
        dcterms:creator example:alice#me;
        bibo:abstract "Really enjoyed Coldplay".

ex:29655 a bibo:Article;
        dcterms:title "Disappointed";
        dcterms:date "2010";
        dcterms:creator example:alice#me;
        bibo:abstract "Not up to the standards".
```

Fig. 6.7 The content of the named graph :alice_reviews containing the reviews authored by Alice

NiceTag ontology which allows to define the relationship among the resources and the tags for each tagging action. Negative access conditions are allowed, where we specify which user cannot access the data. This is expressed, as shown in *Cond4*, by means of the FILTER clause, and access is granted to every user except *bob*. *Cond5* expresses an access condition where the user can access the data only if he is a minimum lucky, e.g., one chance out of two. *Cond6* provides a positive exception where only a specific user can access the data, it is the contrary of *Cond4*. *Cond7* grants the access to those users who are members of a particular group to which also the provider belongs.

The Access Control Manager

The ACM, visualized in Fig. 6.8, is the core module which allows the data providers to define and check the Access Conditions.

The framework is developed in the following way:

1. The data consumer queries the SPARQL endpoint to access the content, and at the same time, the social platform sends the user information coupled with the query. This data is sent as an INSERT DATA statement to build the named graph representing the user's data. Summarizing, the user sends two SPARQL queries to the endpoint, the first one for accessing the datastore, and the second one for providing her personal information. A caching mechanism can be introduced here to avoid sending the personal information every time a query is performed.
2. The query of the consumer is not directly processed by the SPARQL endpoint, but it is filtered by the ACM.
3. The ACM selects the policies concerning the consumer's query, and after their evaluation, it returns the set of named graphs the consumer is granted access to.
4. The query of the consumer is processed only on the accessible named graphs.
5. The result of the query is sent to the consumer.

The core of our framework is the Access Enforcement Module. The aim of this component is twofold: first, the module selects the Access Policies to assess, and

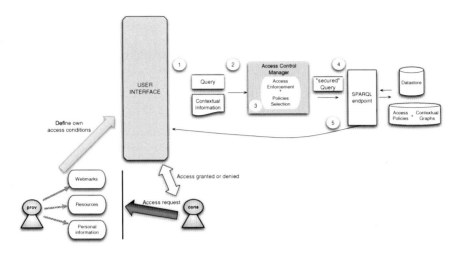

Fig. 6.8 The Access Control Manager

second, it verifies the set of Access Conditions included in the selected policies to allow or not the access. In the following, we describe the two algorithms used to decide whether the consumer is allowed to access or not the data.

Algorithm 6.1 in Fig. 6.11 is the overall algorithm for the query execution with the access enforcement. The input of the algorithm is the consumer's query Q and the RDF graph G_{ctx} modeling the information of the consumer. We assume the existence of a repository of access policies *APS*. The algorithm starts by saving the consumer's graph in a local cache (line 1). The set of selected accessible named graph *NGS* is empty at the beginning of the algorithm execution (line 3). The selection of the Access Policies is addressed by the sub-routine Access Policies Selection (line 4), which returns the set of Access Policies the query is concerned by. Then, the algorithm runs all the Access Conditions composing the selected policies (lines 7–10). For each policy, depending on the kind of Access Conditions Set, i.e., conjunctive or disjunctive, if the policy is verified, then the named graph to which the policy allows the access is added to the set of accessible named graphs (lines 11–12). Finally, after the execution of all the policies, the query of the consumer is sent to the protected SPARQL endpoint with the addition of the FROM and FROM NAMED clauses (line 16). This allows the enforcement module to execute the query only on those named graphs which are accessible, given the user information. Adding the FROM clause is not enough because, in case the client query includes a GRAPH clause, we need to specify the set of named graphs to be queried in a FROM NAMED clause; otherwise, the query will be executed on all the named graphs of the store. USING and USING NAMED describe a dataset in the same way as FROM and FROM NAMED clauses. The keyword USING instead of FROM in update requests has been chosen to avoid possible ambiguities which could arise from writing "DELETE FROM." The algorithm outputs the triples resulting from Q (line 18).

```
ASK{?context a isicil:Context.
     ?context isicil:user ?u.
     ?u foaf:knows ex:alice#me.}
     BINDINGS ?context {(example:actualCtx1)}

ASK {?context a isicil:Context.
     ?context isicil:user ?u.
     ?u rel:hasFriend ?f.
     FILTER (!(?f=ex:ACME_boss#me))}
     BINDINGS ?context {(example:actualCtx1)}
```

Fig. 6.9 The Access Conditions bound to the actual user context with the BINDINGS clause

a
```
DELETE {ex:article dcterms:subject
        <http://dbpedia.org/page/Category: Concert_tours>. }
INSERT {ex:article dcterms:subject
        <http://dbpedia.org/page/Category: Music_performance>. }
WHERE {ex:article a bibo:Article}
```

b
```
DELETE {ex:article dcterms:subject
        <http://dbpedia.org/page/Category: Concert_tours>. }
INSERT {ex:article dcterms:subject
        <http://dbpedia.org/page/Category: Music_performance>. }
```
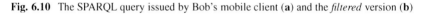
```
USING :peter_data          THE NAMED GRAPH ACCESSIBLE
USING NAMED :peter_data            BY THE CONSUMER

WHERE {ex:article a bibo:Article}
```

Fig. 6.10 The SPARQL query issued by Bob's mobile client (**a**) and the *filtered* version (**b**)

Example 4. An example of client query is shown in Fig. 6.10a, where Bob wants to update the rock festival's reviews.[25] When the query is received by the ACM, it selects the Access Policies concerning this query (Fig. 6.6a). The Access Conditions composing the policy are then coupled with a BINDINGS clause, as shown in Fig. 6.9, where the *?context* variable is bound to the actual Bob's information. Suppose Bob knows Alice, but it is also a friend of Alice's boss (note that these information are retrieved from Bob's FOAF profile): the Access Policy protecting Alice's named graph does not grant access to Bob. After the identification of the named graph(s) accessible by Bob (for instance, the named graph :peter_reviews), the ACM adds the USING and USING NAMED clauses[26] to constrain the execution of the client query only on the allowed named graphs. The "secured" client query is shown in Fig. 6.10b.

---

[25]Notice that the client query can be every kind of query defined by the SPARQL 1.1 Query and Update language, e.g., CONSTRUCT, SELECT.

[26]http://www.w3.org/TR/sparql11-update/#deleteInsert

Algorithm 0.1: Query Execution with Access Enforcement

Input: a SPARQL query Q, an RDF graph G_{ctx}, Access Policy Set APS

Output: the SPARQL query result R

1 save G_{ctx} in local contextual cache;
2 **if** G_{ctx} *has changed* **then**
3 $\quad$ $NGS = \emptyset$;
4 $\quad$ $APS \leftarrow \text{APSelection}(Q,APS)$;
5 $\quad$ **forall the** $AP_i \in APS$ **do**
6 $\quad\quad$ $ACcount_{false} = 0$;
7 $\quad\quad$ **forall the** $AC_j \in ACS_i$ **do**
8 $\quad\quad\quad$ append G_{ctx} to AC_j as BINDINGS clause;
9 $\quad\quad\quad$ **if** ASK_{AC_j} *execution returns false* **then**
10 $\quad\quad\quad\quad$ $ACcount_{false} + +$;
11 $\quad\quad\quad$ **if** $(ACS_{AP_i}$ *is DACS and* $ACcount_{false} < |ACS_{AP_i}|)||(ACS_{AP_i}$ *is CACS and* $ACcount_{false} = 0$ **then**
12 $\quad\quad\quad\quad$ $NGS \leftarrow NGS \cup NG_{AP_i}$;
13 **else**
14 $\quad$ $NGS \leftarrow NGS_{cached}$;
15 **forall the** $NG_i \in NGS$ **do**
16 $\quad$ append $\text{FROM } <NG_i>$, $\text{FROM NAMED}<NG_i>$ to Q;
17 $\quad$ append $\text{USING } <NG_i>$, $\text{USING NAMED}<NG_i>$ to Q;
18 $R \leftarrow$ run Q;
19 **return** R;

Algorithm 0.2: Access Policies Selection

Input: SPARQL client query Q, APS

Output: a reduced set of Access Policies APS_r

1 $AccPrv_Q \leftarrow$ map Q type to CRUD operation;
2 $APS_r = \emptyset$;
3 **forall the** $AP_i \in APS$ **do**
4 $\quad$ **if** $AccPrv_{AP_i} \equiv AccPrv_Q$ **then**
5 $\quad\quad$ $APS_r \leftarrow APS_r \cup AP_i$;
6 **return** APS_r;

Fig. 6.11 SPARQL query execution procedure

Algorithm 6.2 in Fig. 6.11 is the Access Policies Selection routine. The aim of this algorithm is to select, starting from the consumer's query, what are the Access Policies the query is concerned with. The input of the algorithm is the query Q and the repository of the policies *APS*. The idea is that we do not want to verify all the Access Policies every time a query is run. Thus, we adopt a selection mechanism to obtain only a subset of Access Policies to execute. In particular, the algorithm maps the consumer's query to one of the four access privileges S4AC defines (line 1). Then, the algorithm selects all the Access Policies which have the identified Access Privilege (lines 3–7). The selected policies are returned to the main algorithm of access enforcement.

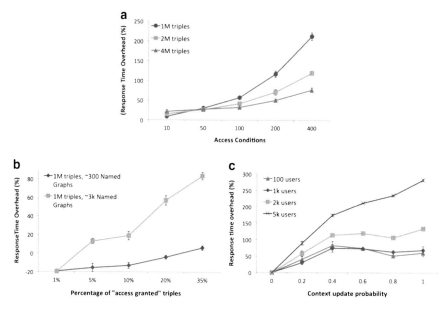

Fig. 6.12 Response time overhead

Evaluation

To assess the impact on response time, we implemented the ACM as a Java EE component and we plugged it to the Corese-KGRAM RDF store[27] and SPARQL 1.1 query engine[28] [10]. We evaluated the prototype on an Intel Xeon E5540, Quad Core 2.53 GHz machine with 48 GB of memory, using the Berlin SPARQL Benchmark (BSBM) dataset 3.1.[29]

In Fig. 6.12, we show the execution of ten independent runs of a test query batch consisting in 50 identical queries of a simple SELECT over bsbm:Review instances (tests are preceded by a warmup run). We measure the response time with and without access control. When executed against the ACM, the test SPARQL query is associated with the static user context. Each Access Policy contains exactly one Access Condition. In Fig. 6.12a, to simulate a worst-case scenario, access is granted to all named graphs defined in the base (i.e., all Access Conditions return true), so that query execution does not benefit from cardinality reduction. Larger datasets are less affected by the delay introduced by our prototype, as datastore

---

[27]Concerning accessing inferred statements, Corese-KGRAM allows to know where are the inferred triples. In this way, we can apply to these inferred triples the same access policies that regulate the access to the triples from which these triples have been inferred.

[28]http://www-sop.inria.fr/edelweiss/software/corese/

[29]http://www4.wiwiss.fu-berlin.de/bizer/BerlinSPARQLBenchmark/spec/Dataset/

	Adopted languages	CRUD	Context awareness	Granularity	Performance evaluation	Role-based	Conflicts verification
WAC	RDF	Read/Write	N/A	RDF document	N/A	N/A	N/A
Sacco and Passant	RDF/SPARQL	Read/Write	N/A	Part of RDF document	N/A	N/A	N/A
Muhleisen et al.	SWRL	Read/Write	N/A	RDF document	YES	YES	N/A
Giunchiglia et al.	DL	N/A	N/A	Resources	N/A	N/A	N/A
Finin et al.	OWL/RDF	N/A	N/A	Resources	N/A	YES	N/A
Hollenbach et al.	RDF	Read/Write/ Control	N/A	RDF document	YES	N/A	N/A
Abel et al.	High level syntax	Read	YES	Triples	YES	N/A	N/A
Cuppens and Cuppens-Boulahia	Datalog	Read/Write	YES but only temporal and spatial	Resources	N/A	N/A	YES
Corradi et al.	RDF	N/A	YES	Resources	YES	N/A	YES
Toninelli et al.	DL	N/A	YES	Resources	YES	N/A	YES
Flouris et al.	High level syntax	Read	N/A	Triples	YES	YES	YES
Carminati et al.	SWRL	Read	N/A	Resources	YES	N/A	N/A
Stroka et al.	RDF	N/A	N/A	Resources	N/A	N/A	N/A
S4AC	RDF/SPARQL	Create/Read/ Update/Delete	YES	Named graphs	YES	N/A	N/A

Fig. 6.13 A summarizing table about the related work

size plays a predominant role in query execution time (e.g., for 4M triples and 100 always-true Access Policies we obtain a 32.6 % response time delay).

In a typical scenario, the ACM restricts the results of a query. In Fig. 6.12b, we assess the impact on performance for various levels of cardinality reduction, using modified versions of the BSBM dataset featuring a larger amount of named graphs (we define a higher number of bsbm:RatingSites, thus obtaining more named graphs). When access is granted to a small fraction of named graphs, the query is executed faster than the case without access control (e.g., if access is granted to only 1 % of named graphs, the query is executed 19 % faster on the 1M triple test dataset). As more named graphs and triples are accessible, performance decreases. In particular, response time is affected by the construction of the active graph, determined by the merge of graphs in the FROM clauses. As shown in Fig. 6.12b, the cost of this operation grows with the number of named graphs returned by the evaluation of the Access Policies.

In Fig. 6.12c, we analyze the overhead introduced on response time by queries executed in dynamic user environments. We execute independent runs of 100 identical SELECT queries, dealing with a range of context change probabilities. In case of a context update, the query is coupled with a SPARQL 1.1 DELETE/INSERT (i.e., update) of the context graph. Not surprisingly, with higher chances of updating the context, the response time of the query grows, since more SPARQL queries need to be executed. The delay of INSERT DATA or DELETE/INSERT operations depends on the size of the triple store and on the number of named graphs (e.g., after a DELETE query, the adopted triple store refreshes internal structures to satisfy RDFS entailment). Performance is therefore affected by the number of active consumers, since each of them is associated with a user context graph.

In Sect. 6.3, we address a qualitative comparison with respect to the related work. On the other hand, addressing a quantitative evaluation is a tricky point: among the list in Fig. 6.13, only few works explicitly designed for the Web come with an evaluation campaign [1, 15, 23, 25, 33]. Moreover, although some of these works provide a response time evaluation, the experimental conditions vary, making the comparison difficult.

6.3 Related Work

6.3.1 XML Languages for Access Control and Digital Rights

Most of the mechanisms of access control implemented in content management systems are based on XML languages dedicated to the description of policies of access control and digital rights management (DRM). These systems exploit the metadata associated with resources to which access must be controlled and these metadata comply with the XML schemas of these dedicated languages. Among these languages, the most famous are XrML[30] (Right eXtensible Markup Language) used as the basic language of expression rights of MPEG-21,[31] ODRL[32] (Open Digital Right Language) implemented by the Open Mobile Alliance (OMA) and XACML[33] developed by OASIS. The ODRL model is based on the concepts of *Asset, Party, Permission, Constraint, Requirement, Condition, Rights holder, Context, Offer, Agreement,* and *Revoking rights.* The XACML model allows to represent access control policies by rules. It is based on the concepts of *Rule, Policy,* and *Policy Set* and these concepts can be refined with those of *Subject, Resource, Action, Environment.* A *Rule* comprises *Conditions* and *Effects* and a *Policy* embeds *Rules* and *Obligations.*

6.3.2 Semantic Approaches to Access Control

With the emergence of the Web of data and people, new approaches to manage access to content have emerged based on semantic Web models and technologies. Notably [2] shows the limitations of solutions using non-semantic description languages for managing access rights. They propose an OWL ontology to describe the access to web services inspired from the XACML model. More generally, in the

---

[30]http://www.xrml.org/

[31]http://www.chiariglione.org/mpeg/standards/mpeg-21/mpeg-21.htm

[32]http://www.w3.org/TR/odrl/

[33]http://www.oasis-open.org/committees/xacml/

few existing semantic models for managing access to content, we recognize some concepts that were already present in the older XML languages.

The W3C initiative is also noticeable: it uses since 2001 an RDF-based system to control access to the files of its servers: W3C ACL System.[34] Hollenbach et al. [25] proposed an evolution of this system to a scalable system that allows for decentralized user authorization via an RDF metadata file containing an ACL. The ontology used in this system is called Basic Access Control Ontology.[35] It is presented as a basis to develop more sophisticated models. Our models extend it for allowing the construction of more fine-grained access control policies. With the AMO ontology we propose a rule-based access control, at document level (rather than directory level). With S4AC-PRISSMA, we provide a fine-grained access control model which grants access to specific RDF data, i.e. the data provider may want to restrict the access to a few named graphs. Moreover, we let consumers submit any SPARQL query, and the user information takes part into the evaluation.

Requirements in terms of access rights management in social platforms like ISICIL are similar to those of digital libraries which [18] propose an overview. However, one of the key issues for digital libraries is not relevant in the context of ISICIL: the respect of the copyrights of available documents and for this purpose the protection of documents by DRM. Indeed the documents handled by the watchers remain in the corporate intranet or are public documents on the web. Among the work on access management in digital libraries, we notice those of [32] on the Fedora architecture for managing digital resources and those of [30] on the semantic Digital Library JeromeDL.

The Fedora authors propose a model called DARS (acronym for Distributed Active Relationships) for associating metadata with objects in a digital library, especially for managing access rights. However, although part of the model of access management is thus in an ontology, the Fedora system also uses XACML metadata associated with resources it handles.

Access management in JeromeDL is based on the EAC ontology[36] (acronym for Extensible Access Control) [29]. EAC enables to associate licenses with resources, for each license corresponding to an Access Policy. For example, a license can specify that only people of a given organization can access some resources of the library. The purpose of EAC is to filter access to resources while that of AMO is to define access rights associated with user roles.

Approaches of access control based on annotations of resources are particularly well suited for social platforms. For example, in [34] end users are able to annotate by tagging both resources and members of their social network. Access control policies are then based on these annotations. For example, a basic policy states that if a resource shares the same tag as a member of the social network, this member has access to the resource. This user-centric approach is more flexible than role-

---

[34]http://www.w3.org/2001/04/20-ACLs

[35]http://www.w3.org/ns/auth/acl

[36]http://www.jeromedl.org/eac/1.0/spec/index.html

centric access control since no real role nor actions need to be defined. It does not require an administrator user having global maintenance access rights to the system. Therefore, it seems more dedicated to the management of personal data rather than public shared data with many contributors (like in wikis). However, we plan to investigate how we could combine such a user-centric approach with our role-centric approach.

Sacco and Passant [35, 36] present a Privacy Preference Ontology (PPO[37]), built on top of WAC, to express fine-grained access control policies to an RDF file. In their approach, the consumer asks to access a particular RDF file, e.g., an FOAF profile. Their ACM selects the part of the file the consumer can access and returns it to the consumer. They do not propose an access control filter for generic SPARQL endpoints, mapping the queries with the access policies. They also specify access queries with SPARQL ASKs, but the PPO vocabulary does not consider contextual information. They rely entirely on the WAC vocabulary without distinguishing different kinds of Write actions, and they cannot specify conjunctive and disjunctive sets of privacy preferences.

Muhleisen et al. [33] present a policy-enabled server for Linked Data called PeLDS, where the access policies are expressed using a descriptive language called PsSF, based on SWRL.[38] They distinguish only Read and Update actions, and they do not consider contextual information. The system is based on an ontology of the actions that can be performed on the datasets, but no further description is provided.

Giunchiglia et al. [24] propose a Relation Based Access Control model (Rel-BAC), a formal model of permissions based on description logic. They require to specify who can access the data, while in our framework and in [36] the provider specifies the attributes the consumer must satisfy.

Finin et al. [22] study how to represent RBAC using the OWL language. The authors show also the representation of policies based on general attributes of an action, similar to what we present in this chapter. The difference is that we specify the policies using SPARQL 1.1 ASK queries, where the Bindings clause is used to specify the values of the variables, and temporal and spatial constraints may be expressed too.

Abel et al. [1] present a model of context-dependent access control at triple level, where also contextual predicates are allowed, e.g., related to time, location, credentials. The policies are not expressed using Semantic Web languages, but they introduce a high level syntax then mapped to existing policy languages. They enforce access control as a layer on top of RDF stores. They pre-evaluate the contextual conditions, then the queries are expanded, and sent to the database.

Flouris et al. [23] present a fine-grained access control framework on top of RDF repositories. As in our approach, the authors underline the need of a fine-grained access control framework while being repository independent. Differently from our

---

[37]http://vocab.deri.ie/ppo

[38]http://www.w3.org/Submission/SWRL/

framework, they do not consider the contextual dimension, and they propose a high level specification language which has to be translated into a SPARQL/SerQL/SQL query to enforce the policy, while we use directly SPARQL 1.1 to specify the policies. Moreover, they focus only on read operations.

Carminati et al. [8] propose a fine-grained online social network access control model based on Semantic Web technologies. Their main idea is to encode social network-related information by means of an ontology. By constructing such an ontology, the authors model the Social Network Knowledge Base. They assume that a centralized reference monitor hosted by the social network manager will enforce the required policies. The access control policies are encoded as SWRL[39] rules. This approach is also based on the specification of who can access the resources, i.e., the access request is a triple (u, p, URI), where the user u requests to execute privilege p on the resource located at URI.

Stroka et al. [40] present a preliminary proposal about securing the collaborative content on the platform KiWi. They consider global permissions, individual content item permissions, and RDF-type-based permission management. They do not specify the kind of access polices they can define.

6.3.3 Social Semantic Web Standards

A key specificity of the Semantic Web based approaches we adopt is to be interoperable with the models of the social Web and semantic Web. Specifically, SweetWiki uses FOAF and SIOC to annotate resources and AMO complements these ontologies to manage access to content. FOAF[40] (acronym for Friend Of A Friend) is an RDF vocabulary used in social networks to describe people and the relations among them. SIOC[41] (acronym for Semantically-Interlinked Online Communities) is another RDF vocabulary that models the concepts of social web applications: forums, blogs, wikis. It reuses some concepts from FOAF and other popular ontologies (Dublin Core, SKOS, etc.) and it has established itself as the standard. It is now integrated into numerous applications such as the WordPress blog engine and its adoption within the *Linked Data*[42] project confirms its popularity.

FOAFRealm is an extension of FOAF proposed to collaboratively filter access to resources based on user profiles and their relationships in a social network. This vocabulary is used, for example, in JeromeDL for filtering based on measures of trust in a social network. Such filtering may be complementary to access control allowed by AMO, based on user roles and types of access to resources.

---

[39]http://www.w3.org/Submission/SWRL/

[40]http://xmlns.com/foaf/spec/

[41]http://sioc-project.org/ontology

[42]http://linkeddata.org/

Finally, the problem of authorizing access to resources which AMO addresses is related to the problem of authentication of agents. Our approaches should be compliant with the FOAF-SSL protocol [39].

6.3.4 Context-Oriented Access Control

A significant number of works in various research areas deal with context-aware access control. We describe first a list of proposals not related to the Web, followed by works targeting the Web and its evolution, the Web of Data.

Hulsebosch et al. [26] propose a context-sensitive access control infrastructure enriched by the verification of user-provided context information. Their work shows that context-sensitive access control improves classic access control by imitating real-world authorization procedures. They provide a comprehensive overview of context verification techniques.

Bertino et al. [3] limit on location awareness and discuss the motivations behind enriching access control with position data. The location verification problem is presented, along with solutions (e.g., authenticator devices for physical location of users). Their approach relies on geographical role-based access control (GEO-RBAC), where users are assigned roles with given spatial validity.

Another contextual role-based approach is presented in Kulkarni and Tripathi [31]. The authors propose a context management layer in charge of authenticating the sensors that produce context data used to evaluate access.

Shen and Cheng [38] propose a model called SCBAC which combines Semantic Web technologies with a context-based access control mechanism. Policies are expressed using SWRL.[43] The authors consider four types of contexts: subject contexts (our User and Device dimensions), object contexts, transaction contexts (our access privilege), and environment contexts (our Environment dimension). They do not apply their model to the Social Semantic Web.

Covington et al. [17] present an approach where the notion of role proposed by RBAC [37] is used to capture the context of the environment in which the access requests are made. The environmental roles are defined using a prolog-like logical language for expressing policies. In a subsequent work, Covington proposes a context-aware attribute-based access control model called CABAC [16]. They heavily rely on contextual attributes.

Cuppens and Cuppens-Boulahia [19] propose an Organization Based Access Control (OrBAC) model where also contextual conditions are expressed. Contextual conditions are considered as extra statements to be satisfied to activate a security constraint and are based on Datalog rules. A context algebra is introduced. The main difference is that we entirely rely on Semantic Web languages. Moreover, we consider additional context data, beyond the temporal and spatial dimensions.

---

[43]http://www.w3.org/Submission/SWRL/

Corradi et al. [13] present UbiCOSM, a security middleware adopting context as a basic concept for policy specification and enforcement. As we do, the authors consider context as a first-class design principle to control access to resources. They distinguish among physical (i.e., physical spaces) and logical contexts (e.g., temporal conditions, user activities). We add further contextual dimensions, e.g., the device. Policies are expressed at a high level of abstraction in terms of RDF metadata. Their approach is not applied to the Web of Data.

Toninelli et al. [41, 42] follow two design guidelines, which inspire also the framework proposed in this chapter: context-awareness to control resource access and semantic technologies for context and policy specification. They adopt spontaneous coalitions as an application scenario, while we deal with the Web of Data. Moreover, the semantic technology adopted differs, i.e., rule-based approach with description logic in their case and SPARQL 1.1 in our proposal. Their contextual information does not include the device dimension. Finally, their solution is not meant to be a pluggable framework for SPARQL endpoints. In a separated work, Toninelli et al. [43] present the Proteus policy framework and discuss the role of the quality of context in access control systems. In this chapter, we do not take into account this problem that is left as future work.

In the table visualized in Fig. 6.13, we use N/A when the feature is not available in the model, YES when the feature is implemented in that model, and when necessary we specify the feature. The last row of the table is dedicated to the *S4AC-Prissma* model. The table considers the following key features of the access control models: adopted languages (the kind of language used to express the policies), CRUD (the kind of access privileges the model is able to manage), context-awareness (if the context dimension is used to grant or not the access), granularity (what is protected by the model), performance evaluation (if an experimental evaluation of the model is presented), role-based (if the model is role-based), conflicts verification (if the model is able to detect possible inconsistencies among the policies).

6.4   Summary

Access control is a fundamental issue in the field of the Social Semantic Web. The users aim at protecting their resources to grant the access to their data only to authorized users. In this chapter, we present two semantic access control frameworks we introduced.

First, we have presented an ontology-based model for access control. This model is grounded on the AMO vocabulary for representing access rights to resources and on a base of inference rules using this vocabulary to represent access control strategies. It has been implemented and experienced to manage access to resources in the SweetWiki engine.

Second, we have introduced a fine-grained access control model for the Social Semantic Web. This model is grounded on the S4AC vocabulary which allows the users of social networks to define Access Conditions for their data. In particular, these Access Conditions are implemented as SPARQL 1.1 ASK queries, and they can be either conjunctively or disjunctively evaluated. Moreover, Access Policies can be constrained w.r.t. the set of tags the resources are tagged with. The Access Evaluation Context provides the mapping, implemented as a BINDINGS clause, between the information about the consumer and the Access Conditions. We have presented our ACM, realized in the context of the ISICIL social platform. The manager grants or denies access to the users. Through an interface which allows also non-experts to interact with the system, users can specify the Access Policies to protect their data. The manager looks for the policies which apply to the resource, and after checking the contextual constraints, if present, and the features of the consumer, it states whether the access is granted or not.

There are several lines to follow for future work. First of all, in this chapter we assume that the user's information is trustworthy. The trustworthiness of the information sent by consumers should not be taken for granted. The User's identity needs to be certified: this is an open research area in the Web, and initiatives such as WebID[44] specifically deal with this issue. Hulsebosch et al. [26] provide a survey of context verification techniques (e.g., heuristics relying on context history, collaborative authenticity checks). At the same time, also the other contextual information like the time of the query or the location of the consumer when she is querying the SPARQL endpoint have to be checked. We plan to follow the example of Toninelli et al. [43]. Second, a user evaluation campaign is needed to evaluate our access control framework. The campaign aims at establishing the understandability of the framework, the definition of the access policies from non-expert users, the explanation of the result after the attempt to access the data, and many other features of our framework. Third, as ubiquitous connectivity grows, access control in the Social Semantic Web must not ignore the mobile context in which data consumption takes place. We are currently working on a mobile access control framework which will introduce also the *device* dimension to our user's information.

The two proposals presented in this chapter highlight the expressive power a semantic-based approach to access control adds to the usual approaches. However, our frameworks represent the first step towards a fully semantic approach to network-based access control. Several issues still need to be addressed in this context, as mentioned above, like privacy preservation, consumer's trustworthiness assessment, and user-friendly interfaces definition.

---

[44]http://www.w3.org/2005/Incubator/webid/spec/

References

1. Abel, F., De Coi, J.L., Henze, N., Koesling, A.W., Krause, D., Olmedilla, D.: Enabling advanced and context-dependent access control in RDF stores. In: 6th International Semantic Web Conference (ISWC). Lecture Notes in Computer Science, vol. 4825, pp. 1–14. Springer, Berlin (2007)

2. Alam, A., Subbiah, G., Thuraisingham, B.M., Khan, L.: Reasoning with semantics-aware access control policies for geospatial web services. In: 3rd ACM Workshop on Secure Web Services (SWS), pp. 69–76. ACM, New York (2006)

3. Bertino, E., Kirkpatrick, M.S.: Location-aware authentication and access control. In: IEEE 23rd International Conference on Advanced Information Networking and Applications (AINA), pp. 10–15. IEEE Computer Society, Bradford (2009)

4. Bolchini, C., Curino, C., Quintarelli, E., Schreiber, F.A., Tanca, L.: A data-oriented survey of context models. SIGMOD Rec. **36**(4), 19–26 (2007)

5. Breslin, J., Passant, A., Decker, S.: The Social Semantic Web. Springer, Berlin (2009)

6. Buffa, M., Faron-Zucker, C.: Ontology-based access rights management. In: Advances in Knowledge Discovery and Management. Studies in Computational Intelligence, vol. 398, pp. 49–61. Springer, Berlin (2012)

7. Buffa, M., Gandon, F.L., Erétéo, G., Sander, P., Faron, C.: SweetWiki: a semantic wiki. J. Web Semant. **6**(1), 84–97 (2008)

8. Carminati, B., Ferrari, E., Heatherly, R., Kantarcioglu, M., Thuraisingham, B.M.: Semantic web-based social network access control. Comput. Secur. **30**(2–3), 108–115 (2011)

9. Carroll, J.J., Bizer, C., Hayes, P.J., Stickler, P.: Named graphs. J. Web Semant. **3**(4), 247–267 (2005)

10. Corby, O., Faron-Zucker, C.: The KGRAM abstract machine for knowledge graph querying. In: Web Intelligence, pp. 338–341. IEEE, Toronto (2010)

11. Corby, O., Dieng-Kuntz, R., Faron-Zucker, C.: Querying the semantic web with Corese search engine. In: 16th European Conference on Artificial Intelligence (ECAI), pp. 705–709. IOS Press, Amsterdam (2004)

12. Corby, O., Dieng-Kuntz, R., Faron-Zucker, C., Gandon, F.L.: Searching the semantic web: approximate query processing based on ontologies. IEEE Intell. Syst. **21**(1), 20–27 (2006)

13. Corradi, A., Montanari, R., Tibaldi, D.: Context-based access control management in ubiquitous environments. In: 3rd IEEE International Symposium on Network Computing and Applications (NCA), pp. 253–260. IEEE Computer Society, Los Alamitos (2004)

14. Costabello, L.: DC Proposal: PRISSMA, towards mobile adaptive presentation of the Web of data. In: Doctoral Consortium, 10th International Semantic Web Conference (ISWC). Lecture Notes in Computer Science, vol. 7032, pp. 269–276. Springer, Berlin (2011)

15. Costabello, L., Villata, S., Delaforge, N., Gandon, F.L.: Ubiquitous Access Control for SPARQL Endpoints: Lessons Learned and Future Challenges. In: WWW (Companion Volume), pp. 487–488. ACM, New York (2012)

16. Covington M.J., Sastry, M.R.: A contextual attribute-based access control model. In: Workshops on the Move to Meaningful Internet Systems (OTM). Lecture Notes in Computer Science, vol. 4278, pp. 1996–2006. Springer, Berlin (2006)

17. Covington, M.J., Long, W., Srinivasan, S., Dey, A.K., Ahamad, M., Abowd, G.D.: Securing context-aware applications using environment roles. In: 6th ACM Symposium on Access Control Models and Technologies (SACMAT), pp. 10–20. ACM, New York (2001)

18. Coyle, K.: Rights management and digital library requirements. Ariadne **40**, 125–137 (2004)

19. Cuppens, F., Cuppens-Boulahia, N.: Modeling contextual security policies. Int. J. Inf. Secur. **7**(4), 285–305 (2008)

20. Cuppens, F., Cuppens-Boulahia, N., Ghorbel, M.B.: High level conflict management strategies in advanced access control models. Electron. Notes Theor. Comput. Sci. **186**, 3–26 (2007)

21. Dey, A.K.: Understanding and using context. Pers. Ubiquitous Comput. **5**(1), 4–7 (2001)

22. Finin, T.W., Joshi, A., Kagal, L., Niu, J., Sandhu, R.S., Winsborough, W.H., Thuraisingham, B.M.: ROWLBAC: representing role based access control in OWL. In: 13th ACM Symposium on Access Control Models and Technologies, pp. 73–82. ACM, New York (2008)

23. Flouris, G., Fundulaki, I., Michou, M., Antoniou, G.: Controlling access to RDF graphs. In: 3rd Future Internet Symposium (FIS). Lecture Notes in Computer Science, vol. 6369, pp. 107–117. Springer, Berlin (2010)

24. Giunchiglia, F., Zhang, R., Crispo, B.: Ontology driven community access control. In: 1st Workshop on Trust and Privacy on the Social and Semantic Web (SPOT). CEUR Workshop Proceedings (2009)

25. Hollenbach, J., Presbrey, J., Berners-Lee, T.: Using RDF metadata to enable access control on the social semantic web. In: Workshop on Collaborative Construction, Management and Linking of Structured Knowledge (CK). CEUR-WS.org (2009)

26. Hulsebosch, R.J., Salden, A.H., Bargh, M.S., Ebben, P.W.G., Reitsma, J.: Context sensitive access control. In: 10th ACM Symposium on Access Control Models and Technologies (SACMAT), pp. 111–119. ACM, New York (2005)

27. Khandelwal, A., Bao, J., Kagal, L., Jacobi, I., Ding, L., Hendler, J.A.: Analyzing the AIR language: a semantic web (production) rule language. In: Web Reasoning and Rule Systems, 4th International Conference (RR). Lecture Notes in Computer Science, vol. 6333, pp. 58–72. Springer, Berlin (2010)

28. Korpipää, P., Mäntyjärvi, J.: An ontology for mobile device sensor-based context awareness. In: Modeling and Using Context, 4th International and Interdisciplinary Conference (CONTEXT). Lecture Notes in Computer Science, vol. 2680, pp. 451–458. Springer, Berlin (2003)

29. Kruk, S.R.: Extensible Access Control (EAC) Ontology Specification. DERI. http://www.jeromedl.org/eac/1.0/spec/index.html/ (2008)

30. Kruk, S.R., Cygan, M., Gzella, A.: JeromeDL—semantic and social technologies for improving user experience in digital libraries. In: World Wide Web Conference, WWW 2008. ACM, New York (2008)

31. Kulkarni, D., Tripathi, A.: Context-aware role-based access control in pervasive computing systems. In: 13th ACM Symposium on Access Control Models and Technologies (SACMAT), pp. 113–122. ACM, New York (2008)

32. Lagoze, C., Payette, S., Shin, E., Wilper, C.: Fedora: an architecture for complex objects and their relationships. Int. J. Digit. Libr. **6**(2), 124–138 (2006)

33. Muhleisen, H., Kost, M., Freytag, J.-C.: SWRL-based access policies for linked data. In: 2nd Workshop on Trust and Privacy on the Social and Semantic Web (SPOT). CEUR-WS.org (2010)

34. Nasirifard, P., Peristeras, V., Hayes, C., Decker, S.: Extracting and utilizing social networks from log files of shared workspaces. In: 10th IFIP Working Conference on Virtual Enterprises, (PRO-VE), pp. 643–650. Springer, Berlin (2009)

35. Sacco, O., Passant, A.: A privacy preference manager for the social semantic web. In: 2nd Workshop on Semantic Personalized Information Management: Retrieval and Recommendation (SPIM). CEUR Workshop Proceedings (2011)

36. Sacco, O., Passant, A.: A privacy preference ontology (PPO) for linked data. In: Linked Data on the Web Workshop (LDOW). CEUR-WS.org (2011)

37. Sandhu, R.S., Coyne, E.J., Feinstein, H.L., Youman, C.E.: Role-based access control models. IEEE Comput. **29**(2), 38–47 (1996)

38. Shen, H., Cheng,Y.: A semantic context-based model for mobile web services access control. Int. J. Comput. Netw. Inf. Secur. **3**(1), 18–25 (2011)

39. Story, H., Harbulot, B., Jacobi, I., Jones, M.: FOAF+TLS: RESTful authentication for distributed social networks. In: 1st Workshop on Trust and Privacy on the Social and Semantic Web (SPOT). CEUR-WS.org (2009)

40. Stroka, S., Schaffert, S., Burger, T.: Access control in the social semantic web—extending the idea of FOAF+SSL in KiWi. In: 2nd Workshop on Trust and Privacy on the Social and Semantic Web (SPOT). CEUR-WS.org (2010)

41. Toninelli, A., Montanari, R., Kagal, L., Lassila, O.:   A semantic context-aware access control framework for secure collaborations in pervasive computing environments. In: 5th International Semantic Web Conference (ISWC). Lecture Notes in Computer Science, vol. 4273, pp. 473–486. Springer, Berlin (2006)

42. Toninelli, A., Montanari, R., Kagal, L., Lassila, O.: Proteus: a semantic context-aware adaptive policy model. In: 8th IEEE International Workshop on Policies for Distributed Systems and Networks (POLICY), pp. 129–140. IEEE Computer Society, Los Alamitos (2007)

43. Toninelli, A., Corradi, A., Montanari, R.: A quality of context-aware approach to access control in pervasive environments. In: 2nd International Conference on Mobile Wireless Middleware, Operating Systems, and Applications (MOBILWARE). Lecture Notes of the Institute for Computer Sciences, Social Informatics and Telecommunications Engineering, vol. 7, pp. 236–251. Springer, Berlin (2009)

44. Villata, S., Delaforge, N., Gandon, F., Gyrard, A.: An access control model for linked data. In: 7th International IFIP Workshop on Semantic Web & Web Semantics (SWWS). Lecture Notes in Computer Science, vol. 7046, pp. 454–463. Springer, Berlin (2011)

45. Villata, S., Delaforge, N., Gandon, F., Gyrard, A.: Social semantic web access control. In: 4th International Workshop Social Data on the Web (SDoW), pp. 48–59. CEUR Workshop Proceedings (2011)

Part III
Security and Privacy in Mobile and P2P Social Networks

Chapter 7
Supporting Data Privacy in P2P Systems[*]

Mohamed Jawad, Patricia Serrano-Alvarado, and Patrick Valduriez

Abstract Peer-to-Peer (P2P) systems have been very successful for large-scale data sharing. However, sharing sensitive data, like in online social networks, without appropriate access control, can have undesirable impact on data privacy. Data can be accessed by everyone (by potentially untrusted peers) and used for everything (e.g., for marketing or activities against the owner's preferences or ethics). Hippocratic databases (HDB) provide an effective solution to this problem, by integrating purpose-based access control for privacy protection. However, the use of HDB has been restricted to centralized systems. This chapter gives an overview of current solutions for supporting data privacy in P2P systems and develops in more detail a complete solution based on HDB.

7.1 Introduction

Data privacy is the right of individuals to determine for themselves when, how, and to what extent information about them is communicated to others [40]. It has been treated by many organizations and legislations that have defined well-accepted principles. According to OECD,[1] data privacy should consider: collection

---

[*]Work partially funded by the DataRing project of the French ANR.

[1]Organization for Economic Co-operation and Development. One of the world's largest and most reliable source of comparable statistics on economic and social data (http://www.oecd.org/).

M. Jawad (✉) · P. Serrano-Alvarado
LINA, University of Nantes, Nantes, France
e-mail: Mohamed.Jawad@univ-nantes.fr

P. Valduriez
INRIA and LIRMM, University of Montpellier, Montpellier, France
e-mail: Patrick.Valduriez@inria.fr

R. Chbeir and B. Al Bouna (eds.), *Security and Privacy Preserving in Social Networks*,   195
Lecture Notes in Social Networks, DOI 10.1007/978-3-7091-0894-9_7,
© Springer-Verlag Wien 2013

limitation, purpose specification, use limitation, data quality, security safeguards, openness, individual participation, and accountability. From these principles, we underline *purpose* specification which states that data owners should be able to specify the data access purposes for which their data will be collected, stored, and used.

With the advent of Online Social Networks (OLSN), data privacy has become a major concern. An OLSN is formed by people having something in common and connected by social relationships, such as friendship, hobbies, or co-working, in order to exchange information [11]. Many communities use OLSNs to share data in both professional and non-professional environments. Examples of professional OLSNs are Shanoir,[2] designed for the neuroscience community to archive, share, search, and visualize neuroimaging data, or medscape,[3] designed for the medical community to share medical experience and medical data. There are also non-professional OLSNs for average citizens and amateurs in different domains such as Carenity,[4] designed for patients and their relatives to share medical information about them in order to help medical research. Another example is DIYbio,[5] dedicated to make biology accessible for citizen scientists, amateur biologists, and biological engineers, who share research results. The most popular OLSN, Facebook, with hundreds millions of users, enables groups of friends to share all kinds of personal information among themselves.

Scalable data sharing among community members is critical for an OLSN system. Two main solutions have emerged for scalable data sharing: cloud computing and Peer-to-Peer (P2P). Cloud computing promises to provide virtually infinite computing resources (e.g., CPU, storage, network) that can be available to users with minimal management efforts [26]. Data are stored in data centers, typically very large clusters of servers, operated by infrastructure providers such as Amazon, Google, and IBM. Among others, cloud computing exhibits the following key characteristics: (a) elasticity, as additional resources can be allocated on the fly to handle increased demands, (b) ease of maintenance, which is managed by cloud providers, and (c) reliability, thanks to multiple redundant sites. These assets make cloud computing suitable for OLSNs.

However, cloud computing proposes a form of centralized storage that implies many effects on sensitive data: (a) users need to trust the providers and their servers; (b) providers can use community information to make profits (e.g., profiling, marketing, advertising); and (c) users may find their data censored by providers. In particular, (a) is very hard to enforce as cloud providers can outsource data storage to other providers, yielding a chain of subcontractors (typically in different countries, each with a different legislation on data privacy) which is difficult to track.

---

[2] www.shanoir.org/.

[3] http://www.medscape.com/connect/.

[4] http://www.carenity.com.

[5] http://diybio.org.

As an alternative to a centralized data sharing solution, like the cloud, P2P provides a fully decentralized infrastructure. Examples of very popular P2P applications can be found in networking (e.g., Skype), search engines (e.g., YaCy), OLSN (e.g., Diaspora), and content sharing (e.g., BitTorrent). For instance, one third of the Internet traffic today is based on BitTorrent. P2P systems for data-centered applications offer valuable characteristics: (a) decentralized storage and control, so there is no need to trust one particular server; (b) data availability and fault tolerance, thanks to data replication; (c) scalability to store large amounts of data and manage high numbers of users; (d) autonomy, as peers can join and leave the network at will.

We claim that a P2P solution is the right solution to support the collaborative nature of OLSN applications as it provides scalability, dynamicity, autonomy, and decentralized control. Peers can be the participants or organizations involved in collaboration and may share data and applications while keeping full control over their (local) data sources.

However, despite their assets, P2P systems offer limited guarantees concerning data privacy. They can be considered as hostile because data, that can be sensitive or confidential, can be accessed by everyone (by potentially untrusted peers) and used for everything (e.g., for marketing, profiling, fraudulence, or for activities against the owner's preferences or ethics). Several P2P systems propose mechanisms to ensure privacy such as OceanStore [22], Past [33], and Freenet [8]. However, these solutions remain insufficient. Data privacy laws have raised the respect of user privacy preferences where purpose-based access is cornerstone. Managing data sharing, with trustworthy peers, for specific purposes and operations, is not possible in current P2P systems without adding new services.

Inspired by the Hippocratic oath and its tenet of preserving privacy, Hippocratic databases (HDB) [2] have incorporated purpose-based privacy protection, which allows users to specify the purpose for which their data are accessed. However, HDB have been proposed for centralized relational database systems.

Applied to P2P systems, HDBs could bring strong privacy support as in the following scenario. Consider an OLSN where patients share their own medical records with doctors and scientists, and scientists share their research results with patients and doctors. Scientists have access to patient medical records if their access purpose is for research on a particular disease. Doctors have access to research results for giving medical treatment to their patients. In this context, Hippocratic P2P data sharing can be useful. Producing new P2P services that prevent peers from disclosing, accessing, or damaging sensitive data encourages patients (resp. scientists) to share their medical records (resp. results) according to their privacy preferences. Thus, the challenge is to propose services to store and share sensitive data in P2P systems taking into account access purposes.

This chapter is organized as follows. Section 7.2 surveys data privacy in P2P systems and gives a comparative analysis of existing solutions. Section 7.3 surveys HDBs. Section 7.4 introduces PriMod, a privacy model that applies HDB principles to data sharing in P2P systems. Section 7.5 describes PriServ, a privacy service that supports PriMod in structured P2P networks. Section 7.6 presents the PriServ prototype. Section 7.7 concludes.

7.2   Data Privacy in P2P Systems

P2P systems operate on application-level networks referred to as overlay networks. The degree of centralization and the topology of overlay networks have significant influence on properties such as performance, scalability, and security. P2P networks are generally classified into two main categories: pure and hybrid [29]. In pure P2P networks, all peers are equal and they can be divided into structured and unstructured overlays. In hybrid P2P networks (also called super-peer P2P networks), some peers act as dedicated servers for some other peers and have particular tasks to perform.

Unstructured P2P overlays are created in an ad hoc fashion (peers can join the network by attaching themselves to any peer) and data placement is completely unrelated to their organization. Each peer knows its neighbors but does not know the resources they have. Many popular applications operate as unstructured networks like Napster, Gnutella, Kazaa, and Freenet [8]. In those systems, content is shared among peers without needing to download it from a centralized server. Those systems vary, among others, in the way data is indexed that implies the way data is searched. There exist two alternatives for indexing: centralized and distributed. In centralized indexes, a peer is responsible for managing the index of the system. This centralization facilitates data searching because requesting peers consult the central peer to obtain the location of the data, and then directly contact the peer where the data is located. Napster is an example of a system that maintains this type of index. In the distributed approach, each peer maintains part of the index, generally the one concerning the data they hold. Data searching is typically done by *flooding*, where requesting peers send the request to all of its neighbors which forward the request to all of their neighbors if they do not have the requested data, and so on. For large networks, a Time to Live (TTL) is defined to avoid contacting all peers at every request. Gnutella is a system that maintains this kind of index.

Structured P2P systems have emerged to improve the performance of data searching by introducing a particular structure into the P2P network. They achieve this goal by controlling the overlay topology, the content placement, and the message routing. Initial research on P2P systems led to solutions based on Distributed Hash Tables (DHTs) where a hash function maps a *key* to each peer (such key is considered as the peer's identifier). Data placement is based on mapping a *key* to each data item and storing the $(key, data)$ pair at the peer which identifier is equal or follows the key. Thus, the distributed lookup protocol supported by these systems efficiently locates the peer that stores a particular data item in $O(\log N)$ messages. Representative examples of DHTs are Chord [36], Pastry [34], and Tapestry [41]. The reader can find an analysis of data sharing in DHT-based systems in [32].

Hybrid P2P networks contain a subset of peers (called super-peers) that provide services to some other peers. These services can be data indexing, query processing, access control, meta-data management, etc. With only one super-peer, the network architecture reduces to client-server. The organization of super-peers follows a P2P approach and they can communicate with each other in sophisticated manners. An example of these systems is Edutella [28].

Peers may be participants that share data, request information, or simply contribute to the storage system. Considering that there are peers that own data and do not necessarily act as servers of those data, we distinguish between three kinds of peers:

- *Requester.* A peer that requests data.
- *Server.* A peer that stores and provides data.
- *Owner.* A peer that owns and shares data.

This section surveys the P2P systems that deal with data privacy. Depending on their main application, the data privacy issues are different. We divide these systems into two main classes: those focusing on distributed data storage (Sect. 7.2.1) and those focusing on massive data sharing (Sect. 7.2.2). We compare them, in Sect. 7.2.3, based on their privacy protection guarantees and the techniques they use.

7.2.1 Privacy in Distributed Data Storage Systems

Distributed data storage systems are mainly used by users who want to benefit from large storage space and possibly share data with some other users. Usually, the users of these systems require the following privacy guarantees:

- Data storage: data are available for owners.
- Data protection against unauthorized reads: servers do not have the ability to read the data they store.
- Data protection against corruption and deletion: servers do not have the ability to corrupt or delete the data they store.
- A peer does not claim the property of data owned by another peer.

Various techniques and protocols should be employed together to ensure such guarantees. For instance, replication can be used to guarantee data availability, but to limit privacy breaches, data replicas must be stored at trusted servers. Thus, trust and access control techniques can be used with replication in order to ensure data privacy. In addition, data digital checksums and encryption can be used to protect ownership rights as well as data from unauthorized reads.

Past [33], OceanStore [22], and Mnemosyne [12] are examples of systems that use such techniques.

Past is a large-scale, Internet-based, global storage utility that provides scalability, high availability, persistence, and security. It relies on Pastry and uses smartcards, self-certifying data, and certified-based trust in order to protect data content from malicious servers.

OceanStore is a cooperative infrastructure that provides a consistent, highly available, durable, and secured storage utility. It relies on Tapestry and uses symmetric cryptography and access control techniques to protect data privacy from malicious peers.

Mnemosyne is a storage service that provides a high level of privacy by using a large amount of shared distributed storage to hide data. It relies on Tapestry and uses steganographic data, data whose presence among random data cannot be detected. This allows to protect data from malicious reads and suppressions.

7.2.2 Privacy in Massive Data Sharing Systems

Massive data sharing systems are mainly used by users who want to (a) share data in the system and (b) request and download data from the system. (b) is probably the main reason why this type of system is so well known and used, in particular, in multimedia file sharing. One important difference with data storage systems is that, in massive data sharing, data are massively duplicated. In this type of system, users search for privacy guarantees related to data and users.

1. The data privacy guarantees are:

   • Data storage: data are available to authorized owners and requesters.
   • Data protection against unauthorized reads: servers do not have the ability to read the data they store.
   • Data protection against corruption and deletion: servers do not have the ability to corrupt or delete the data they store.
   • A peer does not claim the property of data owned by another peer.
   • Limited disclosure: data are not provided to unauthorized requesters.

2. The user privacy guarantees (usually referred to as anonymity guarantees) are:

   • Users are not monitored in the system by other peers.
   • Users freedom of behavior is not limited by the system.
   • Users are protected against identity theft.

Data privacy guarantees in massive data sharing systems are similar to those of distributed data storage systems, the only difference is limited disclosure. As said before, data privacy can be protected by techniques such as access control, trust management, data encryption, and digital checksums. The difference here is the potential number of requesters. On the other hand, user privacy can be protected by using different anonymity techniques. However, ensuring user privacy may cause undesired effects on data privacy. Users want to protect their data privacy while remaining anonymous to behave freely, which increases the risk of violating the data privacy of others. This loop is probably the main reason we did not find in the literature systems that guarantee both data and user privacy.

Protecting Data Privacy. Data privacy in massive data sharing systems can be illustrated with the following systems.

Office SharePoint Workspace,[6] previously known as Office Groove, is a desktop application designed for document collaboration within teams (i.e., workspaces). It is based on a partially centralized P2P system. Each user has a private editable copy of the workspace. Workspace copies are synchronized via the network in a P2P manner. Office SharePoint Workspace uses access control, trust, and encryption techniques in order to protect data privacy.

Piazza [38] is a data management system that enables sharing of XML documents in a distributed and scalable way. It is based on an unstructured P2P system. Although the goal and emphasis of Piazza is data sharing and not data privacy, the creators of Piazza proposed in [27] new techniques for publishing a single data instance in a protected form, thus enforcing data privacy.

OneSwarm [14] is a P2P service that provides users with explicit control over their data privacy by letting them determine how data are shared. It relies on BitTorrent and was designed to provide privacy-preserving data sharing. OneSwarm uses asymmetric cryptography, access control, trust, and communication anonymity in order to protect data privacy.

Other systems address the censorship problem which can have effects on data privacy since censorship can be a reason for data suppression. In some systems such as Usenet news,[7] anyone who sees a message can post a cancel message to delete it, allowing censorship. Many systems have been proposed to resist to censorship. *Dagster* [37] is a censorship-resistant publishing scheme that intertwines legitimate and illegitimate data from web pages, so that a censor cannot remove *objectionable content* without simultaneously removing legally *protected content. Tangler* [39] is another censorship-resistant publishing scheme based on the idea of intertwining data. Newly published documents are dependent on previous published blocks. This dependency, called entanglement, provides a user some incentive to replicate and store the blocks of other documents. Thus, data blocks are resistant to censorship and suppression.

Censorship-resistant schemes protect data only from suppression. Data privacy is not fully protected since any user can access these data.

Protecting User Privacy. User privacy in massive data sharing systems can be illustrated with the following systems. They mostly use anonymity techniques to guarantee anonymous publishing and sharing.

Freenet [8] is a free P2P system that ensures anonymous file sharing, browsing, and publishing. It provides users with freedom of behavior by ensuring their anonymity. Freenet has its own key-based routing protocol (similar to that of a DHT), uses symmetric cryptography, and user and communication anonymity in order to guarantee user privacy.

SwarmScreen [7] is a privacy preserving layer for P2P systems that disrupts community identification by obfuscating users' network behavior. SwarmScreen

---

[6]http://office.microsoft.com/en-us/sharepoint-workspace.

[7]http://usenet-news.net/.

relies on BitTorrent and was designed to provide user privacy through plausible deniability. Since a user behavior can be deduced by her interests, SwarmScreen connects the user to other users outside of her community of interest, which can disguise her interests and thus her behavior.

Many systems such as *ANts P2P*[8] and *MUTE*[9] have been proposed as anonymous P2P file sharing softwares. They use anonymity in order to make the user untrackable, hide her identity and encrypt everything she is sending/receiving from others.

Censorship-resistant systems such as *Dagster* and *Tangler* usually use anonymity techniques to hide users identities. Thus, it is not possible to enforce censorship on data belonging to a specific user.

7.2.3 Evaluation

In this section, we evaluate the P2P systems summarized in Table 7.1. We compare them in terms of privacy properties guaranteed and techniques used.

Privacy Properties. The data and user privacy properties we analyze are the following:

- Data protection against unauthorized reads.
- Data protection against corruption and deletion.
- Limited disclosure.
- Anonymity.
- Denial of linkability. Peers have the ability to deny the links they have with other peers.
- Content deniability. Peers have the ability to deny their knowledge on data content.

The two last properties are taken from [31].

Tables 7.2 and 7.3 provide a comparison of the privacy properties guaranteed by the P2P systems we evaluate. In these tables, a cell is kept blank when a privacy property is not guaranteed by the P2P system. *N/A* is used when information about a privacy property is not available in the literature.

Data Protection Against Unauthorized Reads. In order to protect data from unauthorized reads, data encryption is used. Data encryption prevents server peers, and possibly malicious eavesdroppers and routing peers, from reading private data.

OceanStore, Mnemosyne, Office SharePoint Workspace, Piazza, OneSwarm, Dagster, Tangler, and *Freenet* use data encryption. In *Mnemo-syne, Dagster, Tangler, Freenet*, and *SwarmScreen*, data are public. In *Past*, servers are not controlled

---

[8]http://antsp2p.sourceforge.net/.

[9]http://mute-net.sourceforge.net.

Table 7.1 Sample of P2P systems

Application	P2P systems	Focus on	Relies on	Goals
Data storage	Past	Protecting data privacy	Pastry	Scalable, highly available, persistent, and secure storage
	OceanStore		Tapestry	Consistent, highly available, durable, and secured storage
	Mnemosyne		Tapestry	Steganographic data storage
Massive data sharing	Office SharePoint Workspace		Partially centralized P2P	Document sharing, team collaboration
	Piazza		Unstructured P2P	Scalable XML sharing, data management system
	OneSwarm		BitTorrent	Privacy preserving data sharing
	Dagster	Protecting data from censorship	Not specified	Censorship resistant data sharing
	Tangler	Own routing protocol	Own routing protocol	Censorship resistant data sharing
	Protecting user privacy	Anonymous file sharing, freedom of speech		
	SwarmScreen		BitTorrent	Privacy preserving data sharing

Table 7.2 Comparison of P2P systems based on the privacy properties guaranteed

| P2P Systems | Privacy properties guaranteed | | Limited disclosure | |
	Protection against unauthorized reads	Protection against corruption and deletion	Owners	Requesters
Past		Yes, due to data digital checksums	Yes, due to access control (smartcards)	N/A
OceanStore	Yes, due to encryption	Yes, due to data digital checksums	Yes, due to access control	N/A
Mnemosyne	Yes, due to encryption			
Office SharePoint Workspace	Yes, due to encryption		Yes, due to access control	Yes, due to access control and trust techniques
Piazza	Yes, due to encryption		Yes, due to access control	Yes, due to access control
OneSwarm	Yes, due to encryption		Yes, due to access control	Yes, due to access control and trust techniques
Dagster	Yes, due to encryption			
Tangler	Yes, due to encryption			
Freenet	Yes, due to encryption	Yes, due to data digital checksums		
SwarmScreen				

Table 7.3 Comparison of P2P systems based on the privacy properties guaranteed, cont.

P2P Systems	Privacy properties guaranteed, cont.			
	Anonymity			
	Authors	Servers	Readers	Denial of linkability	Content deniability
Past	Yes, due to Pseudonymity	Yes, due to Pseudonymity	Yes, due to Pseudonymity	Yes, due to Pseudonymity	
OceanStore					Yes, due to encryption
Mnemosyne					Yes, due to encryption
Office SharePoint Workspace	Yes, due to workspaces anonymity	Yes, due to workspaces anonymity	Yes, due to workspaces anonymity	Yes, due to workspaces anonymity	Yes, due to encryption
Piazza					Yes, due to encryption
OneSwarm	Only for third-party monitoring		Only for third-party monitoring	Yes, due to communication anonymity	Yes, due to encryption
Dagster	Yes, due to anonymous communication	Yes, due to anonymous communication	Yes, due to anonymous communication	Yes, due to anonymous communication	Yes, due to encryption
Tangler	Yes, due to anonymous identities	Yes, due to anonymous identities	Yes, due to anonymous identities	Yes, due to anonymous communication	Yes, due to encryption
Freenet	Yes, due to anonymous communication	Yes, due to anonymous identities	Yes, due to anonymous identities	Yes, due to anonymous communication	Yes, due to encryption
SwarmScreen	Yes, due to anonymous communication	Yes, due to anonymous communication	Yes, due to anonymous communication	Yes, due to anonymous communication	

and data are not encrypted, so data are not protected against unauthorized server reads.

Data Protection Against Corruption and Deletion. It is hard to prevent data from being corrupted or deleted. However, data integrity techniques and digital checksums can be used to help users to verify if data have suffered unauthorized changes.

OceanStore and *Freenet* use digital checksums to verify that data content has not been tampered with. *Past* uses smarcards to sign data in order to authenticate them and verifies if they have been modified or corrupted. Thus, although private data are not protected against corruption and deletion, malicious changes can be detected. Countermeasures can be taken against malicious peers in order to demotivate any future corruption. In other systems, data checksums are not used, so data changes cannot be detected.

Limited Disclosure. In order to limit data disclosure, access control is used to prevent unauthorized requesters from accessing data.

In addition to access control, encryption is used to prevent unauthorized disclosure due to collusion between servers and requesters. Even if an unauthorized requester receives encrypted data, it cannot access data content without having the corresponding decryption keys.

In addition, trust techniques can be used to make owners feel more comfortable about the use of their data as they can verify the trustworthiness of the requester.

Past and *OceanStore* use access control to limit disclosure only for authorized data owners.[10] *Freenet*, *Dagster*, and *Tangler* do not use access control since the peers are anonymous. *Piazza* uses access control to limit disclosure for authorized users. *Office SharePoint Workspace* and *OneSwarm* not only use access control but also trust techniques, thus, data disclosure is limited not only to authorized peers but also to trustworthy ones.

User Anonymity. Four types of anonymity guarantees are defined in [10]:

1. Author (i.e., owner) anonymity: which users created which documents?
2. Server anonymity: which peers store a given document?
3. Reader (i.e., requester) anonymity: which users access which documents?
4. Document anonymity: which documents are stored at a given peer?

Past guarantees author and server anonymity due to pseudonymity techniques. Each user holds an initially unlinkable pseudonym in the form of a public key. The pseudonym is not easily linked to the user's real identity. If desired, a user may have several pseudonyms to hide that certain operations were initiated by the same user. *Past* users do not need to reveal their identity, nor the data they are retrieving, inserting, or storing.

---

[10]These systems are not meant for massive data sharing, thus information about data disclosure for requesters is not available.

Office SharePoint Workspace guarantees author, server, and reader anonymity, due to anonymous workspaces. Let us recall that inside workspaces, anonymity is not preserved.

OneSwarm and *SwarmScreen* guarantee author and reader anonymity due to anonymous communication but only from third-party monitoring. In OneSwarm, users' identities are known by servers in order to perform access control, so their anonymity is not guaranteed. Server anonymity is also not guaranteed because they must be easily located when publishing or requesting data.

Dagster and *Tangler* guarantee author, server, and reader anonymity, due to anonymous communication and anonymous identities. Because all connections between the server and the owner/requester are over an anonymous channel, there is no correlation between their identities and the documents they are publishing or requesting.

Freenet guarantees author, server, and reader anonymity due to anonymous communication. For reader and server anonymity, while a peer can get some indication on how early the request message is on the forwarding chain by using the limit on the number of hops (hop-to-live), the true reader and server are kept private due to anonymous communication. Author anonymity is protected by occasional resetting of the data source field in response messages. The peer appearing as the data source does not imply that it actually supplies that data.

Document anonymity is discussed later.

Denial of Linkability. Linkability refers to identifying the correlation between users. Knowledge on linkability can be denied by using anonymous communication. Denial of linkability can protect peers from third-party monitoring.

Past uses pseudonymity techniques, thus knowledge on linkability may be denied by peers. *Office SharePoint Workspace* guarantees denial of linkability outside workspaces due to anonymous workspaces. *OneSwarm, Dagster, Tangler, Freenet,* and *SwarmScreen* guarantee denial of linkability, due to anonymous communication.

Content Deniability. Content deniability refers to whether peers can deny the knowledge on the content stored or transmitted (document anonymity). Knowledge on content can be denied if the content is not readable by the peer that holds it.

In all systems that use data encryption, knowledge on data content can be denied. This is the case of *OceanStore, Mnemosyne, Office SharePoint Workspace, Piazza, OneSwarm, Dagster, Tangler,* and *Freenet*. In *Past* and *SwarmScreen*, servers can access data, so they cannot deny the knowledge on the data content they store.

Privacy Techniques. The privacy techniques we analyze are the following:

- Access control. Data access can be controlled with respect to the identity of the user, her role, and the access purpose of the requester.
- Anonymity techniques. A user (resp. a data) is made indistinguishable from other users (resp. data), thus providing her anonymity among a group of users (resp. data set).

- Trust techniques. The behavior of users is predicted.
- Cryptography techniques. Data can be made "unreadable" by converting ordinary information (plain text) into unintelligible cipher text.

Table 7.4 provides a summary of the techniques used to ensure privacy by the systems we evaluate.

Access Control. Access control is essential to guarantee that data will not be read or shared with unauthorized peers.

In *Past*, access control is based on the use of smartcards which generate and verify various certificates. Users may access data or not within the access rights related to their certificate.

In *OceanStore*, access control is based on two types of restrictions: *reader* and *writer* restrictions. In the *reader* restriction, to prevent unauthorized reads, the data decryption keys are distributed by the data owner to users with read permissions. To revoke read permissions, the data owner requests users to delete replicas or re-encrypt them with new encryption keys. A recently revoked reader is able to read old data from cached copies or from misbehaving servers that fail to delete or re-encrypt. This problem is not specific to *OceanStore*, even in conventional systems, as there is no way to force readers to forget what has been read. To prevent unauthorized writes, they must be signed so that well-behaved servers and clients can verify them against an Access Control List (ACL). The data owner can define an ACL by datum by providing a signed certificate. ACLs are public so that servers can check whether a write is allowed. Thus, writes are restricted at servers by ignoring unauthorized updates.

In *Office SharePoint Workspace*, access control is based on the use of member-ship lists and workspace rules. Users are identified by accounts and passwords that allow them to log in workspaces. If they can log in a workspace W, they are listed in the membership list of W. A user can access or remove data from W as long as she is a member of W and, in addition, respects the workspace usage rules.

In *Piazza*, the access to a published XML document is restricted to parts of the document in accordance with the data owner preferences. Data owners in *Piazza* can specify access control policies declaratively and generate data instances that enforce them. By granting decryption keys to users, the data owner enforces an access control policy. Once published, the data owner relinquishes all control over who downloads and processes the data. Requesters can access the data conditionally, depending on the keys they possess.

In *OneSwarm*, persistent identities allow users to define per-file permissions. These permissions (i.e., capabilities) restrict access to protected data. For example, OneSwarm can be used to restrict the distribution of a photo file to friends and family only.

Anonymity Techniques. Anonymity can enable censorship resistance and freedom of behavior without fear of persecution. Anonymity is mostly used to hide user identity. If user identity is hidden, access control cannot be deployed. On the other hand, if anonymous communication channels are used, a channel listener is not able to understand the messages sent on the channel or who has sent it.

Table 7.4 Comparison of P2P systems based on used privacy techniques

P2P Systems	Privacy techniques		Trust techniques	Cryptography techniques	Data integrity protection
	Access control	Anonymity			
Past	Use of smartcards	Pseudonymity using smartcards	Certificate-based Trust		Use of smartcards for data certification
OceanStore	Use of two types of restriction (reader and writer)			Symmetric encryption	Use of content hashing as a checksum
Mnemosyne				Symmetric block encryption	
Office SharePoint Workspace	Use of workspace and membership lists	Anonymous workspaces and groups	Use of trust colors	Symmetric encryption	
Piazza	Use of access policies defined by the owner			Symmetric encryption	
OneSwarm	Use of permissions defined by the owner	Anonymous communication	Use of community trust levels	Hybrid encryption	
Dagster		Anonymous communication		Symmetric block encryption	
Tangler		Anonymous communication and identities		Symmetric block encryption	
Freenet		Anonymous communication		Symmetric encryption	Use of signature-verifying keys
SwarmScreen		Random communication			

A way to provide anonymity is to give users fake identities. Fake identities can be ensured by smartcard techniques where the real identity of the user is only known by the authority which distributes the smartcards. In this case, the authority must be considered as a TTP. In *Past*, smartcards are used to allow users to obtain necessary credentials to join the system in an anonymous fashion.

Systems like *Office SharePoint Workspace* organize users into anonymous groups called workspaces. Users are known within their workspace and they are anonymous for users in other workspaces. This choice can be explained by the fact that users do not need to be anonymous to their friends or to co-workers who are authenticated in order to access their workspace.

In *OneSwarm*, anonymity is only used to protect a user identity from a third-party monitoring. Users in OneSwarm perform their queries by using anonymous routes. However, a server has the complete knowledge of the query initiator identity, so access control is possible.

Dagster, Freenet, and Tangler maintain privacy by using anonymous communication.

In *Dagster*, an anonymous channel between owners or requesters and servers is created by using Anonymizer,[11] a trustworthy third party (TTP). Instead of requesting web data directly, a user sends the request to Anonymizer that forwards the request appropriately. The content is then delivered to Anonymizer that returns it to the requesting user. Anonymizer can only be used to retrieve data content and the user is required to trust that it will not reveal her identity and the requested data content.

In *Freenet*, rather than moving directly from sender to recipient, messages travel through peer to peer chains, in which each link is individually encrypted, until the message finally reaches its recipient. Each peer knows only about its immediate neighbors, so the end points could be anywhere in the network. Not even the peer immediately after the sender can tell whether its predecessor was the message's originator or was merely forwarding a message from another peer. Similarly, the peer immediately before the receiver cannot tell whether its successor is the final recipient or will continue to forward it.

In *Tangler*, privacy is maintained by using not only anonymous communication but also identities. Tangler ensures that a user can retrieve data without revealing their identity. around the world, run by volunteers who wants to be protected from censorship. Users publish documents by anonymously submitting blocks to servers. Servers can communicate with each other both directly and anonymously (by using other servers as a mixed network [6]).

In *SwarmScreen*, privacy is maintained by using random connections. They propose a privacy-preserving layer for P2P systems that disrupts community identification by obfuscating users' network behavior. Users can achieve plausible

---

[11]Anonymizer is an online service that attempts to make activity on the Internet untraceable. It accesses the Internet on the user's behalf, protecting personal information by hiding the source identifying information. http://www.anonymizer.com/.

deniability by simply adding some percentage (between 25 and 50%) of additional random connections that are statistically indistinguishable from natural ones.

Trust Techniques. Trust techniques are used in P2P systems in order to reduce the probability of data privacy violation. The right to access data can be given to peers who are trustworthy and forbidden to peers who are untrustworthy.

Mainly, P2P systems that preserve privacy in distributed data storage do not trust data servers. The potential malicious behavior of peers that store data (i.e., servers) can be prevented with cryptography techniques. The systems analyzed here use trust techniques to verify trustworthiness of peers who want to access data stored on servers.

In *Past*, server peers trust owner peers thanks to a smartcard held by each peer wanting to publish data in the system. Smartcards are given by TTPs called brokers who are fully trusted by owner and server peers. A smartcard ensures the integrity of identifiers and trustworthiness assignment of the user which held it. Without a TTP, it is difficult to prevent attackers from misbehaving in the system.

On the other hand, in systems that provide massive data sharing, data owners are usually considered trustworthy and cryptography techniques are used to prevent the malicious behavior of servers. These systems are thus interested in verifying trustworthiness of requesters who want to access private data.

In *Office SharePoint Workspace*, peers can determine how much can they trust other peers through their authentication status. A peer A can optionally organize its contacts by how they were authenticated or check their authentication status by the color of their name. The names of directly authenticated contacts, which are trustworthy, are displayed in green. Other contacts in A's workspace, which are also trustworthy, are displayed in teal. Contacts in other workspaces trusted by the A's domain administrator are displayed in blue. Contacts that are not authenticated are displayed in black, and duplicated names that conflict are displayed in red. The color of the name can be used in the verification of trustworthiness of the requester. Requesters who have their name in black or red are considered untrustworthy and thus they may not gain data access.

In *OneSwarm*, data are located and transferred through a mesh of untrusted and trusted peers populated from user social networks. Peers explicitly define a trust level for a persistent set of peers. This requires some notion of identity to allow peers to relate real-world trust relationships to overlay connections. Public keys can be used as identities in order to verify trustworthiness of the peers. These public keys can be exchanged in three ways. First, requesters discover and exchange keys with owners over the local area network. Second, peers can rely on existing social networks, e.g., Google Talk or Facebook, to distribute public keys. Third, peers can email invitations to friends. Invitations include a one-time use capability that authenticates the recipient during an initial connection, during which public key exchange occurs. OneSwarm also supports key management within a group. It allows peers to subscribe to one or more community servers. A community server maintains a list of registered peers and can delegate trust regarding a subset of their peers.

Cryptography Techniques. Cryptography is largely used by P2P systems in order to protect private data from unauthorized access. Encryption techniques are used to prevent malicious servers from reading private data, while digital checksums are used to detect if malicious peers are modifying or corrupting private data.

Usually, symmetric-key encryption is used to protect data content. Symmetric key generation is less expensive than asymmetric key generation. Since a large number of keys are needed to encrypt data, P2P systems have found more interest in symmetric-key encryption.

In *OceanStore*, data are encrypted using symmetric keys. Encryption keys are distributed to users who are allowed to access data.

In *Piazza*, published data are encrypted with symmetric keys in order to restrict peers from accessing data in accordance with the owners' preferences.

In *Freenet*, all data are encrypted with symmetric keys before publication. This is done mainly for political or legal reasons where servers might wish to ignore the content of the data stored. Data encryption keys are not included in network messages. Owners distribute them directly to end users[12] at the same time as the corresponding data identifiers. Thus, servers cannot read their own files, but users can decrypt them after retrieval.

In *Office SharePoint Workspace*, data are encrypted on the communication channels. Data that may be temporarily stored on servers are also encrypted using symmetric keys kept by owners, thus preventing potentially malicious servers from reading data. However, a user has the choice to delegate her identity management to servers hosted by Microsoft[13] or a TTP. If so, this one will have access to the encryption keys.

Other systems like Mnemosyne, Dagster, and Tangler use block encryption.

In *Mnemosyne*, data are divided into blocks. In order to store data, each block is encrypted using the cryptographic hash function SHA-256 and the Advanced Encryption Standar (AES), and written to a pseudo-randomly chosen location. With a good enough cipher code and key, the encrypted blocks will be indistinguishable from the random substrate, so an attacker cannot even identify the data. On the other hand, users who have the data name and key can reconstruct the pseudo-random sequence, retrieve the encrypted blocks, and decrypt them.

In *Tangler*, data are broken into a number of small blocks (shares). Each of these blocks is treated independently and stored on a subset of the participating servers. Blocks are then entangled with other random blocks, which obscures the real content of the block and makes it unreadable. In order to reconstruct a data block, users have to retrieve a minimum number of blocks of the appropriate shares. By simply stripping away the random value, users can find the original data block.

---

[12] Freenet does not use access control techniques thus key distribution is not restricted.

[13] Microsoft kept their right to collect some information about the use of the Office Share-Point Workspace software and other activities "outside" of workspaces, as explained in their privacy statement at http://office.microsoft.com/en-us/help/privacy-supplement-for-microsoft-office-groove-2007-HA010085213.aspx.

In *Dagster*, data are separated into blocks, then the user generates a symmetric key for each block. Each block is encrypted with the corresponding key and sent to servers using anonymous channels. In order to reconstruct data, a number of blocks are needed along with the decryption keys.

Other systems, such as *OneSwarm*, combine public key encryption with symmetric encryption. While symmetric keys are used to encrypt data, public keys are used to share symmetric keys in a secure manner.

Having private data encrypted prevents unauthorized peers from reading their content. This contributes to protect data content privacy, although it does not protect data from being corrupted or deleted. To protect data from suppression and corruption, cryptographic hash functions (digital checksums) can be used.

In *OceanStore*, the data are named using a secure hash over the data content, giving them globally unique checksums. This provides data integrity, by ensuring that requested data have not been corrupted, since the checksum of corrupted data will be different than the globally unique checksum.

In *Freenet*, when a user publishes data which she later intends to update, she first generates a public–private key pair and signs the data with the private key. Data are published under a pseudo-unique binary key (i.e., hash key), but instead of using the hash of the data contents, the data identifier itself is used (a signature-verifying key). Signature-verifying keys can be used to verify that data content has not been tampered with.

In *Past*, the user smartcard generates reclaim certificates, containing the data identifiers and included in the user request. When processing a request, the smartcard of a server peer first verifies that the signature in the reclaim certificate matches the one in the data certificate stored with the data. This prevents unauthorized users from reclaiming the ownership of data.

To summarize our evaluation, we can see that depending on the target application, the majority of the compared P2P systems guarantee:

- Protection against unauthorized reads, by using data encryption.
- Protection against corruption and deletion, by using data checksums.
- Limited disclosure, by using access control and trust techniques.
- Anonymity and denial of linkability, by using anonymity techniques.
- Content deniability, by using data encryption.

However, these P2P systems do not support purposes. Purpose specification is essential for privacy protection as recommended by the OECD guidelines and should be taken into account in the complete data management cycle.

7.3 Hippocratic Databases

Inspired by the Hippocratic oath and its tenet of preserving privacy, Hippocratic databases (HDB) aim at incorporating privacy protection within relational database systems [2]. The important concepts of HDBs are the following.

- Privacy policies. A privacy policy defines for each column, row, or cell of a table (a) the usage purpose(s), (b) the external recipients, and (c) the retention period.
- Privacy authorizations. A privacy authorization defines which purposes each user is authorized to use on which data.

HDBs define ten founding principles to protect data privacy according to users preferences.

1. Purpose Specification. For personal information stored in the database, the purposes, for which the information has been collected, shall be associated with that information.
2. Consent. The purposes associated with personal information shall have the consent of the owner of the personal information.
3. Limited Collection. The personal information collected shall be limited to the minimum necessary for accomplishing the specified purposes.
4. Limited Use. The database shall run only those queries that are consistent with the purposes for which the information has been collected.
5. Limited Disclosure. The personal information stored in the database shall not be communicated outside the database for purposes other than those for which there is a consent of the owner of the information.
6. Limited Retention. Personal information shall be retained only as long as necessary for the fulfillment of the purposes for which it has been collected.
7. Accuracy. Personal information stored in the database shall be accurate and up-to-date.
8. Safety. Personal information shall be protected by security safeguards against theft and other missappropriations.
9. Openness. An owner shall be able to access all information about her stored in the database.
10. Compliance. An owner shall be able to verify compliance with the above principles. Similarly, the database shall be able to address a challenge concerning compliance.

In an HDB, queries are submitted along with their intended purpose. Query execution preserves privacy by using query rewriting and restrictions by column, row, or cell.

Purpose Specification. Purpose specification is the cornerstone of an HDB. It states that purposes should be specified and attached to data items to control their usage. In order to do this, simple specification language such as Platform for Privacy Preferences (P3P) [9] can be used as a starting point. P3P is a standard developed by the World Wide Web Consortium. Its goal is to enable users to gain more control over the use of their personal information on web sites they visit. P3P provides a way for a Web site to encode its data-collection practices in a machine-readable XML format, known as a P3P policy [9], which can be programmatically compared against a user's privacy preferences [23]. In [20, 21] authors propose ideas for reducing the complexity of the policy language which include arranging purposes in a hierarchy. Subsumption relationships may also be defined for retention periods and recipients.

Limited Disclosure. Limited disclosure is another vital component of an HDB system. It states that the private data shall not be disclosed for purposes other than those defined by the data owner. A scalable architecture for enforcing limited disclosure rules and conditions at the database level is proposed in [25]. For enforcing privacy policies in data disclosure, privacy policies can be stored and managed in the database. These policies are expressed in high-level privacy specification languages (e.g., P3P). Enforcing privacy policies does not require any modification to existing database applications. Authors provide techniques for enforcing a broad class of privacy policies by automatically modifying all queries that access the database in a way that the desired disclosure semantics is ensured. They examine several implementation issues, including privacy metadata storage, query modification algorithms, and structures for storing conditions and individual choices.

HDB Implementation. Subsequent works have proposed solutions for implementing HDBs. In [1], authors address the problem of how current relational DBMS can be transformed into their privacy-preserving equivalents. From specifications of privacy policies, they propose an algorithm that defines restrictions (on columns, rows, and cells) to limit data access. In [4], authors propose query modification techniques and access control to ensure data privacy based on purposes. They propose to organize purposes in a tree hierarchy where the root is the most general purpose and the leafs the more specific ones. In this way, if data access is allowed for a purpose x, all descendant purposes of x are also allowed. They also propose data labeling (with allowed purposes) at different granularity levels (table, column, row, or cell). In addition, they propose some SQL modifications to include purposes, for instance *Select* column-name *From* table-name *For* purpose-name.

HDBs are the first privacy techniques that include the notion of purpose in relational databases. They are essential to users who would like to know for which purpose their data are used. However, enforcing HDBs in P2P systems is a challenge which we address in the next sections.

7.4 PriMod

P2P systems analyzed in Sect. 7.2 cannot be used in our illustrative example on the medical OLSN (introduced in Sect. 7.1), where participants are allowed to use medical data depending on the purpose specification made by data owners. Thus, a new privacy solution for P2P-based OLSNs is necessary that:

- Allows to specify access purposes by using HDB principles.
- Limits data disclosure by using purpose-based access control.
- Protects data against unauthorized reads.

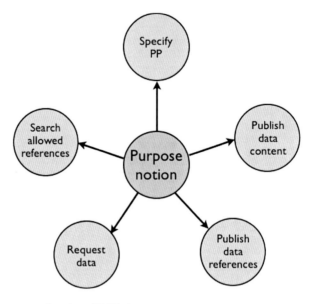

Fig. 7.1 Purpose as mainspring of PriMod

- Allows to detect corruption or unauthorized deletion.

PriMod [15, 17, 18], a data privacy model for P2P systems, is proposed to answer the need of data owners to share their sensitive data in a P2P system while preserving data privacy. It makes no assumptions about the P2P system organization. The unique important hypothesis is that each peer has a unique identifier in the system for all its connections.[14] Figure 7.1 shows how the purpose notion is mainspring of PriMod functionalities. PriMod allows owners to specify their privacy policies (PPs) and to publish data for specific purposes and operations. They can choose between publishing only their data references (e.g., filenames and primary keys) or publishing encrypted data content. Requesters can search for sensitive data but must specify the access purpose and operation in their requests, thus they are committed to their intended and expressed use of data. Requesters can also ask which sensitive data they can access for a particular purpose.

To summarize, the PriMod assets are the following:

- It benefits from P2P assets in data publishing and sharing while offering data privacy protection based on access purposes.
- It can be easily integrated to any P2P system.

---

[14]In [5, 35], authors have treated peer identification. We are fully aware of the impact of identification on user privacy. However, peer identities do not necessarily reveal users real identities, thus user privacy can be somehow protected.

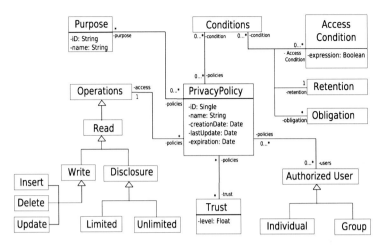

Fig. 7.2 Privacy policy (PP) model

- It proposes/uses privacy policies concepts and defines models for trust and data management.
- It offers operations for publishing data content, publishing references, requesting, and purpose-based searching.
- Data owners can define their privacy preferences in privacy policies.
- Sensitive data are associated with privacy policies. This association creates private data ready to be published in the system.
- Requests are always made for particular purposes and operations.
- Trust techniques are used to verify trustworthiness of requester peers.

In the following, Sect. 7.4.1 presents the privacy policy model of PriMod. Section 7.4.2 presents the data model. Section 7.4.3 introduces the functions of PriMod.

7.4.1 Privacy Policy Model

In PriMod, each data owner should define her privacy preferences. Those privacy preferences are registered in PPs independently of data. Once defined, they are attached to appropriate data. PPs are dynamic because they can vary with time. For instance, at the end of the medical treatment, a doctor will only allow reading access to other doctors for analyzing the patient medical record. Updating diagnosis will not be allowed anymore. We consider that each owner is responsible for defining and maintaining her PPs in an independent way.

Inspired from the Platform for Privacy Preferences (P3P) [9], Fig. 7.2 shows a PP model. This model does not claim to be exhaustive, but shows information about PPs that can include:

Authorized Users. It is a list of users who are authorized to access data, a kind of ACL. A user can be an individual or a group.

Purpose. An access purpose states the data access objective. With this concept, an owner is able to specify the purposes for which its data can be accessed by users.

Operations. An operation determines what a peer can do with data. We use three basic operations, read, write, and disclose, but others can be defined.

- *Read.* A peer can read the data content.
- *Write.* A peer can modify the data content with the following operations: insert, update, delete.
- *Disclose.* A peer is able to disclose shared data for other peers. Disclosure can be limited or unlimited. If a peer disclosure right is limited, it cannot give disclosure rights on the data to other peers.

Conditions. Conditions state the access conditions a user should respect, the obligations a user should accomplish after accessing data, and the limited time for data retention.

- *Access condition.* Conditions state under which semantic condition data can be accessed. This may concern data values, for example age > 10.
- *Obligation.* Obligations state the obligation a user must accomplish after the data access. For example, researcher Ri should return research results after using the patient records.
- *Retention time.* The retention time states the time limit of retention of the data. For example, the local copy obtained by a requester of a patient record should be destroyed after 1 year of use.

Minimal Trust Level. It is the minimal trust level a requester peer should have in order to gain access to data.

7.4.2 Data Model

In order to respect PPs, they should be associated with data. In the following, we use relational tables; however, any type of data can be considered (files, XML documents, rich text files, etc.).

Data Table. Each owner peer stores locally the data it wants to share (see Table 7.5). Those data can be stored in relational tables called *data tables*. The unique restriction about data tables is that primary keys should be generic and impersonal to respect privacy and to not disclose any information. If considered data are files, their identifiers or names should be impersonal.

Privacy Policy Table. Data contained in PPs are stored in a table named *privacy policies table* (see Table 7.6). To simplify, all elements of Fig. 7.2 are not included.

Table 7.5 Data table of doctor Dj

Data table DTj							
Id (PK)	SS	Name	Country	Birthdate	Gender	Smoker	Diagnosis
Pat1	001044001001	Alex	France	2000	Male	No	NO cardio-vascular disease
Pat2	900344001001	Bea	France	1990	Female	No	NO cardio-vascular disease

Table 7.6 Privacy policy table of doctor Dj

Privacy policy table PPTj					
Id (PK)	Operation	User	Purpose	Condition	Minimal trust level
PP1	Read	Pharmacists, Doctors	Consulting record	Birthdate < 2000	0.5
PP2	Read	Researchers	Researching on cardiovascular disease	—	0.6

In this table, one tuple corresponds to one PP. The same PP can be used with different data. Each policy contains operations (read, write, or disclose), allowed users, access purposes, conditions (if they exist), and the required minimal trust level of allowed users.

Purpose Table. Information about the available purposes are stored in a table named *purpose table*. A tuple of the purpose table contains the purpose identifier, the purpose name, and the purpose description. We recall that in HDB, purposes can be organized in a hierarchy. To simplify, in PriMod, we make abstraction of such hierarchy.

Trust Table. Each peer maintains a local *trust table* that contains the trust level of some peers in the system. A tuple of the trust table contains the identifier of a peer, its trust level, and a cell defining if this peer is considered as a *friend* or not locally.

Private Data Table. This table joins data to privacy policies. It allows fine-grained access control by specifying which table, column, line, or cell can be accessed by preserving which privacy policy. For instance, in PD1 of Table 7.7, only some columns of the data table DTj (those who do not disclose patients identities) are concerned by the privacy policy PP1 where pharmacists and doctors can read records of patients who were born before 2000.

Purpose-Based Data Reference Table. To ease data searching, a purpose-based index is necessary. Information about the references of data allowed for particular purposes and operations for particular requesters are stored in a table named *purpose-based data reference table* (PBDRT for short). This purpose-based index allows requesters to know which data they can access for a particular purpose and

Table 7.7 Private data table of doctor Dj

Private data table j					
	Data				Privacy
Id (PK)	Table	Column		Id	Policy
PD1	DTj	Birthdate, Gender, Smoker, Diagnoses		–	PP1
PD2	DTj	Country, Birthdate, Gender, Smoker, Diagnosis		–	PP2

Table 7.8 Purpose-based data reference table (PBDRT) of doctor Dj

Purpose-based data reference table j (PBDRT j)		
Key (PK)	requesterID	DataRefList
hash(diagnosis, write)	Doctor1	{DataRef1}
	Doctor2	{DataRef1, DataRef2}
hash(research, read)	Doctor1	{DataRef1}
	Scientist1	{DataRef1}
	Scientist2	{DataRef1, DataRef3, DataRef5}
hash(accessing, read)	Patient1	{DataRef1}
	Patient2	{DataRef2}
	Patient3	{DataRef3, DataRef5}

operation. Each tuple of this table is identified by a *key*, obtained, for instance, by hashing the couple (*purpose, operation*) (see Table 7.8). Besides the key, a tuple contains the identifiers of requesters and the list of data references that are allowed to access.

7.4.3 PriMod Functions

PriMod proposes the next set of functions.

Publishing Data. PriMod provides two ways of publishing sensitive data indicating the PP that users should respect. An owner may choose to publish her data content or only data references. In the first case, data storage is protected from malicious servers by using cryptography techniques. In the second case, there is no need of data encryption since references do not show any private information about the data if they are well chosen (i.e., personal information such as security numbers and addresses should not be used in references).

Boolean publishData(data, PPId). Owner peers use this function to publish data content in the system. The second parameter is the privacy policy that dictates the usage conditions and access restrictions of the published data. This function returns true if data content is successfully distributed, false otherwise. It is similar to a traditional publishing function. To protect data privacy against potential malicious servers, before distribution, data content is encrypted (by using symmetric cryptography) and digital checksums are used to verify data integrity. Symmetric keys are stored locally by the owner. Requesters must contact owners to retrieve keys and decrypt requested data.

Boolean publishReference(data, PPId).   Owner peers use this function to publish data references in the system while data content are stored locally. This function returns true if data references are successfully distributed, false otherwise. Servers store data references and help requesters to find data owners to obtain data content. Publishing only data references allows owners to publish private data while being sure that data content will be provided to right requesters. This hypothesis cannot be guaranteed in the previous function because malicious servers may misbehave by returning encrypted data to unauthorized peers.

Requesting Data.   For requesters, how data have been published is transparent and a unique function to request data is proposed by PriMod.

Data request(dataRef, purpose, operation).   Requester peers use this function to request data (*dataRef*) for a specific purpose (e.g., researching, diagnosis, or analysis) to perform a specific operation (i.e., read, write, or disclose). This function returns the requested data if the requester has corresponding rights, otherwise it returns null. This function compels requesters to specify the access purposes and the operations that they have the intention to apply to requested data.

   This explicit request is the cornerstone of this work, it commits requesters to use data only for the specified purposes and to perform only specified operations. Legally, this commitment may be used against malicious requesters if it is turned out that obtained data have been used differently.

TrustLevel searchTrustLevel(requesterID).   Owner peers use this function to search the trust level of the requester *requesterID*. This function returns the trust level of the requester if it is found else it returns null. This trust level is used in the requesting process to verify the trustworthiness of the requester in order to give him access rights.

Purpose-Based Reference Searching.   PriMod provides users with a function for purpose-based reference searching based on the PBDRT. This function allows requesters to know which data they are authorized to request for a particular purpose and operation. This prevents users from denying knowledge about their access rights. The allowed data reference lists (contained in the PBDRT) can be created transparently while publishing data. These lists can be published periodically in the system.

DataRefList dataRefSearch(purpose, operation).   Requester peers use this function to know the data they are authorized to access for a specific purpose and operation. This function returns a list of data references of data the requester is authorized to access (*dataRefList*). If the list is empty, the function returns null. This request by peer protects privacy because it avoids that all users know which are the access rights of other users and know the complete list of available data (global schema).

Table 7.9 PriMod: used privacy techniques and guaranteed properties

P2P Model	Guaranteed properties				
	Protection against unauthorized reads	Protection against corruption and deletion	Limited disclosure		Content deniability
			Owners	Requesters	
PriMod	Yes, due to encryption	Yes, due to data digital checksums		Yes, due to access control	Yes, due to encryption
	Privacy Techniques				
	Access control	Anonymity	Trust techniques	Data encryption	Data integrity protection
	Use of purpose-based access control		Use of trust levels	Symmetric encryption	Use of content hashing as a checksum

7.4.4 Analysis

Table 7.9 compares PriMod to works shown in Sect. 7.2.

- Private data in PriMod are protected against malicious reads, corruption, and deletion by using encryption and data checksums.
- Data disclosure is limited by using access control only for requesters.
- Unlike all the models presented, the notion of purpose is omnipresent in PriMod.
- PriMod does not guarantee anonymity since it is not designed to protect user behavior but only data privacy.
- Trust levels are used to prevent malicious behavior of requesters.

7.5 PriServ

PriServ is a privacy service that implements PriMod. Figure 7.3 shows the Priserv's architecture, which is on top of a DHT layer. This DHT layer has two functional components: one is in charge of the routing mechanism that supports the *lookup()* function as well as the dynamicity of peers (join/leave of peers); the other ensures key-based data searching and data distribution by implementing the *put()* and *get()* functions. These two layers provide an abstraction from the DHT.

Conceptually, PriServ is an APPA (Atlas Peer-to-Peer Architecture) service [3]. APPA is a data management system for large-scale P2P and Grid applications. The PriServ implementation uses Chord for its efficiency and simplicity; however, any DHT can be used. PriServ uses the traditional get() and put() functions of DHTs to locate and publish data, each incurring $O(\log N)$ messages.

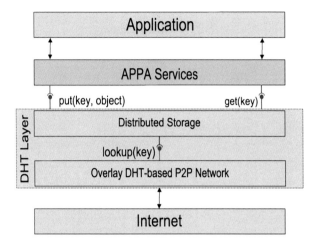

Fig. 7.3 Global architecture

- *put*(*key, data*) stores a key k and its associated data object in the DHT.
- *get*(*key*) retrieves the data object associated with k in the DHT.

Data keys in PriServ are created by hashing the triplet (*dataRef, purpose, operation*). We consider that *dataRef* is a unique data reference, *purpose* is the data access purpose and *operation* is the operation that can be executed on requested data with respect to the corresponding privacy policy. Thus, the same data with different access purposes and different operations have different keys.

7.5.1 PriServ Architecture

Figure 7.4 shows the component-based architecture of PriServ. It provides five interfaces to the application layer: publishReference(), publishData(), request(), dataRefSearch(), and dataRefPut(). The first four correspond to PriMod operations. The last one allows users to construct a distributed purpose-based index used in the dataRefSearch() function. PriServ also provides two retrieve functions necessary for the interaction between requesters and owners. As said before, it uses the two traditional interfaces of the DHT layer (put() and get()).

Inside PriServ, several components are gathered around an orchestrator which organizes PriServ activities. The orchestrator may execute three different workflows depending on the role of the peer (owner, requester, or server/storer).

Storage Manager. Its role is to manage data storage. Data storage is made locally before data are stored in the P2P system. To store data in the P2P system, this component invokes the put() and get() functions of the DHT layer.

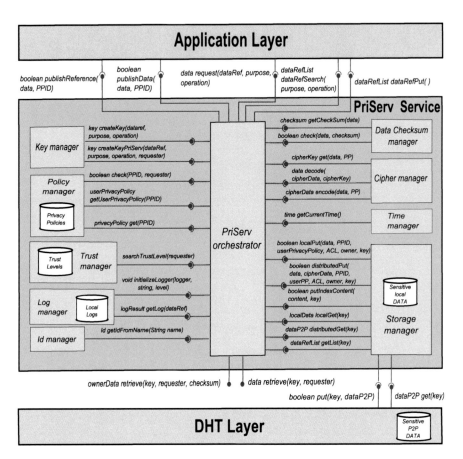

Fig. 7.4 PriServ architecture

Policy Manager. Its role is to manage owner PPs. Data owners organize their privacy preferences in PPs. From PPs, this component creates "user privacy policies" that should be respected by requesters. From PPs, this component extracts the list of authorized users. This list is a kind of access control list (ACL) that contains only allowed users. To simplify, we consider that users may belong to groups, so, this ACL contains mainly a list of groups of users and maybe some individual users known in advance.

Trust Manager. Its role is to manage trust levels. We consider that each peer stores locally a list of peers and their corresponding trust levels. If a required trust level does not exist locally, the trust manager asks for it to other peers. In PriServ, three ways of obtaining trust levels are used, namely, with-friends, without-friends, and with-or-without-friends. Section 7.5.2 explains those algorithms.

Cipher Manager. Its role is to manage cryptography in PriServ. It offers a function that creates a symmetric cipher key for each pair (data, PP). It also offers two functions to encrypt and decrypt data. In this work, symmetric-key algorithms can be used because they are generally much less computationally intensive than other cryptography algorithms. However, PriServ is independent of the encryption technique used.

Data Checksum Manager. Its role is to check the integrity of data. This component offers a function to calculate digital data checksums (e.g., by using MD5). Digital checksums can be used by owners to verify if servers have tampered with the data. PriServ is independent of the techniques used to create data checksums.

Key Manager. Its role is to generate data keys used in the put() and get() functions. It creates data keys by hashing the triplet *(data references, purposes, operations)*. This component offers two functions, the first one, used during the put() process, returns the created data key and the second, used during the get() process, returns the created data key and adds it the identifier of the requester.

Id Manager. Its role is to identify peers from their names (e.g., URI). Peer identifiers are used in access control to authorize or prohibit access to data.

Log Manager. Its role is to manage logs. It stores logs in a dedicated database on each peer. These logs can be used for recovery and auditing process.

Time Manager. Its role is to give the current time. It is used for synchronizing clocks of all connected peers.

PriServ Orchestrator. It is the central component of PriServ. According to the role of the peer, the orchestrator executes a different workflow by using the components introduced before.

- *Owner orchestrator.* Its role is to orchestrate the owner functionalities. It is responsible for publishing in the P2P system references or data depending on the called function (publishData() or publishReference()). It is also responsible for directly returning data or symmetric keys during the requesting process (retrieve()). It interacts with the application layer for publishing and with the requester orchestrator for retrieving.
- *Requester orchestrator.* Its role is to orchestrate data requesting. It interacts with the application layer for requesting and with the owner orchestrator for retrieving.
- *Server orchestrator.* Its role is to orchestrate the server functionalities. For that, it interacts with the DHT layer to store data for the P2P system and to return stored data.

7.5.2 PriServ Mean Functions

PriServ implements the PriMod functions so it offers to the application layer two ways for publishing and allows searching data and data references for a particular

```
Owner orchestrator
0:    publishReference(data, PPID)
1:    begin
2:       privacyPolicy = policyManager.get(PPID);
3:       userPP = policyManager.getUserPrivacyPolicy(PPID);
4:       key = keyManager.createKey(data.dataref, privacyPolicy.purpose,
                                       privacyPolicy.operation);
5:       storageManager.localPut(data, PPID, userPP, privacyPolicy.ACL,
                                    owner, key);
6:       key2 = keyManager.createKey(null, privacyPolicy.purpose,
                                        privacyPolicy.operation);
7:       storageManager.addRef(key2, data.getDataref(), privacyPolicy.ACL);
8:    end;

Owner storage manager
9:    localPut(data, PPID, userPP, ACL, owner, key)
10:    begin
11:       dataTable.localSave(data);
12:       privateDataTable.localSave(data.dataRef, PPID);
13:       dataP2P = createDataP2P(userPP, ACL, owner);
14:       DHT.put(key,dataP2P);
15:    end;
```

Fig. 7.5 Algorithm of the publishReference() function

purpose and operation. The main procedures are publishReference(), publishData(), request(), dataRefSearch(), dataRefPut(), and searchTrustLevel(). All but the last function use the DHT organization. The searchTrustLevel() function uses instead an unstructured P2P approach as we will see latter.

In the following, consider 6 peers with identifiers P23,[15] P25, P31, P33, P51, and P60. Consider also that:

- P23 is an owner peer.
- P31 and P25 are requester peers.
- P33, P51, and P60 are server peers.

Boolean publishReference(data, PPId). Owners use this function to publish data references under a particular PP. Publishing only data references and storing data locally allow owners to provide themselves their data to right requesters.

When the owner orchestrator receives the publishReference() call from the application, it uses the algorithm shown in Fig. 7.5 to publish data references in the P2P system and to store data locally. The object sent to the P2P system contains the *conditions* that data requesters should respect (userPP) when using the data (see Fig. 7.2), the ACL that servers should verify, and the owner id of the data reference.

---

[15]To distinguish data keys from peer keys, we prefix peer keys with letter P.

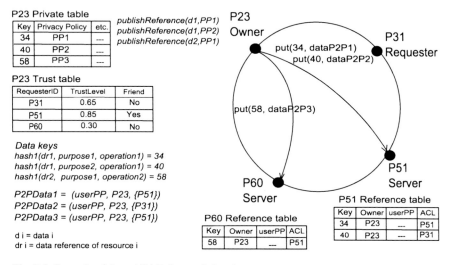

Fig. 7.6 Example of the publishReference() function

Figure 7.6 illustrates this algorithm. P23 shares data with keys 34, 40, and 58. *hash1* is used to produce these keys from data identifiers (IDri), purposes, and operations. From the DHT principle, P51 is the server peer responsible for key 34 and 40 and P60 for 58. Only data references of P23 are published with its identifier by using the put function.

Boolean publishData(data, PPId). Owners use this function to publish data content under a particular PP (PPId). To protect data privacy against potential untrusted servers, before distribution, data content is encrypted (by using symmetric cryptography), and digital checksums are used to protect data integrity from servers.

When the owner orchestrator receives the publishData() call from the application, it uses the algorithm shown in Fig. 7.7 to publish the data content in the P2P system and to store locally a copy of the data.

Figure 7.8 illustrates this algorithm. P23 shares data with keys 36, 42, and 59. *hash1* is used to produce keys from data identifiers (IDri), purposes, and operations. P51 is the server peer responsible for key 36 and 42, and P60 for key 59. P2P data are created by P23 by encapsulating encrypted data, the P23 identifier, and the corresponding ACL that contains requesters' identifiers allowed to access this data. Then, P2P data are published by using the put function.

Data request(dataRef, purpose, operation). Requesters use this function to request data (*dataRef*) for a specific *purpose* to perform a specific *operation*. This function compels requesters to specify the access purposes and the operation that they will apply to requested data. When a requester orchestrator receives the request() call from the application, it uses the algorithm shown in Fig. 7.9. For requesters, the way data have been published is transparent (publishData or publishReference), so they always use this request function.

```
Owner orchestrator
0:    publishData(data, PPID)
1:    begin
2:        privacyPolicy = policyManager.get(PPID);
3:        userPP = policyManager.getUserPrivacyPolicy(PPID);
4:        key = keyManager.createKey(data.dataref, privacyPolicy.purpose,
                                          privacyPolicy.operation);
5:        cipherData = cipherManager.encode(data, privacyPolicy);
6:        storageManager.distributedPut(data, cipherData, PPID,
                          userPP, privacyPolicy.ACL, owner, key);
7:        key2 = keyManager.createKey(null, privacyPolicy.purpose,
                                          privacyPolicy.operation);
8:        storageManager.addRef(key2, data.getDataref(), privacyPolicy.ACL);
9:    end;


Owner storage Manager
10:    distributedPut(data, cipherData, PPID, userPP, ACL, owner, key);
11:    begin
12:        dataTable.localSave(data);
13:        privateDataTable.localSave(data.dataRef, PPID);
14:        dataP2P = createDataP2P(cipherData, userPP, ACL, owner);
15:        DHT.put(key, dataP2P);
16:    end;
```

Fig. 7.7 Algorithm of the publishData() function

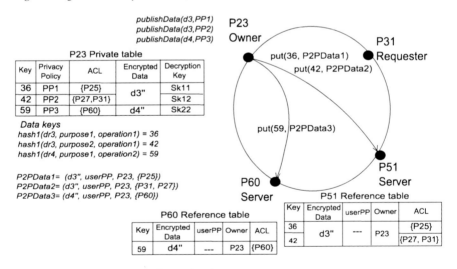

Fig. 7.8 Example of the publishData() function

Figure 7.10 illustrates the data requesting algorithm on the example where encrypted data are published (see Fig. 7.8). P31 requests data for purpose2 and operation1 that corresponds to key 42, so the storage manager does a *get*(42). The peer which identifier is equal to or follows 42 is P51. P51 returns P23 which is

```
Requester orchestrator
0:    data request(dataRef, purpose, operation)
1:    begin
2:        data = null;
3:        key = keyManager.createKeyPriServ(dataRef, purpose,
                                            operation, requester);
4:        dataP2P = storageManager.distributedGet(key);
5:        if (dataP2P contains cipher data) then
6:            checksum = dataCheckSumManager.getCheckSum(dataP2P.cipherData);
7:            ownerData = dataP2P.owner.retrieve(key, requester, checksum);
8:            if (ownerData contains a cipherKey) then
9:                data = cipherManager.decode(dataP2P.cipherData,
                                            ownerData.cipherKey);
10:           else
11:             if (ownerData contains data) then
12:                data = ownerData.data;
13:              end if;
14:           end if;
15:         end if;
16:       if (dataP2P contains a data reference) then
17:         data = dataP2P.Owner.retrieve(key, requester);
18:       end if;
19:       return data;
20:   end;

Requester storage manager
21:    dataP2P distributedGet(key)
22:    begin
23:        dataP2P = DHT.get(key);
24:        return dataP2P;
25:    end;
```

Fig. 7.9 Algorithm of the request() function

the owner peer of 42 and the encrypted data corresponding to 42. In this example, we consider that P51 misbehaves and returns a corrupted data $cr'3$. P31 calculates a checksum of the received data, then it contacts directly P23 to retrieve the decryption key corresponding to 42 ($retrieve(42, P31, checksum)$). P23 verifies the information contained in the privacy policy of 42 (PP2). In this example, consider that the trust table of P23 contains the trust level of P31 that we suppose is 0.7. As the trust level of P31 is higher than the level required in PP2 that is 0.6 (see Table 7.6), the access is granted. P23 calculates the checksum of the data corresponding to 42 and compares it to the checksum value sent by P31. P23 finds out that the checksums are not equal and deduces that someone has misbehaved. Since P31 has a corrupted data, P23 sends the encrypted data with the encryption key.

During the request process, the servers do an access control based on the ACL sent by the owner during the publishing process. The data owner is always contacted by the requester to request either the data content (if only references

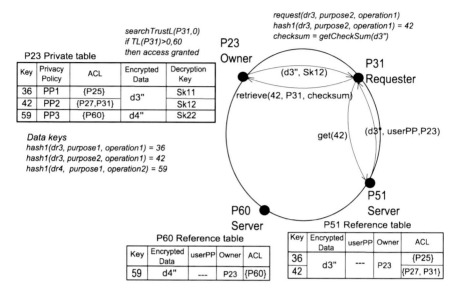

Fig. 7.10 Example of the request() function

have been published) or the decryption key (when encrypted data content have been published). Before retrieving data, owners check the trust level of requesters as you will see in the function searchTrustLevel().trust level does not exists locally, they make a limited flooding among some friend peers (or known peers). contacting the owner, requesters generate the digital checksum of the encrypted data sent by the server (by using a known hash function for example). This signature is sent to the owner which can verify if the server has attacked the integrity of the data.

To use this request function, it is necessary to know the references of available data in the system. This information is maintained in an index that can be centralized or distributed (see Sect. 7.2). A centralized index represents a point of failure and potential bottlenecks. Besides, it implies to trust one single peer (or server) that has the control and the responsibility of maintaining this important meta-information. We argue that when preserving privacy, the distribution of control is essential, which is why PriServ implements a distributed index.

DataRefList dataRefSearch(purpose, operation). PriServ implements the purpose-based reference searching function of PriMod. The index represented by PBDRT (see Table 7.8) in PriServ is implemented in a distributed way by using the DHT organization. The couple (*purpose, operation*) is hashed to create the keys of the index. Keys are assigned to peers that are responsible for maintaining information about the peers that can request data for the purpose and operation represented by the key. The hash function used to produce these keys may be different from the one used to publish data. Thus, each peer maintains a PBDRT of all the keys for which its id is the closest in the DHT organization, which gives a partial view of the global index.

```
             Owner storage manager
             0:    addRef(key, dataRef, ACL)
             1:    begin
             2:        dataP2P=createDataP2P(dataRef, ACL);
             3:        DHT.put(key, dataP2P);
             4:    end;
```

Fig. 7.11 Algorithm of the addRef() function

```
Requester orchestrator
0:    dataRefList dataRefSearch(purpose, operation)
1:    begin
2:        dataRefList = null;
3:        key = keyManager.createKeyPriServ(null, purpose, operation,
                                                requesterID);
4:        dataRefList = storageManager.getList(key);
5:        return dataRefList;
6:    end;

Requester storage manager
7:    dataRefList getList(key)
8:    begin
9:        dataRefList = DHT.get(key);
10:       return dataRefList;
11:   end;
```

Fig. 7.12 Algorithm of the dataRefSearch() function

During the publishing process, the key for the purpose and operation is created and a new data reference is added to the PBDRT index (a) when publishing references (lines 6–7 of Fig. 7.5) and (b) when publishing encrypted data (lines 7–8 of Fig. 7.7). The addRef() function used by owners during publishing follows the algorithm of Fig. 7.11.

When the requester orchestrator receives the dataRefSearch() call from the application, it uses the algorithm shown in Fig. 7.12. The requester orchestrator asks the key manager to create a key by hashing the purpose and the operation. Then, it asks its storage manager to obtain the data references it has access rights for the specified purpose and operation from the P2P system. The storage manager gets the data reference list corresponding to the key by invoking the get() function of the DHT.

Figure 7.13 illustrates the dataRefSearch() function. P25 requests data references it has right access for *purpose1* and *read* that corresponds to key 37 (*get(37)*). The peer whose identifier is equal to or follows 37 is P51. P51 returns the data reference list {*dr1, dr8*} corresponding to key 37 and requester P25. You can see that P25 can request not only dr1 for which owner is P23 but also dr8 for which owner is not shown in the figure.

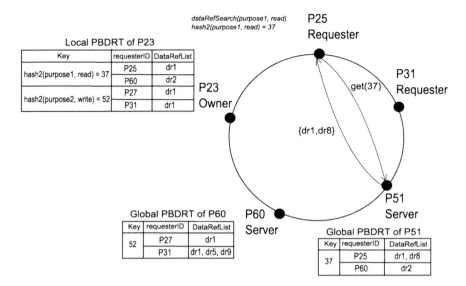

Fig. 7.13 Example of the dataRefSearch() function

```
Owner orchestrator
0:    dataRefPut()
1:    begin
2:        for each key i contained in the localPBDRT do
3:            contentKey = localPBDRT.getContent(key);
4:            storageManager.putIndexContent(contentKey, key);
5:        end for;
6:    end;

Owner storage manager
7:    putIndexContent(contentKey, key)
8:    begin
9:        dataP2P=createDataP2P(contentKey);
10:       DHT.put(key, dataP2P);
11:   end;
```

Fig. 7.14 Algorithm of the dataRefPut() function

To optimize the update of the index, a periodic publication of references can be done by using the dataRefPut() algorithm shown in Fig. 7.14. The idea is that if an owner publishes many data for the same (*purpose, operation*), only one update is done. For that, a local PBDRT is maintained and *flushed* periodically. With this periodical update, lines 6–7 of Fig. 7.5 (resp. lines 7–8 of Fig. 7.7) should be suppressed. See the local PBDRT of P23 in Fig. 7.13.

The last function that PriServ implements focuses on searching the trust level of requesters. PriServ uses trust levels to make the final decision of sharing or not data. The trust level reflects a peer reputation wrt other peers. A peer can have different trust levels at different peers. The peer reputation influences its trust level. Peers which are suspicious have lower trust level than peers considered as honest. A peer can have locally the trust levels of some well-known peers or peers it has interacted with. If a peer P does not have a particular trust level it can ask for it to its *friends*. A friend is a peer considered as honest from the point of view of P and the number of friends can vary from one peer to another.

The implementation of the searchTrustLevel() function does not use the DHT organization but uses an unstructured overlay. Thus, this function has been redefined to take into account a Time To Live (TTL).

TrustLevel searchTrustLevel(requesterID, nestedLevel). Trust levels are considered in the range [0..1]. A peer with a trust level of 1 is completely trustworthy. A peer with a trust level of 0 has very bad reputation. During requesting, if the trust manager of the data owner has the trust level of the requester it does not have to contact other peers. Otherwise, PriServ defines three methods for searching the requester trust level. Choosing one of them depends on the number of friends of the owner. Briefly, the three methods are explained below, and only the algorithm of the first one is presented. More details can be found in [16].

With-Friends Algorithm. This version of the searchTrustLevel function considers that each peer has at least one friend (Fig. 7.15). With this assumption, the data owner asks its friends for the trust level of the requester. Each received trust level (*RTL*) is weighted with the trust level (*FTL*) of the sending friend. The final trust level is computed from the received trust levels as the average, the maximum, the minimum, etc. This searching is recursive. If a friend does not have the requested trust level, it asks for it to its friends and the number of nested levels (*nestedLevel*) is incremented. Recursion is limited to *maxDepth*. The maximum number of contacted friends can also be limited to a predefined number.

Without-Friends Algorithm. In this algorithm, we consider that peers have no friends. In this case, data owners ask for the trust level of the requester to the subset of known peers from the DHT, i.e., their finger table.

With-or-Without-Friends Algorithm. Here, peers may have friends or not and priority is given to ask for trust levels to friends. If a data owner has some friends, it asks them for the trust level by using the with-friends algorithm, else it asks the peers in its finger table by using the without-friends algorithm.

7.5.3 PriServ Validation

We validated PriServ in three steps, with costs analysis, simulations, and implementation of a Java prototype. The costs analysis and results of simulation with SimJava are presented next, the prototype is presented in Sect. 7.6.

```
0:   trustLevel searchTrustLevel(requesterID, nestedLevel)
1:   begin
2:      requesterTrustLevel = 0;
3:      if (nestedLevel has reached maxDepth) then
4:         if (trustLevel of requesterID in trustTable) then
5:            requesterTrustLevel=trustLevel of requesterID;
6:         else
7:             return -1;
8:         end if;
9:      else
10:         if (trustLevel of requesterID in trustTable) then
11:            requesterTrustLevel=trustLevel of requesterID;
12:         else
13:            nestedLevel is incremented;
14:            nbPeersContacted = 0;
15:            for each friend do
16:               FTL = trustLevel of friendID;
17:               RTL = friendTrustManager.searchTrustLevel(requesterID,
                                                            nestedLevel);
18:               if (RTL != -1) then
19:                   FTL*RTL is added to requesterTrustLevel;
20:                   nbPeersContacted is incremented;
21:               end if;
22:            end for;
23:            requesterTrustLevel = requesterTrustLevel/
                                         nbPeersContacted;
24:         end if;
25:      end if;
26:      return requesterTrustLevel;
27:   end;
```

Fig. 7.15 Algorithm of the searchTrustLevel() function: with-friends version

Publishing Costs. Publishing data in the system conserves the logarithmic cost of the traditional put function. By using the DHT, $O(\log N)$ messages are needed to publish each key. In PriServ, the number of keys is equal to the number of entries (ept) of the private data table. Additional costs induced by the cipher key generation and the data encryption are negligible wrt the network costs. Thus, the publishing cost is:

$$C_{\text{Publish}} = \sum_{i=1}^{ept} O(\log N) = O(ept * \log N)$$

The maximum value of ept is equal to the number of shared data ($nbData$) multiplied by the number of purposes ($nbPurpose$) multiplied by the number of operation (*nbOperation*). At worst, each data item is shared for all purposes and all operations:

$$C Max_{Publish} = O(nbData * nbPurpose * nbOperation * \log N)$$

We can see that the number of purposes and operations affects the publishing cost. Previous studies have shown that considering ten purposes allows to cover a large number of applications [24, 30]. Used with ten purposes (by data item) and three operations (read, write, and disclosure), PriServ incurs a small overhead. Overall, the publishing cost remains logarithmic.

Requesting Costs. Concerning the requesting cost, it is the addition of two costs: get() cost and the retrieving cost. We disregard access control, checksum calculation, and decryption costs, which are negligible wrt network costs. The get() cost is in $O(\log N)$ and the server returns its answer in one message. For data retrieval, a requester needs one additional message to contact the data owner that answers in another message.

To summarize, the requesting cost is:

$$C_{Requesting} = C_{DHTGet} + C_{Retrieving}$$

$$= O(\log N) + 1 + 2$$

$$= O(\log N) + 3$$

$$= O(\log N)$$

Trust Level Searching Cost. The trust level searching cost (C_{STL}) depends on the trust searching algorithm:

- *With-friends algorithm.* In this case, the owner sends a message to each of its friends that in turn do the same in a nested search. This cost depends on the number of friends (NF) and the maximum depth of the nested search (MaxDepth).

$$C_{STL_{WF}} = \sum_{i=1}^{MaxDepth} NF^i = O(NF^{MaxDepth})$$

- *Without-friends algorithm.* In this case, the owner sends a message to each of the peers in its finger table, which in turn do the same in a nested search. This cost depends on the number of fingers, which is $\log N$, and the maximum depth of the nested search (MaxDepth).

$$C_{STL_{WOF}} = \sum_{i=1}^{MaxDepth} (\log N)^i = O((\log N)^{MaxDepth})$$

- *With-or-without-friends algorithm.* In this case, if the owner has friends, it sends a message to each of its friends. Otherwise, it sends a message to each of the peers in its finger table. A peer contacted by an owner does the same in a nested search. The trust level searching cost depends on the number of friends (NF), the number of fingers, which is $\log N$, and the maximum depth of the nested search (MaxDepth).

$$C_{\text{STL}_{\text{WWF}}} = O((\max(\log N, NF))^{\text{MaxDepth}})$$

The trust level searching cost C_{STL} can be one of the three costs $C_{\text{STL}_{\text{WF}}}$, $C_{\text{STL}_{\text{WOF}}}$, or $C_{\text{STL}_{\text{WWF}}}$. Note that if $NF > \log N$, $C_{\text{STL}_{\text{WWF}}}$ is equal to $C_{\text{STL}_{\text{WF}}}$, else it is equal to $C_{\text{STL}_{\text{WOF}}}$. In all cases $C_{\text{STL}_{\text{WWF}}}$ can be used for C_{STL}:

$$C_{\text{STL}} = O((\max(\log N, NF))^{\text{MaxDepth}})$$

To summarize,

$$C_{\text{publishing}} = O(nbData * nbPurpose * nbOperation * \log N)$$

$$C_{\text{Requesting}} = C_{\text{Request}} + C_{\text{Retrieve}} = O(\log N)$$

$$C_{\text{STL}} = O((\max(\log N, NF))^{\text{MaxDepth}})$$

For the simulation, we used SimJava [13] and the Chord protocol was simulated with some modifications in the put() and get() functions. Tests consider N peers, peer keys are selected randomly between 0 and 2^n. N is set to 11, which corresponds to 2^{11} peers. This number of peers is enough to simulate collaborative applications like the medical one. MaxDepth is set to 11 and the number of friends is set to 2.

Trust level searching introduces a large overhead because of flooding in the unstructured network. Figure 7.16 compares the three algorithms seen above. The with-friends case introduces the smallest cost while the without-friends case introduces the highest cost. However, intuitively, the probability to find the trust level is higher in the without-friends algorithm than in the with-friends algorithm. This is due to the fact that the number of contacted peers is higher in the without-friends algorithm, which increases the probability to find the trust level. We estimate that the with-or-without-friends algorithm is the most optimized because it is a trade-off between the probability to find the requester trust level and the trust level searching cost.

Figure 7.17 shows that the trust level searching cost decreases with the number of requests and stabilizes. When peers ask for a trust level, answers are returned in the requesting order and the trust tables are updated with the missing trust level. Thus, the trust tables evolve with the number of searches. After a while, these tables stabilize. Thus, the number of messages for searching trust levels is reduced to a

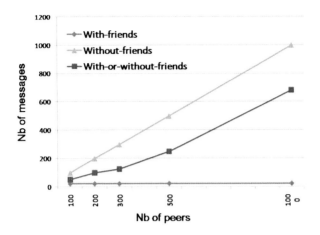

Fig. 7.16 Comparison of the three algorithms of trust level searching

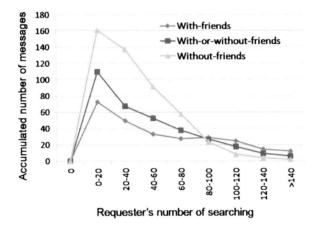

Fig. 7.17 Stabilization of the cost of trust level searching

stable value. This value is not null because of the dynamicity of peers. Simulations consider that the number of peers joining the system is equal to those leaving the system. Thus, there are always new peers which do not know the requester trust level. We also observe in the figure that the trust level searching cost in the without-friends algorithm stabilizes first. This is due to the fact that a larger number of peers are contacted. The with-or-without-friends algorithm comes in second place, and the with-friends algorithm comes last. As can be seen in the comparison of the three algorithms, we find again that the with-or-without-friends algorithm is the most optimized because it is a trade-off between the time to stabilization and the trust level searching cost.

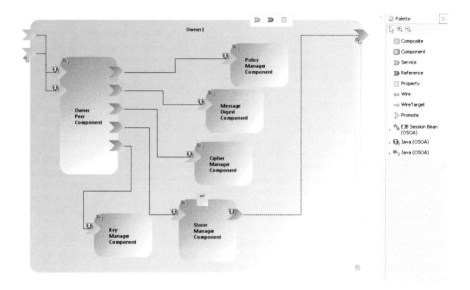

Fig. 7.18 Owner peer modeled with SCA

7.6 PriServ Prototype

A prototype of PriServ for privacy-preserving data sharing applications for online communities was developed [19]. The prototype uses the Java language, SCA (Service Component Architecture) tools,[16] and RMI (Remote Method Invocation). Figure 7.18 shows the owner peer implementation with SCA.

PriServ was tested and validated by PeerUnit[17] on Grid5000.[18] Grid5000 is a scientific instrument for the study of large-scale parallel and distributed systems. The tests were done on a population of 180 peers on 42 Grid5000 nodes. Several results have validated the performance of the prototype concerning:

- Respect of privacy policies of owners. PriServ limits data access if a requester does not intend to respect the owner privacy policy. The test results show the failure of the request each time the access purpose or operation is different from the intended purpose and operation specified while publishing data.
- Limited access only for authorized requester. The test results show that 100 % of queries from an unauthorized requesters are systematically rejected by servers.

---

[16]http://www.oasis-opencsa.org/sca
http://www.obeo.fr/pages/sca/.

[17]http://peerunit.gforge.inria.fr/.

[18]http://www.grid5000.fr/.

- Traceability of the data distribution. The test results show that it is possible to review, automatically, all the requesters who had access to data, and to check the list of those who have tried to access them. This allows the data owner to be sure that its data privacy requirements are respected.

7.6.1 Medical PPA

Privacy-Preserving Applications (PPA) manage sensitive data (financial projects, unpublished research results, patients records, etc.). PriServ is illustrated with the collaborative medical research application of Sect. 7.1. The participants of this applicaiton are scientists, doctors, and patients. In order to control disclosure of sensitive information (e.g., medical records owned by doctors, research results owned by scientists) without violating privacy, data access should respect the privacy preferences of their owners. In this medical PPA:

- Patients and doctors who own and manage private medical records can be considered as owners. Scientists who may use medical records for scientific research can be considered as requesters. Servers are peers of the storage system.
- Doctors may define the privacy preferences of patients in privacy policies and attach them to their medical records. For instance, a doctor may allow writing access on her information to scientists for adding comments on her diagnosis.
- Scientists may define their own privacy preferences and attach them to their research results. For instance, a scientist may allow reading access on her results to doctors for giving diagnosis.

  PriServ is used for these scenarios:

- After defining their privacy preferences on their medical records, doctors and patients can publish their data in the system while preserving their privacy.
- Scientists can search for data by using procedures that respect the privacy preferences of the data owners.

  Through a GUI, we show scenarios that exhibit important aspects of private data management: privacy policy management, data publishing, data searching, and data reference searching.

  Figure 7.19 shows the interface through which the DHT is launched. Once launched, the DHT creates server peers whose number can be specified. Then the main user creates owners and requesters by specifying their names and clicking on *create user*. The interface of each user newly created will appear automatically. For instance, Fig. 7.20 shows the user interface of DoctorHouse.

  Scenarios. The key features of the prototype, are demonstrated through the following scenarios of the medical PPA:

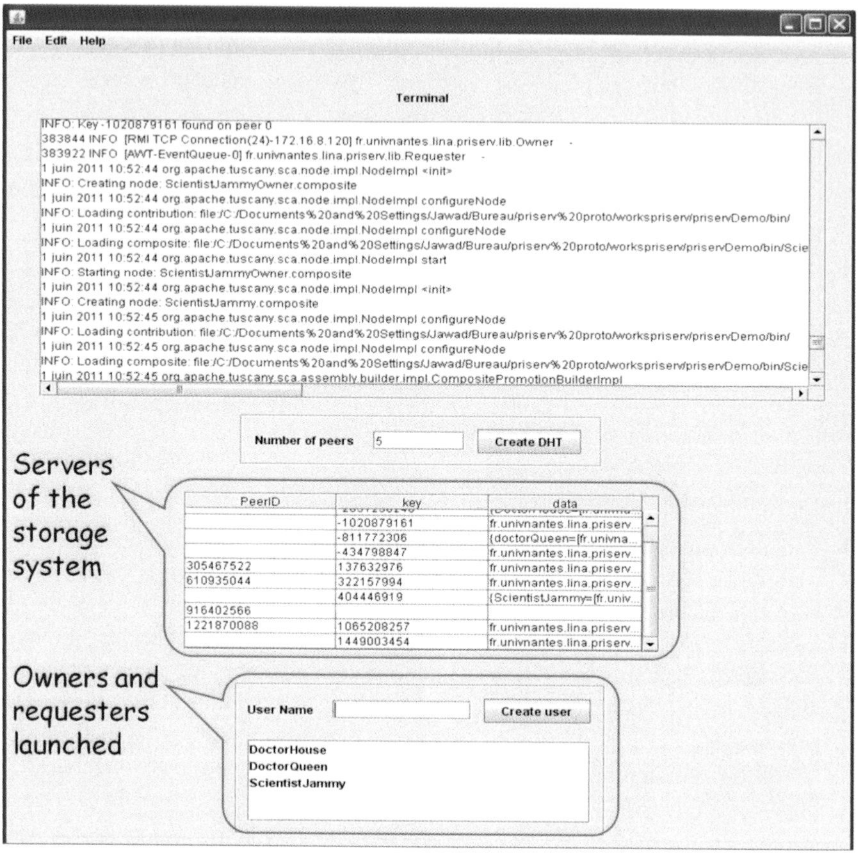

Fig. 7.19 Dashboard of the storage system (DHT)

Privacy Policies Management. This scenario is used to show how DoctorHouse (an owner) can specify his privacy preferences by defining his own privacy policies (see Fig. 7.21). He is also able to attach different policies to a datum in order to control the access to his data. He also has an interface that shows information about his privacy policies.

Data Publishing. This scenario shows how DoctorHouse can publish sensitive data in the P2P system (see Fig. 7.22a). He can specify for which privacy policy his data will be published. He also has the choice between publishing encrypted data or data references. He can also have a view of his published data and the policies attached to them.

Data Searching. This scenario shows how ScientistJammy can search for data (see Fig. 7.22b). He has a choice between: (a) a local search on his own local data, (b) a P2P data requesting. Searching is made only if he selects the purpose and the operation for the requested data.

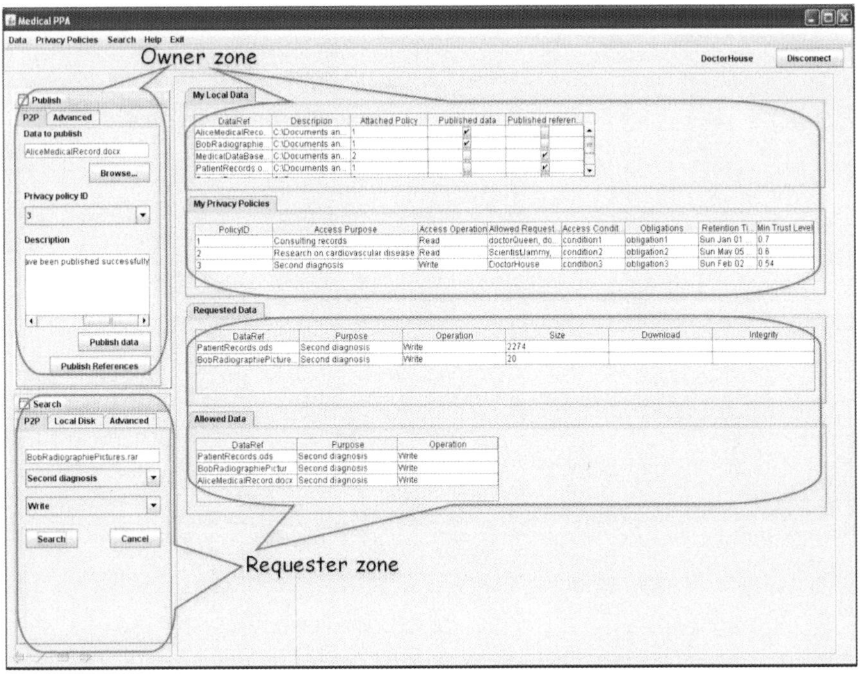

Fig. 7.20 GUI of the medical PPA that uses PriServ

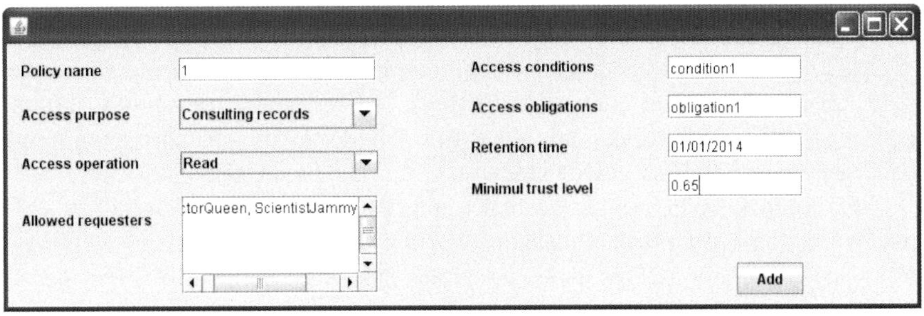

Fig. 7.21 Privacy policy specification interface

Data Reference Searching. This scenario shows how ScientistJammy can search for data references for particular purpose and operation (see Fig. 7.22c). For this, we show that he will be able to have a list of references for data which he can access for particular purpose and operation. Then, he can access data content by specifying the data reference that he has gotten, the purpose and the operation in the data searching interface.

The PriServ web site is https://sites.google.com/site/gddlina/priserv and the prototype code can be found at http://sourceforge.net/projects/priserv/.

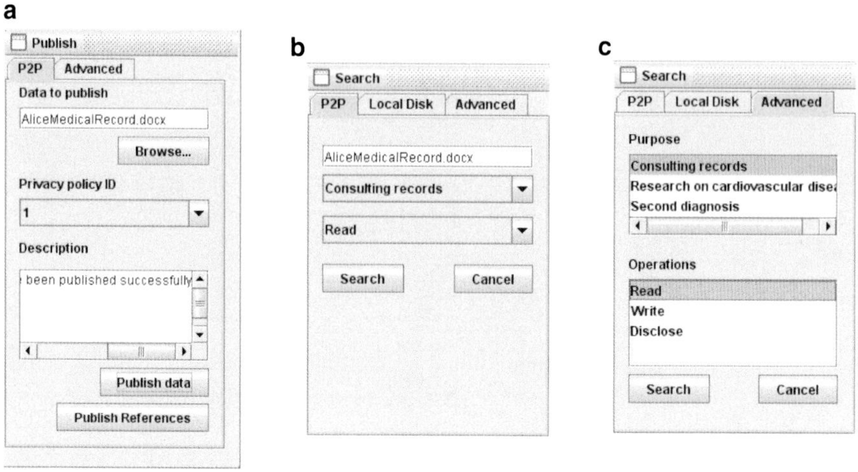

Fig. 7.22 Publishing and requesting interfaces. (**a**) Data publishing interface (**b**) Data requesting interface (**c**) Data reference searching interface

7.7 Conclusion

This chapter gave an overview of current solutions for supporting data privacy in P2P systems. Our evaluation of existing solutions is made based on the techniques used to protect privacy of data and users (i.e., access control, anonymity, trust, and cryptography) and the guaranteed privacy properties (i.e., protection against unauthorized reads, corruption and deletion, limited disclosure, anonymity guarantees, denial of linkability, and content deniability). This analysis showed that while the notion of purpose of Hippocratic databases (HDB) is gaining more attention, in particular, because of the OECD guidelines, it has not been used in P2P systems.

Then we developed in more detail a complete solution (Primod and Priserv) for data privacy in P2P systems that supports the notion of purpose of HDB. PriMod, a privacy model for P2P systems, integrates the purposes notion as mainspring. Purposes are omnipresent in several process of sensitive data management. Data owners specify, through personal privacy policies, the access purpose for their data. Data publication attaches the allowed access purposes. Data requesters specify the access purpose in their requests, thus they are committed to their intended and expressed use of data.

PriServ is a privacy service that implements PriMod. The PriServ prototype combines purpose and operation-based access control, trust techniques, cryptography techniques, and digital checksums. A privacy-preserving data sharing application for online social networks illustrates this approach.

Several improvements can be made to PriMod and PriServ. The purpose-based index should be anonymized to avoid servers to know the partial view of the index they store. A more semantically rich query language may also be proposed. But above all, auditing solutions should be proposed to verify compliance of data use with the specified privacy preferences. This is still an open and challenging issue.

References

1. Agrawal, R., Bird, P., Grandison, T., Kiernan, J., Logan, S., Rjaibi, W.: Extending relational database systems to automatically enforce privacy policies. In: IEEE Conference on Data Engineering (ICDE), Tokyo, Japan (2005)
2. Agrawal, R., Kiernan, J., Srikant, R., Xu, Y.: Hippocratic databases. In: Very Large Databases (VLDB), Hong Kong, China (2002)
3. Akbarinia, R., Martins, V., Pacitti, E., Valduriez, P.: Design and implementation of APPA. In: Baldoni, R., Cortese, G., Davide, F. (eds.) Global Data Management. IOS Press, pp. 98–123 (2006)
4. Byun, J.W., Li:, N.: Purpose based access control for privacy protection in relational database systems. Very Large Databases (VLDB) J. 17(4) (2008)
5. Castro, M., Druschel, P., Ganesh, A., Rowstron, A., Wallach, D.S.: Secure routing for structured peer-to-peer overlay networks. In: Operating Systems Design and Implementation (OSDI), Boston, MA (2002)
6. Chaum, D.L.: Untraceable electronic mail, return addresses, and digital pseudonyms. Comm. ACM 24(2) (1981)
7. Choffnes, D.R., Duch, J., Malmgren, D., Guierma, R., Bustamante, F.E., Amaral, L.: Swarm-Screen: privacy through plausible deniability in P2P systems. Tech. rep., Northwestern EECS University (March 2009)
8. Clarke, I., Miller, S.G., Hong, T.W., Sandberg, O., Wiley, B.: Protecting free expression online with freenet. IEEE Internet Comput. 6(1) (2002)
9. Cranor, L., Langheinrich, M., Marchiori, M., Presler-Marshall, M., Reagle, J.: The Platform for Privacy Preferences 1.0 (P3P1.0) Specification (2002)
10. Daswani, N., Garcia-Molina, H., Yang, B.: Open problems in data-sharing peer-to-peer systems. In: International Conference on Database Theory (ICDT), Siena, Italy (2003)
11. Garton, L., Haythornthwaite, C., Wellman, B.: Studying online social networks. J. Comput. Mediat. Comm. 3(1) (1997)
12. Hand, S., Roscoe, T.: Mnemosyne: peer-to-peer steganographic storage. In: International Peer To Peer Systems Workshop (IPTPS), Cambridge, MA (2002)
13. Howell, F., McNab, R.: Simjava: a discrete event simulation library for Java. In: International Conference on Web-Based Modeling and Simulation, San Diego, CA (1998)
14. Isdal, T., Piatek, M., Krishnamurthy, A., Anderson, T.: Privacy-preserving P2P data sharing with oneswarm. Tech. rep., University of Washington (2009)
15. Jawad, M.: Data privacy in P2P systems. Ph.D. thesis, Université de Nantes (2011)
16. Jawad, M., Serrano-Alvarado, P., Valduriez, P.: Protecting data privacy in structured P2P networks. In: Data Management in Grid and P2P Systems (Globe), Linz, Austria (2009)
17. Jawad, M., Serrano-Alvarado, P., Valduriez, P., Drapeau, S.: A data privacy service for structured P2P systems. In: Mexican International Conference in Computer Science (ENC), México D.F., México (2009)
18. Jawad, M., Serrano-Alvarado, P., Valduriez, P., Drapeau, S.: Data privacy in structured P2P systems with PriServ. In: Bases de Données Avancées (BDA), Namur, Begium (2009)
19. Jawad, M., Serrano-Alvarado, P., Valduriez, P., Drapeau, S.: Privacy support for sensitive data sharing in P2P systems. In: Bases de Données Avancées (BDA), demonstration paper, Rabat, Morocco (2011)
20. Karjoth, G., Schunter, M., Waidner, M.: Platform for enterprise privacy practices: privacy-enabled management of customer data. In: Workshop on Privacy Enhancing Technologies, San Francisco, CA (2002)

21. Kleinberg, J., Papadimitriou, C.H., Raghavan, P.: On the value of private information. In: Theoretical Aspects of Rationality and Knowledge (TARK), Siena, Italy (2001)
22. Kubiatowicz, J., Bindel, D., Chen, Y., Czerwinski, S.E., Eaton, P.R., Geels, D., Gummadi, R., Rhea, S.C., Weatherspoon, H., Weimer, W., Wells, C., Zhao, B.Y.: OceanStore: An architecture for global-scale persistent storage. In: Architectural Support for Programming Languages and Operating Systems (ASPLOS), Cambridge, MA (2000)
23. Langheinrich, M.: A P3P Preference Exchange Language (APPEL1.0) Specification (2001)
24. Liberty Alliance Project, Privacy Preference Expression Languages (PPELs). http://projectliberty.org/liberty/content/download/371/2670/file/Final_PPEL_White_Paper.pdf
25. LeFevre, K., Agrawal, R., Ercegovac, V., Ramakrishnan, R., Xu, Y., DeWitt, D.J.: Limiting disclosure in hippocratic databases. In: Very Large Databases (VLDB), Toronto, Canada (2004)
26. Mell, P., Grance, T.: The NIST definition of cloud computing. Natl. Inst. Stand. Tech. **53**(6) (2009)
27. Miklau, G., Suciu, D.: Controlling access to published data using cryptography. In: Very Large Databases (VLDB), Berlin, Germany (2003)
28. Nejdl, W., Wolf, B., Qu, C., Decker, S., Sintek, M., Naeve, A., Nilsson, M., Palmér, M., Risch, T.: Edutella: A P2P networking infrastructure based on RDF. In: ACM World Wide Web Conference (WWW), Hawaii, USA (2002)
29. Ozsu, M.T., Valduriez, P.: Principles of Distributed Database Systems, 3rd edn. Springer, New York (2011)
30. 1.0 P3P Purposes of Data Collection Elements. http://p3pwriter.com/LRN_041.asp
31. Pfitzmann, A., Hansen, M.: A terminology for talking about privacy by data minimization: anonymity, unlinkability, undetectability, unobservability, pseudonymity, and identity management. Tech. rep., Dresden University of Technology (2009)
32. Roncancio, C., del Pilar Villamil, M., Labbé, C., Serrano-Alvarado, P.: Data sharing in DHT based P2P systems. Trans. Large Scale Data Knowl. Centered Syst. I **5740** (2009)
33. Rowstron, A., House, G.: Storage management and caching in PAST, a large-scale, persistent peer-to-peer storage utility. In: Symposium on Operating Systems Principles (SOSP), Banff, Alberta, Canada (2001)
34. Rowstron, A.I.T., Druschel, P.: Pastry: scalable, decentralized object location, and routing for large-scale peer-to-peer systems. In: ACM/IFIP/USENIX Middleware Conference (MIDDLEWARE), Heidelberg, Germany (2001)
35. Ryu, S., Butler, K., Traynor, P., McDaniel, P.: Leveraging identity-based cryptography for node ID assignment in structured P2P systems. In: Advanced Information Networking and Applications Workshops (AINA), Niagara Falls, Canada (2007)
36. Stoica, I., Morris, R., Karger, D.R., Kaashoek, M.F., Balakrishnan, H.: Chord: a scalable peer-to-peer lookup service for internet applications. In: ACM Conference on Applications, Technologies, Architectures, and Protocols for Computer Communication (SIGCOMM), San Diego, CA (2001)
37. Stubblefield, A., S.Wallach, D.: Dagster: Censorship-Resistant Publishing without Replication. Tech. rep., Rice University (2001)
38. Tatarinov, I., Ives, Z.G., Madhavan, J., Halevy, A.Y., Suciu, D., Dalvi, N.N., Dong, X., Kadiyska, Y., Miklau, G., Mork, P.: The piazza peer data management project. ACM Spec. Interest Group Manag. Data (SIGMOD) Rec. **32**(3) (2003)
39. Waldman, M., Mazières, D.: Tangler: a censorship-resistant publishing system based on document entanglements. In: Computer and Communications Security (CCS), Philadelphia, PA (2001)
40. Westin, A.F.: Privacy and Freedom. Atheneum, New York (1967)
41. Zhao, B.Y., Huang, L., Stribling, J., Rhea, S.C., Joseph, A.D., Kubiatowicz, J.: Tapestry: a resilient global-scale overlay for service deployment. IEEE J. Sel. Areas Comm. **22**(1) (2004)

Chapter 8
Privacy Preserving Reputation Management in Social Networks

Omar Hasan and Lionel Brunie

Abstract Reputation management is a powerful security tool that helps establish the trustworthiness of users in online applications. One of the most successful uses of reputation systems is on e-commerce web sites such as eBay.com and Amazon.com, which use reputation systems to root out fraudulent sellers. Reputation systems can also play an important role in social networks to enforce various security requirements. For example, a reputation system can help filter fake user profiles. However, a major challenge in developing reputation systems for social networks is that users often hesitate to publicly rate fellow users or friends due to the fear of retaliation. This trend prevents a reputation system from accurately computing reputation scores. Privacy preserving reputation systems hide the individual ratings of users about others and only reveal the aggregated community reputation score thus allowing users to rate without the fear of retaliation. In this chapter, we describe privacy preserving reputation management in social networks and the associated challenges. In particular, we look at privacy preserving reputation management in decentralized social networks, where there is no central authority or trusted third parties, thus making the task of preserving privacy particularly challenging.

8.1 Social Networks and Relationships

We take a look at the key social concepts of social networks and social relationships. In particular, we discuss the nature of social relationships by identifying the various attributes that characterize them.

O. Hasan (✉) · L. Brunie
University of Lyon, CNRS, INSA-Lyon, LIRIS, UMR5205, F-69621, France
e-mail: omar.hasan@insa-lyon.fr; lionel.brunie@insa-lyon.fr

R. Chbeir and B. Al Bouna (eds.), *Security and Privacy Preserving in Social Networks,*
Lecture Notes in Social Networks, DOI 10.1007/978-3-7091-0894-9_8,
© Springer-Verlag Wien 2013

8.1.1 Social Networks

A social network is a composition of nodes and the relationships between them. The nodes in a social network may be individuals or collectives of individuals. The relationships between nodes are founded on human ties such as friendship, membership in the same family or organization, mutual interests, common beliefs, trade, exchange of knowledge and geographical proximity.

8.1.2 Characteristics of Social Relationships

The most commonly discussed characteristics of social relationships include *roles*, *valence*, *provenance*, *history*, and *strength* [43].

Roles.   A social relationship is defined by the roles that are associated with it. For example, the roles of employer and employee define the relationship of employer–employee in a professional setting. The same pair of nodes may take on different roles in a parallel relationship. For example, an employer–employee relationship may be complemented by a neighbor–neighbor relationship.

Valence.   A social relationship can have positive, negative, or neutral sentiments associated with it. For example, an individual may like, dislike, or be apathetic towards another individual.

Provenance.   Some attributes of a social relationship may be asymmetric, that is, perceived differently by the individual participants of the relationship. For example, a sentiment of *like* from one node may not be reciprocated by the other node in the relationship.

Relationship history.   Social relationships have a temporal dimension. A social relationship may evolve with time through interactions or the absence thereof. The history of a social relationship can be considered as an indicator of the current and future status of the relationship. For example, a long positive relationship in the past is likely to be followed by a positive relationship in the present and in the near future.

Strength.   Strength of a tie (or social relationship) is a quantifiable property that characterizes the link between two nodes [50]. The notion of tie strength was first introduced by sociologist Mark Granovetter in his influential paper "The strength of weak ties" [27] published in 1973. He defined the strength of a tie as a "combination of the amount of time, the emotional intensity, the intimacy (mutual confiding), and the reciprocal services which characterize the tie" [27]. The strength of a social relationship is a complex construct, which is itself composed of several properties of social relationships. We discuss the strength of social relationships in detail in the following section.

8.1.3 Strength of Social Relationships

Granovetter proposed four dimensions of tie strength: *amount of time, intimacy, intensity,* and *reciprocal services* [25, 27]. A number of researchers (including Burt [9], Wellman and Wortley [58], Lin et al. [40], Marsden [41]) have since studied the dimensions of tie strength and have refined and expanded the original list of four. The existing literature suggests seven dimensions of tie strength: *intensity, intimacy, duration, reciprocal services, structural factors, emotional support,* and *social distance* [25].

In a study on predicting tie strength between individuals based on their exchanges on social networking sites [25], Gilbert and Karahalios have identified a number of indicators that predict tie strength belonging to each of the seven dimensions. In a study with similar goals, Petroczi et al. [50] have developed a set of questions that they pose the members of a virtual social network in order to establish the strength of ties between them. In the following list, we discuss each of the seven dimensions of tie strength as well as some associated indicators and questions that yield tie strength.

Intensity. The indicators of the intensity of a tie strength include the *frequency of contact* and the *amount of information exchanged* between two nodes.

Homans presented the argument in his 1950 book "The Human Group" that "the more frequently the persons interact with one another, the stronger their sentiments of friendship for one another are apt to be" [27, 34].

Gilbert and Karahalios [25] use the amount of information exchanged (for example, the number of words and messages exchanged) on a social networking site as an indicator of the intensity of the tie strength between individuals.

Intimacy. Mutual confiding (or trust) is an indicator of the intimacy and the strength of a social tie [27, 41, 50]. Sociologist Diego Gambetta [24] characterizes trust as contextual and quantifiable as subjective probability.

Petroczi et al. [50] ask the members of an online discussion forum the following question in order to determine the trust and consequently the tie strength between them: "Which participants do you trust (for example, they know your real name, email address, password to your introduction sheet)?".

Gilbert and Karahalios [25] use the variable "Relationship status," with the possible values of *single, in relationship, engaged,* and *married,* as an indicator of the intimacy of two individuals. Other variables that they use as indicators of intimacy include "Distance between hometowns," "Appearances together in photos," and "Days since last communication."

Duration. The duration or the span of the relationship is considered as an indicator of the strength of the relationship.

Gilbert and Karahalios [25] use the variable "Days since first communication" on social networking sites as a proxy for the length of the relationship between two individuals.

Reciprocal services. A social relationship is stronger if it is reciprocated by both participants. For example, a sentiment of *like* shared by both nodes would result in a strong social relationship.

Gilbert and Karahalios [25] use *the number of links and applications mutually shared* between friends as variables quantifying reciprocal services on social networking sites.

Structural factors. Ronald Burt proposed that structural factors shape tie strength, factors like network topology and informal social circles [9, 25].

A structural factor that Gilbert and Karahalios [25] use to predict tie strength is the "Number of mutual friends." They also use structural factors such as membership in common interests groups, and association with the same institutions, organizations, or geographical locations (for example, graduation from the same university, employment in the same company, or residence in a common city).

Emotional support. Wellman and Wortley argue that providing emotional support, such as offering advice on family problems, indicates a stronger tie [25, 58].

To determine the emotional support between the members of a virtual social network, Petroczi et al. [50] ask the members which other members they have requested or they feel they could request for a favor or help. Gilbert and Karahalios [25] monitor emotion words (as identified by the Linguistic Inquiry and Word Count (LIWC) dictionary [49], for example, *birthday, congrats, sweetheart*) exchanged between the members of a social networking site as indicators of emotional support.

Social distance. Lin et al. show that social distance, embodied by factors such as socioeconomic status, education level, political affiliation, race and gender, influences tie strength [25, 40].

Gilbert and Karahalios [25] measure social distance by considering parity in age, occupation, education, political, and religious views of the individuals.

8.2   Trust

Trust is an important indicator of the strength of a social relationship. It inherently takes into account a number of other aspects of a social relationship.

8.2.1   *Modeling Trust*

Sociologist Diego Gambetta [24] proposes the following definition of trust:

> Trust (or, symmetrically, distrust) is a particular level of the subjective probability with which an agent assesses that another agent or group of agents will perform a particular action, both before he can monitor such action (or independently of his capacity ever to be able to monitor it) and in a context in which it affects his own action.

This is one of the seminal definitions that describe trust as a quantifiable construct. Gambetta observes that trust is an agent's degree of belief (the level of subjective probability) that another entity will perform an expected action. An additional important aspect of this definition is the recognition that trust is contextual.

The advantage of Gambetta's model of trust is its quantification of trust as subjective probability. It allows trust to be modeled as a mathematical construct and to be manipulated using the wide range of tools available in probability theory. Moreover, trust modeled with subjective probability is more intuitive than trust modeled with other theories such as subjective logic and fuzzy logic.

8.2.2 Characteristics of Trust

From Gambetta's definition, we can infer that trust has the following characteristics:

Binary-Relational and Directional. According to the definition, "Trust ... is a particular level of the subjective probability with which *an agent* assesses that *another agent or group of agents* will perform a particular action . . . ." From this excerpt, it is evident that trust is a relationship between two entities. Moreover, it is also clear that trust is directional. The first entity is an agent who has trust in a second entity which may be another agent or a group of agents.

Contextual. As given in the definition, "Trust ... is a particular level of the subjective probability with which an agent assesses that another agent or group of agents will perform *a particular action* . . . ." We infer that trust is in the context of "a particular action" that the second entity may perform.

Quantifiable as Subjective Probability. "Trust ... is *a particular level of the subjective probability* with which an agent assesses that another agent or group of agents will perform a particular action, *both before he can monitor such action (or independently of his capacity ever to be able to monitor it)* . . . ." From this excerpt of the definition, we deduce that trust is quantifiable as subjective probability.

We discuss below some other characteristics of trust which are not evident from Gambetta's definition. We provide examples to support their validity as characteristics of trust. These characteristics have been previously identified by several authors (such as Capra [12]).

Non-Reflexive. An agent may or may not trust herself. For example, a patient Alice may trust her doctor to prescribe her the correct medicine, whereas she might not trust herself to do so.

Asymmetric. If an agent Alice trusts an agent Bob, then Bob may or may not trust Alice. For example, in the context of car repair, a car owner Alice may trust her mechanic Bob; however, Bob may not necessarily trust Alice.

Non-Transitive.   If an agent Alice trusts an agent Bob who in turn trusts an agent Carol, then Alice may or may not trust Carol. For example, an email server A might trust an email server B to not send spam. If B trusts an email server C in the same context, then A may or may not trust C depending on various factors such as its strength of trust in B, the availability of additional evidence, etc.

Dynamic.   Trust may change with time. For example, let's say that an online shopper Alice has so far had good experiences with an online vendor Bob and therefore she has high trust in him. However, if her latest transaction with Bob is less than satisfactory, then her trust in Bob is likely to decrease instead of staying constant.

8.2.3   Inferring Trust

There are a number of techniques that enable inferring trust between entities. The first technique that we describe is *direct interaction* that requires explicit input from nodes. The other three methods that we discuss aim to infer trust from existing information.

Direct Interaction

The primitive method of establishing trust in an unknown entity is to directly interact with it and observe its behavior in the desired context. However, this method requires that the entity be trusted at least once without any prior background on that entity. This approach is perhaps suitable for low-risk transactions and in situations when no other recourse is available. However, when reliance on an unknown entity may lead to substantial damage, the other approaches for trust establishment are clearly preferable, since they allow the truster to base his trust on some prior knowledge provided by others.

McKnight et al. [42] introduce the notion of *initial trust*, which is described as the trust in an unfamiliar trustee—a relationship in which the actors do not yet have credible information about, or affective bonds with each other [7].

Trust Recommendation and Propagation

Establishing trust in an unknown entity through trust recommendation and propagation takes advantage of the possible transitivity of trust. Let's say that Alice wishes to establish trust in an unknown individual, Carol. If another individual Bob trusts Carol, then he could give a recommendation to Alice about Carol's trustworthiness. Taking Bob's trust recommendation and her own trust in Bob into account, Alice may establish a trust relationship with Carol. Thus a transitive path of trust that leads from Alice to Bob to Carol enables Alice to develop trust in Carol. If Alice wishes to establish trust in Carol through Bob's recommendation, we say that Bob's trust in Carol has propagated to Alice.

Guha et al. [29] term the above-described one-step propagation as *atomic propagation*. The term stems from the observation that the conclusion is reached based on a single argument, rather than a possibly lengthy chain of arguments. Guha et al. identify four types of atomic propagations: *direct propagation*, *co-citation*, *transpose trust*, and *trust coupling*. We briefly elaborate each of these types of atomic trust propagation:

Direct Propagation. The example given in the first paragraph represents direct propagation. If i trusts j, and j trusts k, then a direct propagation allows us to infer that i trusts k. Guha et al. refer to this particular atomic propagation as direct propagation since the trust propagates directly along an edge.

Co-Citation. Let's consider that i_1 trusts j_1 and j_2, and i_2 trusts j_2. Under co-citation, it is concluded that i_2 also trusts j_1.

Transpose Trust. In transpose trust, i's trust in j causes j to develop some level of trust towards i. Let's say that i trusts j, then transpose trust implies that j should also trust i.

Trust Coupling. Let's suppose that i and j both trust k, then trust coupling leads us to infer that i and j should trust each other since they both trust k.

Iterative propagation builds upon multiple atomic propagations to help establish trust in an unknown entity. Let's extend the example presented in the first paragraph: Alice trusts Bob and Bob trusts Carol. We further assume that Carol trusts Dave. Alice may establish trust in Dave as a result of the following two atomic propagations: (1) the first atomic propagation builds Bob's direct trust in Dave and (2) now since Bob trusts Dave, Alice can establish trust in Dave through a second atomic propagation. This sequence of atomic propagations is referred to as iterative propagation.

Trust Negotiation

Trust negotiation is an approach that can enable strangers to electronically share sensitive data and services. Trust negotiation establishes trust between entities based not on their identities but their properties. For example, in the case of an individual, the properties that may be considered include their place of employment, age, membership in certain organizations, etc. With trust negotiation, the trust between two entities is acquired through iterative requests for credentials and their disclosure.

An example from Bertino et al. [6]: CARS is an online car rental agency, which has an agreement with a company called CORRIER to provide rental vehicles free of charge to their employees, provided that they prove their employment status (which also implies that they are authorized to drive). Other customers (who are not employees of CORRIER) can rent a vehicle by showing a valid driving license and by providing a credit card for payment. Thus, CARS establishes trust in customers to be legitimate drivers through the exchange of multiple possible credentials.

```
Customer: Request a vehicle
CARS: Show digital employment ID from CORRIER
Customer: Not available
CARS: Show digital driving license
Customer: Digital driving license
CARS: Provide digital credit card
Customer: Digital credit card
CARS: Vehicle granted (vehicle info, pickup info, etc.)
```

Reputation

Reputation is the general opinion of the community about the trustworthiness of an individual or an entity. A person who needs to interact with a stranger may analyze her reputation to determine the amount of trust that he can place in her. In the physical world, reputation often comes from word of mouth, media coverage, physical infrastructure, etc. However, the reputation of a stranger is often difficult to observe in online communities, primarily due to their global scale, the cheap availability of anonymous identities, and the relative ease of acquiring high quality digital presence.

A reputation system computes the reputation of an entity based on the feedback (quantified trust) provided by fellow entities. Reputation systems make certain that users are able to gauge the trustworthiness of an entity based on the history of its behavior. The expectation that people will consider one another's pasts in future interactions constrains their behavior in the present [54].

Hoffman et al. [33] provide the following description of reputation:

> In general, reputation is the opinion of the public toward a person, a group of people, or an organization. In the context of collaborative applications such as peer-to-peer systems, reputation represents the opinions nodes in the system have about their peers. Reputation allows parties to build trust, or the degree to which one party has confidence in another within the context of a given purpose or decision. By harnessing the community knowledge in the form of feedback, reputation-based trust systems help participants decide who to trust, encourage trustworthy behavior, and deter dishonest participation by providing a means through which reputation and ultimately trust can be quantified and disseminated.

8.3 Privacy Preserving Reputation Systems

An accurate reputation score is possible only if the feedback is accurate. However, it has been observed that the users of a reputation system may avoid providing honest feedback [53]. The reasons for such behavior include fear of retaliation from the target entity or mutual understanding that a feedback value would be reciprocated.

A solution to the problem of fear of retaliation is computing reputation scores in a privacy preserving manner. A privacy preserving protocol for computing reputation scores does not reveal the individual feedback of any entity. Private feedback ensures that there are no consequences for the feedback provider and thus he is uninhibited to provide honest feedback.

Slandering is the act of sabotaging an honest user's reputation by assigning them unwarranted low feedback. A trade-off of private feedback is that it creates the opportunity for slandering without consequences. However, we draw attention to the processes of voting and election, where the privacy of the voters is often guaranteed to allow them complete freedom of opinion. Since feedback providers in reputation systems are similarly entitled to personal opinion, it can be argued that their privacy should also be preserved. Slandering is most effective when it is carried out by a collusion of users. An important challenge to be addressed by future work is the detection of collusions in privacy preserving reputation systems.

8.3.1 Architecture

The architecture of a reputation system is one of the key factors in determining how the following activities are conducted:

- Feedback collection
- Feedback aggregation (reputation computation)
- Reputation dissemination

The two common architectures are: centralized and decentralized.

Centralized Reputation Systems.   Centralized reputation systems are characterized by the existence of a trusted central authority. The central authority receives feedback from users, aggregates it to compute the reputation, and disseminates the reputation scores.

One of the benefits of a centralized solution is that it is straightforward to implement. Moreover, a centralized reputation system is often less vulnerable to certain attacks, such as the sybil attack, since the central authority can monitor and correlate all activities in the reputation system. Additionally, the central authority is universally trusted; therefore, users can be assured that the feedback collection, aggregation, and dissemination are being done correctly.

Unfortunately, the requirement of universal trustworthiness of the central authority is also a liability. If the central authority fails or becomes compromised, then the whole reputation system crashes. Thus, the central authority is a single point of failure and a high-value target for attackers. As with any other centralized system, another major disadvantage of centralized reputation systems is that they are very expensive to deploy and maintain, particularly for large numbers of users. Centralized reputation systems are also unable to cater for decentralized environments, particularly decentralized social networks.

Decentralized Reputation Systems.   Decentralized environments are characterized by the absence of a central authority. Advantages of such networks include: lack of a single point of failure, no need to deploy and maintain an expensive central authority, a more democratic environment, scalability, etc.

Decentralized reputation systems are suitable for decentralized environments such as decentralized social networks as they do not assume the presence of a central entity. In decentralized reputation systems, a central location for submitting and aggregating feedback, and disseminating reputation does not exist. Feedback is commonly stored locally by the node who generates it, for example in response to his experiences with another party. Computing reputation of an entity in the system requires finding all or a portion of the nodes who carry feedback about that entity. Once the feedback providers have been located, the aggregation may be done at a single location after receiving all feedback, or a more sophisticated protocol may be employed to aggregate the feedback in a distributed manner.

8.3.2   Adversarial Models

Semi-Honest.   In the semi-honest model, the agents do not deviate from the specified protocol. In other words, they always execute the protocol according to the specifications. The adversary abstains from wiretapping and tampering of the communication channels. However, within these constraints, the adversary passively attempts to learn the inputs of honest agents by using intermediate information received during the protocol and any other information that it can gain through other legitimate means.

Disruptive Malicious.   Disruptive malicious agents are not bound to conform to the protocol. Agents under the malicious model may deviate from the protocol as and when they deem necessary. They actively attempt to achieve their objectives. They may participate in extra-protocol activities, devise sophisticated strategies, and exhibit arbitrary behavior. Specifically, malicious agents may (1) refuse to participate in the protocol, (2) provide out of range values as their inputs, (3) selectively drop messages that they are supposed to send, (4) prematurely abort the protocol, (5) distort information, and (6) wiretap and tamper with all communication channels. A malicious adversary may have one or both of the following objectives: (1) learn the inputs of honest agents and (2) disrupt the protocol for honest agents. The reasons for disrupting the protocol may range from gaining illegitimate advantage over honest agents to completely denying the service of the protocol to honest agents.

8.4   Centralized Privacy Preserving Reputation Systems

We review some reputation systems that can be utilized for privacy preserving reputation management in centralized social networks.

8.4.1   Androulaki et al. [2]: A Reputation System for Anonymous Networks

Androulaki et al. [2] propose a reputation scheme for pseudonymous peer-to-peer systems in anonymous networks. Users in such systems interact only through disposable pseudonyms such that their true identity is not revealed. Reputation systems are particularly important for such environments since otherwise there is little incentive for good conduct. However, reputation systems are hard to implement for these environments. One of the reasons is that a user must keep his reputation even if he cycles through many pseudonyms. Moreover, the pseudonyms must be unlinkable to the user as well as to each other even though they share the same reputation score. Another issue that arises in reputation systems for anonymous networks is that a user may lend his good reputation to less reputable users through anonymous pseudonyms.

The proposed system employs the following cryptographic building blocks: anonymous credential systems, e-cash, and blind signatures. Reputation is exchanged in the form of e-coins called *repcoins*. The higher the amount of repcoins received from other users, the higher is the reputation of the user.

The system requires the presence of a *bank*, which is a centralized entity. Additionally, the system also requires that all communication takes place over an anonymous network, such as Mixnet [13] or a network using Onion routing [20]. This requirement makes the solution inaccessible to applications in non-anonymous networks.

The security goals of reputation systems for anonymous networks are different than those of privacy preserving reputation systems. The reputation systems for anonymous networks aim to hide the identity of a user who interacts and assigns feedback to others whereas, in privacy preserving reputation systems, the goal is to hide the feedback value assigned but not the identity of the user who assigned it. The choice between the two kinds of reputation systems depends on the security objectives of the application.

Security Model

Some of the security requirements of the reputation system are as follows:

Unlinkability.   An adversary, controlling the bank and a number of corrupted users, is unable to link a pseudonym with the identity of its non-corrupted user any better than by making a random guess. Moreover, the adversary has no advantage in telling whether two pseudonyms belong to the same non-corrupted user or not.

No Over-Awarding.   A user who tries to double-award (forge) a repcoin, using one or even two different pseudonyms, gets detected and his identity is revealed.

Exculpability.   Any coalition of corrupted users (including the bank) is unable to falsely accuse a user of forgery in order to expose his identity.

Reputation Unforgeability, Non-Transferability. A user cannot forge better reputation. In particular, a user U_1 cannot borrow reputation from another user U_2, unless U_2 reveals his master secret key to U_1.

Cryptographic Building Blocks

The following cryptographic building blocks are used for the construction of the scheme:

Anonymous Credential Systems. In anonymous credential systems (for example, [5, 10]), organizations grant credentials to pseudonymous identities of users. Verifiers are able to verify the authenticity of credentials in possession of users. However, neither an organization nor a verifier is able to link a credential to the true identity of a user.

E-Cash. E-cash [14, 15] is a digital currency that offers the following properties: anonymity, unforgeability (or identification of double-spenders), and fungibility. Please see section "E-Cash and Endorsed E-Cash" for further detail. A centralized *bank* is a key player in an e-cash system.

Blind Signatures. In a blind signature scheme (for example, [14]), an entity signs a message for a user, however the entity does not learn the content of the message.

A Reputation System for Anonymous Networks

The system assumes the presence of a central entity called the bank, which is needed for implementing the above listed cryptographic schemes. The system also requires that all communication takes place over an anonymous network, for example, a Mixnet, or a network using Onion routing. The users interact with each other in a peer-to-peer manner. However, the users must also communicate with the central bank to withdraw and deposit repcoins.

From the above-listed building blocks, Androulaki et al. build a reputation system in which each user has a reputation that he cannot lie about or shed. However, a user may generate as many one time pseudonyms as he needs for his transactions. All pseudonyms of a user share the same reputation. The system is robust against self-promotion attacks. Reputation is updated and demonstrated in a way such that anonymity is not compromised. The system maintains unlinkability between the identity of a user and his pseudonyms, and unlinkability among pseudonyms of the same user.

The system by Androulaki et al. follows upon the work by Dingledine et al. [17–19] on reputations systems and anonymous networks.

The reputation system is summarized in Table 8.1.

Table 8.1 Androulaki et al. [2]: A reputation system for anonymous networks

Architecture	Centralized
Target environment	Peer-to-peer systems
Adversarial model	Malicious (Disruptive)
Key security mechanisms	Anonymous credential systems, E-cash (bank), Blind signatures, Mixnets/Onion routing
Privacy guarantee	Satisfies unlinkability, no over-awarding, exculpability, and reputation unforgeability if the underlying primitives (anonymous credential system, e-cash system, and blind signatures) are secure
Complexity (Messages)	$O(1)$

8.4.2 Steinbrecher [56]: Privacy-Respecting Reputation Systems within Centralized Internet Communities

Steinbrecher [56] argues that traditional cryptographic techniques such as encryption and digital signatures can provide only "technical" security guarantees. For example, encryption and digital signatures can guarantee the confidentiality and integrity of the *text* of a reply sent by an expert to a user on a self-help forum. However, these techniques cannot guarantee the misbehavior of the users themselves. For example, the user might violate confidentiality by relaying the *content* of the text to a third party, or the expert may violate integrity by giving *false advice*. It is argued that trust can mitigate these risks and that reputation systems are a suitable technology for acquiring trust.

However, the author contests that the design of current reputation systems (such as the eBay reputation system) allows open access to the interests and behavior profiles of users. A third-party may acquire information such as the time and frequency of participation, interests in specific items, and feedback provided. Moreover, it is easy to associate the pseudonym of a user with their real identity, for example, through a mailing address.

To counter this issue, Steinbrecher presents a privacy-respecting reputation system for centralized Internet communities. The system relies on simultaneous use of multiple pseudonyms and changing them frequently to achieve anonymity and unlinkability.

A Generalized Model for Centralized Reputation Systems

The paper presents a generalized model for centralized reputation systems. Users use global pseudonyms tied to global reputations. The set of global pseudonyms at time t is considered as $P_t = \{p_{t,1}, \ldots, p_{t,m}\}$. The set of possible reputations that might be associated with a pseudonym is given as R. $(R, +)$ is a commutative group and $+$ an operator to combine elements from R independently of t. At time t_1, each

Table 8.2 Steinbrecher [56]: A centralized privacy preserving reputation system

Architecture	Centralized
Target environment	E-commerce, Self-help forums, etc.
Adversarial model	Malicious (Disruptive)
Key security mechanisms	Pseudonym/Identity management
Privacy guarantee	Unlinkability and anonymity are satisfied if the provider (central server) is honest and secure
Complexity (Messages)	$O(1)$

pseudonym $p_{t_1,l}$ has the reputation $rep(t_1, p_{t_1,l}) \in R$, where $l \in 1 \dots m$. After $p_{t_1,i}$ receives a rating r_{j,i,t_1} from $p_{t_1,j}$, the reputation of $p_{t_1,i}$ at time t_2 is computed as:

$$rep(t_2, p_{t_1,i}) = rep(t_1, p_{t_1,i}) + r_{j,i,t_1} \tag{8.1}$$

where $t_2 \geq t_1$, and $p_{t_1,i}$ does not receive any rating other than r_{j,i,t_1} between t_1 and t_2.

Using Pseudonyms for Unlinkability and Anonymity

The system proposes simultaneous use of multiple pseudonyms by a user. The idea is to have a separate pseudonym for each context (for example, the context of a seller on an auction site, the context of an expert on a self-help forum, etc.). It is suggested that this design leads to unlinkability between the different roles of a user on the Internet.

The system permits users to regularly change their pseudonyms to achieve anonymity. A new and an old pseudonym are unlinkable from the perspective of third parties; however, the provider (central server) is able to link the two pseudonyms. The unlinkability also assumes that a large number of pseudonyms have the same reputation.

To prevent the provider from linking new and old pseudonyms, the system suggests using a set of non-colluding trustworthy third parties who make incremental changes to the pseudonym of the user.

Steinbrecher's work on reputation and privacy also includes [51, 55]. These proposals are oriented for centralized environments as well.

An adversary may compromise unlinkability by monitoring all pseudonyms with the same reputation. The adversary can deduce that a new pseudonym with the same reputation as a recently deleted pseudonym belong to the same user.

The reputation system is summarized in Table 8.2.

8.5 Decentralized Privacy Preserving Reputation Systems

In the following sections, we discuss reputation systems that can be deployed in decentralized social networks for privacy preserving reputation management.

8.5.1 Clifton et al. [16]: Secure Sum

Secure multi-party computation is the study of protocols that take inputs from distributed entities and aggregate them to produce outputs, while preserving the privacy of the inputs.

One of the well-known secure multi-party computation protocols is secure sum [16], which takes inputs from entities and computes their sum. This protocol is clearly a natural fit for the problem at hand. The protocol may be used directly to compute reputation in the form of sum or mean.

Secure Sum

The secure sum protocol assumes that there are three or more sites and there is no collusion between them. It is also assumed that the value to be computed, $v = \sum_{l=1}^{s} v_l$ lies in the range $[0..m]$. The sites are numbered as $1 \ldots s$. Site 1 generates a random number R uniformly chosen from $[0..m]$. It then sends $R + v_1$ mod m to site 2, where v_1 is site 1s local input. Site 2 does not learn any information about v_1 since $R + v_1$ mod m is distributed uniformly across the range $[0..m]$ due to R. For sites $l = 2 \ldots s - 1$, the protocol proceeds as follows: Site l receives:

$$V = R + \sum_{j=1}^{l-1} v_j \quad \text{mod } m \tag{8.2}$$

Site l learns nothing since the value is distributed uniformly across $[0..m]$. Site l computes:

$$R + \sum_{j=1}^{l} v_j \quad \text{mod } m = (v_l + V) \quad \text{mod } m \tag{8.3}$$

Site l then sends this value to site $l + 1$. Eventually, site s also performs the above step. Site s sends the result back to site 1, who subtracts R from it to obtain the sum. Site 1 does not learn any of the private values due to the uniform distribution of the received result over the range $[0..m]$.

The protocol may be used to compute reputation as the sum of the feedback values provided as private inputs by the participants of the protocol.

The security of the secure sum protocol breaks down if the sites collude. Any two sites $l - 1$ and $l + 1$ can use the values that they send and receive, respectively, to compute the private input v_l of site l.

The reputation system is summarized in Table 8.3.

Other Secure Multi-Party Computation Protocols

Other secure multi-party computation protocols include: secure product [1, 3, 16, 35], secure set union [16, 39], secure set intersection [16, 39], and secure multiset

Table 8.3 Clifton et al. [16]: Secure sum

Architecture	Decentralized
Target environment	Distributed environments
Adversarial model	Semi-Honest + Agents do not collude
Key security mechanisms	Secure multi-party computation
Privacy guarantee	The chances that the adversary will learn private information are no better than making a random guess over the range $[0..m]$. Probability: $\frac{1}{m+1}$
Complexity (Messages)	$O(n)$, where n = number of sites

operations [39]. The doctoral thesis of Wenliang Du [21] describes several secure two-party computation protocols for problems in linear programming, geometry, and statistical analysis. A seminal work in secure multi-party computation is the study of the Millionaire's problem [59], in which two parties must determine whose number is larger without disclosing their numbers. We refer the reader to [26] for a comprehensive study of secure multi-party computation.

8.5.2 Pavlov et al. [46]: Decentralized Additive Reputation Systems

Pavlov et al. [46] propose several protocols for decentralized additive reputation systems. Two of their protocols are secure under the semi-honest and the malicious adversarial models, respectively. The protocols draw their strength from witness (feedback provider) selection schemes, which guarantee the inclusion of a certain number of honest witnesses as participants. The security mechanisms used in the protocols include secure multi-party computation, secret sharing, and discrete log commitment.

Problem Setting

A querying agent consults a group of n witnesses to compute the reputation of a target agent, where $0 < n < N$, and $N > 1$ is the number of potential witnesses. $b < N$ is the number of dishonest agents in N.

Decentralized Additive Reputation Systems

A decentralized additive reputation system is described in the article as a reputation system that satisfies the following two requirements: (1) feedback collection, combination, and propagation are implemented in a decentralized way; (2) combination of feedbacks provided by agents is calculated in an additive manner. The Beta reputation system [36] is cited as an example. The eBay reputation system is additive, however, not decentralized.

Impossibility of Perfect Privacy

The paper argues that it is impossible to guarantee perfect privacy for an honest feedback provider in a decentralized additive reputation protocol. The argument is that a dishonest agent may deterministically create a set of n feedback providers, with $n-1$ dishonest agents and the one honest agent under attack. Given the inputs of the $n - 1$ dishonest agents and the output (the reputation score), the secret feedback of the honest agent is easily obtained.

The impossibility argument does not apply to protocols in which an honest agent may choose not to contribute his feedback. The argument also does not apply to protocols in which the set of feedback providers cannot be created deterministically.

Witness Selection Scheme 1 (WSS-1)

A witness selection scheme for a reputation protocol is a process that results in the creation of a set of witnesses. The witnesses in the set contribute their feedback towards computing the reputation of the target agent.

The first scheme [46, Lemma 2] guarantees that if honest agents are uniformly distributed over N, then at least two honest witnesses will be selected with probability greater than $\left(1 - \frac{1}{n}\right)\left(\frac{N-b-1}{N-1}\right)$. The scheme is secure under the semi-honest adversarial model, in which all agents follow the protocol correctly.

According to our analysis, the complexity of the number of messages exchanged is linear in terms of the number of potential witnesses: $O(N)$. After each witness is selected, it is probabilistically decided whether to add more witnesses, therefore the count may run up to N. If each agent sends its successor the current set of witnesses, the total bandwidth utilized is $O(N^2)$.

The complexity of the scheme is a function of the population size of the potential witnesses (N) instead of the witnesses who contribute their feedback (n). The scheme also has the potential of leaving out many honest witnesses from the reputation protocol. Moreover, the scheme works only if $b < n - 1$, because otherwise $n - 1$ dishonest witnesses can select themselves into the set if the first witness selected is dishonest. Even then the scheme might fail since the number of witnesses selected is probabilistic and it may be the case that the actual number of selected witnesses is less than n.

Witness Selection Scheme 2 (WSS-2)

The second scheme [46, Lemma 3] guarantees under the malicious adversarial model that if honest agents are uniformly distributed over N, then at least $n\left(\frac{N-b-n}{N}\right)$ honest witnesses would be selected. A coin flipping scheme is utilized to grow the set of witnesses by selecting the next witness randomly from the available pool of witnesses. According to the paper, the scheme requires $O(n^3)$ messages among the n selected witnesses.

Table 8.4 Pavlov et al. [46]: A reputation protocol based on WSS-1

Architecture	Decentralized
Target environment	Distributed environments
Adversarial model	Semi-honest
Key security mechanisms	Secure multi-party computation, secret sharing
Privacy guarantee	$\left(1 - \frac{1}{n}\right)\left(\frac{N-b-1}{N-1}\right)$
Complexity (Messages)	$O(N) + O(n^2)$, where N = number of potential witnesses, and n = number of selected witnesses

A Reputation Protocol Based on WSS-1

In this reputation protocol, the set of source agents is created using the first witness selection scheme, which guarantees that at least two source agents are honest. Agent q chooses a random number as its secret. Each agent splits its secret into $n+1$ shares such that they all add up to the secret. Each agent keeps the $n + 1$th share and sends its other n shares to the other n agents in the protocol such that each agent receives a unique share. Each agent then adds all shares received along with his $n + 1$th share and sends it to the querying agent. The querying agent adds all sums received and subtracts the random number to obtain the reputation score.

The protocol guarantees the privacy of an honest source agent under the semi-honest model as long as all the other $n - 1$ source agents do not collude. The probability that all other source agents will not collude is greater than $\left(1 - \frac{1}{n}\right)\left(\frac{N-b-1}{N-1}\right)$. The number of messages exchanged is analyzed as $O(n^2)$. We estimate that the size of the messages exchanged is as follows: $O(n^2)$ IDs and $O(n^2)$ numbers.

The complexity is claimed to be $O(n^2)$; however, we believe it to be $O(N) + O(n^2)$ due to the utilization of the witness selection scheme.

The reputation system is summarized in Table 8.4.

A Reputation Protocol Based on WSS-2

This protocol uses the Pedersen verifiable secret sharing scheme [48] and a discrete log commitment method. The Pedersen scheme is resilient up to $n/2$ malicious agents. The set of source agents is created using the second witness selection scheme. It guarantees the presence of less than $n/2$ malicious agents, if $b < \frac{N}{2} - n$.

The protocol is secure under the malicious model as long as $b < \frac{N}{2} - n$. The number of messages exchanged is $O(n^3)$, due to the second witness selection scheme.

The reputation system is summarized in Table 8.5.

Table 8.5 Pavlov et al. [46]: A reputation protocol based on WSS-2

Architecture	Decentralized
Target environment	Distributed environments
Adversarial model	Malicious (Disruptive)
Key security mechanisms	Verifiable secret sharing, discrete log commitment
Privacy guarantee	If $b < \frac{N}{2} - n$, then the adversary does not learn any more information about the private feedback of an honest witness
Complexity (Messages)	$O(n^3)$, where $n = $ number of witnesses

8.5.3 Gudes et al. [28]: The Knots Reputation System

Gudes et al. [28] present several schemes that augment their Knots reputation system [23] with privacy preserving features. A defining characteristic of the Knots reputation model is the notion of subjective reputation. The reputation of a target member is computed by each querying member using a different set of feedback, thus the reputation is subjective for each querying member. The feedback that a querying member uses for computing reputation comes exclusively from the members in which he has a certain amount of pre-existing trust. An advantage of this approach is that the querying member has confidence in each of the feedback values that are used for computing reputation.

The disadvantage is that the opinion of the members whom the querying agent does not know is not taken into account. The notion of subjective reputation tends to be non-conformant with the idea of reputation, which is generally considered to be the aggregate of feedback of the community at large. The concept of subjective reputation seems closer to trust propagation than reputation.

The Knots Model

The Knots model differentiates between two types of users in the system. The *experts* in the system are the users who provide services and the *members* are users who consume those services. The reputation system is concerned with computing the reputation of the experts through the feedback provided by the members. Members have trust relationships among themselves in the context of providing reliable feedback about the experts.

$TrustSet_x(A)$ is defined as the set of members whom member A trusts to provide feedback about expert x. $TM(A, B)$ represents the amount of direct trust that a member A has in another member B. $DTE(A, x)$ is defined as the amount of direct trust that a member A has in an expert x. The subjective reputation of an expert x by a member A is computed as follows:

$$TE(A, x) = \frac{\sum_{\forall B \in TrustSet_x(A)} DTE(B, x) \cdot TM(A, B)}{\sum_{\forall B \in TrustSet_x(A)} TM(A, B)} \tag{8.4}$$

Table 8.6 Gudes et al. [28]: Reputation Scheme 1

Architecture	Decentralized
Target environment	Distributed environments
Adversarial model	Semi-honest
Key security mechanisms	TTP, Public-key cryptography
Privacy guarantee	If the TTP is honest, the chances that A will learn $DTE(B, x)$ are no higher than making a random guess across $|TrustSet_x(A)|$ given values. $B \in TrustSet_x(A)$
Complexity (Messages)	$O(n)$, where $n = |TrustSet_x(A)|$

In the privacy preserving version of the Knots model, the challenge is to compute $TE(A, x)$, such that the privacy of each $DTE(B, x)$ is maintained, where $B \in TrustSet_x(A)$. The three decentralized privacy preserving schemes presented in the paper compute $\rho(A, x)$ (the numerator of the fraction in Eq. (8.4)), such that A cannot learn any of the $DTE(B, x)$ values.

The privacy goal does not include preserving the privacy of the trust between the members (the TM values). It is limited to preserving the privacy of the feedback about the experts (the DTE values).

Reputation Scheme 1

Each member $B \in TrustSet_x(A)$ receives $TM(A, B)$ from A and then computes $E_A(DTE(B, x) \cdot TM(A, B))$ and sends it to a Trusted Third Party (TTP), Z (where $E_A(.)$ is an encryption with the public key of member A). The TTP Z relays each message to A without revealing the source member. A decrypts the messages and obtains $\rho(A, x)$.

Since A does not know the source of a message, it cannot reverse a received value to reveal the private feedback. The messages are encrypted; therefore, the TTP does not learn any information either. The scheme requires $O(n)$ messages to be exchanged, where n is the cardinality of $TrustSet_x(A)$.

The scheme requires disclosure of the trust that A has in each member B. Moreover, there is heavy reliance on the TTP. If the TTP and A collude, then they can easily determine each $TM(B, x)$.

The reputation system is summarized in Table 8.6.

Reputation Scheme 2

Each member $B \in TrustSet_x(A)$ generates $E_A(DTE(B, x))$ and sends it to a TTP, Z. The TTP sends a randomly permuted vector of the messages to A, who decrypts the messages and obtains a vector (vector 1) of the DTE values. A then sends a vector of all values $TM(A, B)$ to Z, where $B \in TrustSet_x(A)$. Z permutes the vector (vector 2) according to the DTE vector (with respect to the order of the members). A and Z compute the scalar product of vectors 1 and 2 using a secure product protocol (such as [1]) to obtain $\rho(A, x)$.

Table 8.7 Gudes et al. [28]: Reputation Scheme 2

Architecture	Decentralized
Target environment	Distributed environments
Adversarial model	Semi-honest
Key security mechanisms	TTP, Public-key cryptography, Secure product
Privacy guarantee	If the TTP is honest, the chances that A will learn $DTE(B, x)$ are no higher than making a random guess across $|TrustSet_x(A)|$ given values. Moreover, B does not learn $TM(A, B)$. $B \in TrustSet_x(A)$
Complexity (Messages)	$O(n)$, where $n = |TrustSet_x(A)|$

Table 8.8 Gudes et al. [28]: Reputation Scheme 3

Architecture	Decentralized
Target environment	Distributed environments
Adversarial model	Semi-honest + Agents do not collude
Key security mechanisms	Secure multi-party computation
Privacy guarantee	A does not learn more information about $DTE(B, x)$, where $B \in TrustSet_x(A)$. The chances of B learning $TM(A, B)$ are no better than its chances of guessing the random number Q from $TM'(A, B)$
Complexity (Messages)	$O(n^2)$, where $n = |TrustSet_x(A)|$

Due to the random permutation generated by the TTP, A is unable to correlate the DTE values with individual members. The TTP does not learn any of the DTE values due to encryption. A key advantage of the scheme is that any member B does not learn $TM(A, B)$.

We analyze that the number of messages exchanged is $O(n)$, whereas the bandwidth utilized is $O(n^2)$ in terms of k-bit numbers transferred, where k is the security parameter (key length).

The privacy of the $TM(A, B)$ values is still not fully preserved since they must be disclosed to the TTP.

The reputation system is summarized in Table 8.7.

Reputation Scheme 3

A executes the reputation protocol for the semi-honest model from Pavlov et al. [46] to obtain $\Sigma_{\forall B \in TrustSet_x(A)} DTE(B, x)$. A sends $TM'(A, B) = TM(A, B) + Q$ to each $B \in TrustSet_x(A)$, where Q is a random number. A executes the secure sum protocol [16] to obtain $\Sigma_{\forall B \in TrustSet_x(A)}(TM'(A, B) \cdot DTE(B, x))$. A calculates:

$$\rho(A, x) = \Sigma_{\forall B \in TrustSet_x(A)}(TM'(A, B) \cdot DTE(B, x))$$

$$-(Q \cdot \Sigma_{\forall B \in TrustSet_x(A)} DTE(B, x)) \tag{8.5}$$

This scheme has the advantage that the privacy of both the $DTE(B, x)$ values and the $TM(A, B)$ values is preserved without the presence of any TTPs. The protocol requires $O(n^2)$ messages due to the inclusion of the protocol from [46].

The reputation system is summarized in Table 8.8.

Proposals for the Malicious Adversarial Model

The work also includes some proposals for augmenting the schemes for the malicious adversarial model. The proposals are largely based on the assumption that a member who provides feedback (member B) would lack the motivation to act maliciously if it does not know the identity of the querying member (member A). However, this assumption does not take into account the case when an attacker may want to attack the system simply to disrupt it, for example, in a denial-of-service attack.

8.5.4   Hasan et al. [32]: The k-Shares Reputation Protocol

The k-shares protocol by Hasan et al. [30–32] offers the following advantages over comparable protocols such as those by Pavlov et al. [46, Sect. 5.2] and Gudes et al. [28]: (1) Lower message complexity of $O(n)$ as opposed to $O(n^2)$ and higher of the protocols in [46] and [28]; (2) The k-Shares protocol allows agents to quantify and maximize the probability that their privacy will be preserved before they submit their feedback.

Framework

An action called "preserve privacy" is defined. Agents are assumed to have trust relationships with some other agents in the context of this action. This assumption is called *trust awareness* and derives from the fact that agents have social relationships and a key component of such relationships is the trust that each other's privacy will be preserved. For example, a user may trust his family members and close friends to help him preserve his privacy.

The adversary is considered as semi-honest and is allowed to collude. Privacy is formalized using the Ideal-Real approach. An ideal protocol for computing reputation is one in which a Trusted Third Party (TTP) receives all inputs and then locally computes the reputation. On the other hand, a real protocol computes reputation without the participation of any TTP. The real protocol is said to preserve privacy if the adversary, with high probability, cannot obtain any more information about the private input of an agent than it can learn in the ideal protocol.

The Protocol

A simplified version of the protocol is outlined below.

1. *Initiate.* The querying agent q retrieves the set of source agents S_t of the target agent t and sends the set to each of the source agents.

Table 8.9 Hasan et al. [32]: The k-shares reputation protocol

Architecture	Decentralized
Target environment	Distributed environments
Adversarial model	Semi-honest
Key security mechanisms	Secure multi-party computation, Trust awareness, Secret sharing
Privacy guarantee	The privacy of an agent a is preserved with high probability if it finds k trustworthy agents in the set of feedback provider agents S_t, such that $k \ll n$ and the probability that all k agents will collude to break agent a's privacy is low
Complexity (Messages)	$O(n)$, where n is the number of feedback providers

2. *Select Trustworthy Agents.* Each source agent selects up to k other agents in S_t. Each agent selects these agents such that the probability that all of them will collude to break his privacy is low. k is a constant, such that $k \ll n$, where n is the number of all source agents. The risk to privacy is thus quantified before submitting the feedback.
3. *Prepare and Send Shares.* Each agent generates k shares such that their sum is equal to the secret feedback value. The secret cannot be revealed until all shares are known. The shares are sent to the selected fellow agents.
4. *Compute Sums and Reputation.* Each agent that receives shares from fellow agents computes the sum of all shares received and sends the sum to the querying agent q. Agent q receives all the sums and computes the grand total and divides it by n to learn the reputation score.

The full version of the protocol takes measures to ensure that a share is not compromised even if it is the only share received by an agent. Moreover, the protocol also takes steps so that the protocol does not reach certain failure states.

The highlights of the protocol are as follows: (1) It requires each source agent to send only $k \ll n$ messages, which implies that the protocol requires only $O(n)$ messages. (2) The risk to privacy can be quantified before submitting feedback. Thus, an agent knows the risk and if that risk is unacceptable it can opt to not participate in the protocol. As a consequence, even up to $n - 1$ dishonest agents in the protocol cannot breach the privacy of one dishonest agent.

The reputation system is summarized in Table 8.9.

Experimental Results

The work comprises of experiments on the web of trust of the Advogato.org social network. The members of Advogato rate each other in the context of being active and responsible members of the open source software developer community. The choice of feedback values are *master*, *journeyer*, *apprentice*, and *observer*, with *master* being the highest level in that order. The result of these ratings is a rich web of trust. The members of Advogato are expected to not post spam, not attack the Advogato trust metric, etc. It is therefore argued that the context "be a responsible member of the open source software developer community" comprises of the context "be honest." The four feedback values of Advogato are substituted as follows: *master* = 0.99, *journeyer* = 0.70, *apprentice* = 0.40, and *observer* = 0.10. For the experiments, the lowest acceptable probability that privacy will be preserved is defined as 0.90. This means that a set of two trustworthy agents must include either one *master* rated agent or two *journeyer* rated agents for this security threshold to be satisfied. The two experiments and their results are as follows:

Experiment 1: In the k-Shares protocol, the following assumption must hold for an agent a's privacy to be preserved: the probability that the agents to whom agent a sends shares, are all dishonest must be low. The experiment determines the percentage of instances of source agents in the Advogato data set for whom this assumption holds true.

Results: Consider the case where there are at least 50 source agents present in the protocol and $k = 2$, that is, only two trustworthy agent can be selected to preserve privacy. It is observed that the assumption holds for 85.8 % of instances of source agents. At $n \geq 5$, the percentage is 72.5 %.

Experiment 2: The experiment observes the effect of increasing k on the percentage of instances of source agents whose privacy is preserved by the k-Shares protocol in the Advogato.org data set.

Results: Consider the case where there are at least 50 source agents present in the protocol and $k = 1$, that is only one trustworthy agent can be selected to preserve privacy. In the percentage of instances of source agents whose privacy is preserved is 75.4 %. At $k = 2$, the percentage is 85.8 %. The rise is due to the possibility with $k = 2$ to rely on two trustworthy agents. Increasing k over 2, even up to 500, does not result in a significant advantage (86.3 % at $k = 500$). These results validate the assumption that the privacy of a large number of agents can be preserved with $k \ll n$.

8.5.5 Belenkiy et al. [4]: A P2P System with Accountability and Privacy

Selfish participants are a major threat to the functionality and the scalability of peer-to-peer systems. Belenkiy et al. [4] propose a content distribution peer-to-peer system that provides accountability, which makes it resilient against selfish participants. The solution is based on e-cash technology. Despite making peers accountable, the system does not compromise the privacy of the peers. The system ensures that transactions between peers remain private. The only exception is the case when there is a dispute between transacting peers.

Although the system is not directly related to reputation, we study it here because it provides insight into designing a privacy preserving system using the e-cash technology. In Sect. 8.4.1, we discuss a privacy preserving reputation system based on e-cash by Androulaki et al. [2].

E-Cash and Endorsed E-Cash

E-cash [14, 15] is a digital currency that offers the following properties:

Anonymity. It is impossible to trace an e-coin (the monetary unit of e-cash) to the user who spent it. This property holds even when the bank (a central entity who issues the e-coins) is the attacker.

Unforgeability. The only exception to the anonymity property is that e-cash does not guarantee the anonymity of a user who tries to double-spend an e-coin. In this case, the bank can learn the identity of the dishonest user. A forged e-coin allows the bank to trace down the user who forged it.

Fungibility. A user can use the e-coins received for services provided as payment for services received from any other user in the system.

Endorsed e-cash [11] adds the following property to e-cash:

Fair Exchange. Fair exchange means that a buyer gets the item only if the seller gets paid and vice versa.

A Currency-Based Model

The authors describe a peer-to-peer content distribution system inspired by BitTorrent [52]. However, the proposed system provides stronger accounting in its protocols that allow nodes to *buy* and *barter* data blocks from their neighbors in a fair manner.

The system requires the participation of two trusted entities: (1) A *bank*, which maintains an endorsed e-cash account for each user. Users are able to make deposits and withdrawals of e-coins. (2) An *arbiter*, which protects the fair exchange of e-cash for data.

A user has two options for acquiring the data blocks that it needs: (1) it can pay e-coins to users who own those data blocks; (2) or it can barter its own data blocks for the ones that it needs. To earn e-cash, a user has to offer data blocks that other users want and exchange them for e-coins. A user is prevented from being selfish since it cannot consume the service provided by the peer-to-peer system unless he contributes as well.

An unendorsed e-coin cannot be deposited into the seller's bank account until the buyer endorses it. Each unendorsed e-coin has a contract associated with it. The fair exchange takes place according to the contract. If the seller fulfills its commitments, then the unendorsed e-coin must be endorsed by the buyer or otherwise by the arbiter.

The Buy and Barter Protocols

The *buy* protocol operates as follows: Alice requests a data block from Bob. Bob encrypts the block with a random key and sends the ciphertext to Alice.

Alice sends an unendorsed e-coin and a contract for the data block. If the unendorsed e-coin and the contract are formed correctly, Bob sends the decryption key for the data block to Alice. If the key decrypts the data block correctly, Alice endorses the sent e-coin, which Bob can then deposit into his account.

The protocol ensures that fair exchange of e-coins and data takes place. If Bob is dishonest and the key is incorrect, Alice does not endorse the e-coin. In case Alice is dishonest and she does not endorse the coin after receiving the key, Bob can present the arbiter with proof of his correct service (in the form of the contract and other credentials received from Alice) and have the arbiter endorse the e-coin for him.

Moreover, the privacy of the transaction is preserved since no third party involvement is required, unless there is a need for arbitration. The e-coin spent by Alice is unlinkable to her due to the anonymity provided by e-cash.

The barter protocol also provides fair exchange and privacy. Alice and Bob initially send each other an unendorsed e-coin as collateral and a contract which lets them have the arbiter endorse the coin in case the key for a bartered data block is incorrect. Alice and Bob then continue to exchange data blocks until the occurrence of fair termination or arbitration.

Endorsed e-cash requires that each received e-coin must be deposited back to the bank before it can be spent. The buy protocol therefore incurs significant overhead due to this requirement. However, the barter protocol is scalable since it does not require any involvement from the bank under normal circumstances.

The bank and the arbiter are centralized entities. This implies that the system is not fully decentralized. The two centralized entities present scalability issues (at least for the buy protocol) as well as single points of failure.

8.5.6   Nin et al. [45]: A Reputation System for Private Collaborative Networks

Nin et al. [45] present a reputation system that computes the reputation of a user based on the access control decisions that he makes. If a user makes good access control decisions, such as granting access to legitimate users and denying access to unauthorized users, then he receives good reputation. In contrast, making dishonest access control decisions leads to bad reputation. The privacy objective of the reputation system is to keep the trust relationships between the users private.

The system operates as follows: A node keeps record of its access control decisions. Other nodes can view anonymized details of those decisions and verify if the decisions were made according to the access control rules or not. The anonymization is derived through the multiplicative homomorphic property of the ElGamal encryption scheme. Private details are not revealed to a third-party due to the anonymization.

Private Collaborative Networks

A private collaborative network is described as a network of users that has the following properties: (1) the users are connected with each other through trust relationships; (2) users own resources that can be accessed by other users if sufficient trust exists; and (3) trust relationships among users remain private.

A private collaborative network is modeled as a directed labeled graph. Edges represent trust relationships between nodes (users). Each edge is labeled with the type of trust relationship as well as the weight of the trust.

Access to each resource in the network is governed by a set of access conditions. An access condition is of the form $ac = (v, rt, d_{max}, t_{min})$, where v is the owner with whom the requester of the resource must have a direct or transitive trust relationship of type rt to gain access. d_{max} and t_{min} are the required maximum depth and minimum trust, respectively, to obtain access.

Each trust relationship also exists in the form of a certificate signed by the truster and the trustee. Since relationships must be kept private, a certificate itself is considered a private resource. To gain access to a resource, a requester must demonstrate to the owner, the existence of a "certificate path" linking the requester to the owner.

The Reputation Model

The reputation system assigns good reputation to a user who performs decisions in accordance with the specified access conditions. In contrast, a user who does not correctly enforce access control rules receives lower reputation. Reputation lies in the interval $[0, 1]$.

A user can act dishonestly in two ways: (1) deny access to a resource to a legitimate requester or (2) allow access to a resource to an unauthorized requester. The access control decision is considered wrong if it violates either of the rt, d_{max}, t_{min} parameters in the access condition. For a wrong decision that violates the trust requirement (t_{min}), the absolute difference between the minimum amount of trust required (t_{min}) and the trust computed over the certificate path is given as wd. The values arising from all such wrong decisions are given as the set $\{wd_1, \ldots, wd_{|WD_{t_A}|}\}$, where $|WD_{t_A}|$ is the number of wrong decisions.

The values in the set $\{wd_1, \ldots, wd_{|WD_{t_A}|}\}$, which represent the wrong decisions made by user A in terms of trust, are aggregated as:

$$AGt_{AC_{SET_A}} = OWA_Q(wd_1, \ldots, wd_{|WD_{t_A}|}) \tag{8.6}$$

where $AGt_{AC_{SET_A}}$ is the aggregated value of the wrong decisions with respect to trust. OWA is an Ordered Weighted Averaging function and Q is a non-decreasing fuzzy quantifier. According to the authors: "The interest of the OWA operators is that they permit the user to aggregate the values giving importance to large (or small) values."

Table 8.10 Nin et al. [45]: A reputation system for private collaborative networks

Architecture	Decentralized
Target environment	Private collaborative networks
Adversarial model	Semi-honest
Key security mechanisms	ElGamal encryption scheme
Privacy guarantee	Trust relationships among users remain private if the underlying encryption scheme is secure
Complexity (Messages)	$O(1)$

The wrong decisions of the user that violate the depth and path requirements are aggregated as $AGd_{AC_{SET_A}}$ and $AGp_{AC_{SET_A}}$, respectively. The reputation of user A is then computed as:

$$R_A = 1 - \frac{1}{3}(AGt_{AC_{SET_A}} + AGd_{AC_{SET_A}} + AGp_{AC_{SET_A}}) \qquad (8.7)$$

which implies that the mean of the aggregates of the three types of wrong decisions is subtracted from the perfect reputation of 1 to arrive at the reduced reputation of the user. The more dishonest decisions a user makes, the lower his reputation.

Anonymized Audit Files

After a user makes an access control decision, an entry about that decision is added into the user's anonymized audit file. The entry includes information such as the identity of the requester of the resource, the certificate path demonstrated by the requester, etc. However, all private information in the entry is encrypted using the ElGamal encryption scheme [22]. Therefore, a third-party who analyzes the entry is unable to acquire any information about these private elements. Due to the multiplicative homomorphic nature of the ElGamal encryption scheme, the encrypted information can be manipulated to compute reputation. A network participant who wishes to learn the reputation of a certain user can analyze the anonymized audit file of that user and derive the reputation score without compromising privacy.

We analyze the number of messages exchanged to compute reputation as constant ($O(1)$), since all required information is provided directly by the target node.

The reputation system has the following advantages: (1) the reputation of a node is not derived from the feedback of other nodes but from objective information about its behavior (its access control decisions) and (2) a node itself manages and furnishes the evidence required for another node to judge its reputation.

The adversarial model is not specified in the paper; however, we estimate that the scheme would be secure only upto the semi-honest model since nodes are assumed to manage their audit files honestly.

The reputation system is summarized in Table 8.10.

8.5.7 Kinateder and Pearson [37]: A Privacy-Enhanced P2P Reputation System

The decentralized reputation system proposed by Kinateder and Pearson [37] requires a Trusted Platform Module (TPM) chip at each agent. The TPM enables an agent to demonstrate that it is a valid agent and a legitimate member of the reputation system without disclosing its true identity. This permits the agent to provide feedback anonymously.

Security Goals

The reputation system sets the security requirements listed below. An attacker must not be able to:

- Provide false feedback on an honest user's behalf.
- Access an honest user's private database and modify data such as feedback, reputation, etc.
- Learn the identity of a feedback provider (which implies that a user should be able to provide feedback anonymously).

Moreover, it is required that:

- The identity of a dishonest user can be revealed if there is sufficient legal justification.

Trusted Platform

The reputation system presented in the paper relies on the Trusted Platform (TP) [44, 47] technology for security. A trusted platform is described as a secure computing platform that preserves the privacy of the user by providing the following three functionalities:

Protected Storage. Data on the TP is protected from unauthorized access.

Integrity. The TP can prove that it is running only the authorized software and no malicious code.

Anonymity. The TP can demonstrate that it is a genuine TP without revealing the identity of the user. The TP uses a pseudonym attested by a PKI Certification Authority (CA).

A Trusted Platform comprises of a Trusted Platform Module (TPM), which is a hardware device with cryptographic functions that enable the various security functionalities of the TP. The TPM is unforgeable and tamper-resistant.

System Model and Functionality

An agent in the system can take up one of following three roles at any given time: *recommender*, *requester*, and *accumulator*.

Recommender.   A recommender agent has interacted directly with other agents and has feedback about them. He regularly announces the availability of feedback to other agents in the system. A recommendation comprises of the target agent's pseudonym, the recommender agent's pseudonym, and the feedback value. The recommendation is digitally signed by the recommender.

Accumulator.   An accumulator agent stores feedback about other agents. However, his feedback is not based on direct experience with the target agent but formed through the feedback that he has received from other agents in the system.

Requester.   A requester agent queries other agents for feedback and then locally aggregates the feedback to determine the reputation of the target agent. A requester agent propagates the query to its peer agents who in turn propagate to their peer agents. Each peer decides when to discontinue further propagation based on whether recommendations are available among its peers. The requester agent receives the feedback from the recommender and accumulator agents queried and then aggregates the feedback to learn the reputation of the target agent.

It is not elaborated how the feedback announcement and feedback query protocols work, for example, if an algorithm such as broadcast or gossip is used. As a consequence, the communication complexity of the protocols is not clear. Moreover, the mechanism for aggregating the feedback is not discussed.

How Security Is Achieved

The security requirements are fulfilled as follows:

- An attacker is unable to provide false feedback on an honest user's behalf since each feedback is digitally signed by the recommender. A requester agent can also verify through the recommender's TP that it has not been compromised by the adversary.
- An attacker is unable to access an honest user's private database and modify data such as feedback and reputation. This is achieved due to the protected data storage functionality of the TP. Therefore, a requester can be certain that the given feedback is not false.
- An attacker does not learn the true identity of a feedback provider since only pseudonyms are used. Thus, a user is able to provide feedback anonymously and without inhibition. The pseudonym is protected by the TP and the CA of the user. Moreover, the use of MIX cascades is suggested to prevent the attacker from correlating the pseudonym with the IP address of the user.
- In case of legal justification, the CA of a user can reveal his true identity.

Table 8.11 Kinateder and Pearson [37]: A privacy-enhanced P2P reputation system

Architecture	Decentralized
Target environment	Peer-to-peer systems
Adversarial model	Malicious (Disruptive)
Key security mechanisms	Trusted platform, MIX cascades, Digital signatures
Privacy guarantee	Security goals are satisfied if the underlying primitives (trusted platform, MIX cascades, digital signatures) are secure
Complexity (Messages)	Not Provided

Voss et al. [57] and Bo et al. [8] also present decentralized systems that are based on similar lines. They both suggest using smart cards as the trusted hardware modules. A later system by Kinateder et al. [38] avoids the hardware modules; however, it requires an anonymous routing infrastructure at the network level.

The reputation systems have some disadvantages. A sale on an e-commerce system may result in the disclosure of the true identities of the seller and the buyer to each other (through mailing addresses, etc.), even if they use anonymous pseudonyms. We must also consider that the privacy of the pseudonym itself may need to be protected. For example, if pseudonym A assigns pseudonym B negative feedback in retaliation, then B's reputation is adversely affected due to the lack of privacy of B's feedback. Better solutions include: preserving the privacy of the feedback, or using disposable pseudonyms, which a user may change after every transaction (such as in the solution by Androulaki et al. [2]).

The reputation system is summarized in Table 8.11.

8.6 Discussion

Tables 8.12 and 8.13 provide a comparison of the reputation systems that aim to preserve privacy under the semi-honest adversarial model and the disruptive malicious adversarial model respectively. Centralized and decentralized architectures are abbreviated as C and D respectively in the tables.

8.6.1 The Semi-Honest Adversarial Model

The Secure Sum protocol is simple and efficient. However, secure sum is secure only under a restricted semi-honest adversarial model where the entities are not allowed to collude. The protocol is therefore not suitable for preserving privacy under the more realistic model where collusion is possible.

The schemes 1 and 2 by Gudes et al. provide security under the full semi-honest model. However, both schemes rely on Trusted Third Parties (TTPs). The issue with TTPs is that if they are not fully honest, they can learn private data with little or no effort.

Table 8.12 Privacy under the semi-honest adversarial model

System/ Protocol	Architecture	Target environment	Key security mechanisms	Privacy guarantee	Complexity (Messages)
Clifton et al. [16]—Secure Sum	D	Distributed environments	Secure multi-party computation	Probability: $\frac{1}{m+1}$, only if nodes don't collude	$O(n)$, where n = number of sites
Pavlov et al. [46]—WSS-1	D	Distributed environments	Secure multi-party computation, secret sharing	$\left(1-\frac{1}{n}\right)\left(\frac{N-b-1}{N-1}\right)$	$O(N) + O(n^2)$, where N = no. of potential witnesses, and n = no. of selected witnesses
Gudes et al. [28]—Scheme 1	D	Distributed environments	TTP, Public-key cryptography	Random guess across $|TrustSet_x(A)|$	$O(n)$, where $n = |TrustSet_x(A)|$
Gudes et al. [28]—Scheme 2	D	Distributed environments	TTP, Public-key cryptography, Secure product	Random guess across $|TrustSet_x(A)|$	$O(n)$, where $n = |TrustSet_x(A)|$
Gudes et al. [28]—Scheme 3	D	Distributed environments	Secure multi-party computation	A does not learn more information about $DTE(B, x)$, where $B \in TrustSet_x(A)$	$O(n^2)$, where $n = |TrustSet_x(A)|$
Hasan et al. [32]	D	Distributed environments	Secure multi-party computation, Trust awareness, Secret sharing	If k trustworthy agents in the set S_t, $k \ll n$	$O(n)$, where n is the number of feedback providers
Nin et al. [45]	D	Private collaborative networks	ElGamal encryption scheme	If the underlying encryption scheme is secure	$O(1)$

The reputation system by Nin et al. is very efficient. It requires exchange of a constant number of messages. However, the system is limited to Private Collaborative Networks, where reputation is computed based on the access control decisions of an entity. The reputation system is not applicable to more general social networks.

The protocol by Pavlov et al. (based on their first witness selection scheme) is secure under the full semi-honest model. Moreover, the protocol is general purpose,

Table 8.13 Privacy under the disruptive malicious adversarial model

System/ Protocol	Architecture	Target environment	Key security mechanisms	Privacy guarantee	Complexity (Messages)
Pavlov et al. [46]—WSS-2	D	Distributed environments	Verifiable secret sharing, discrete log commitment	If $b < \frac{N}{2} - n$	$O(n^3)$, where $n =$ number of witnesses
Androulaki et al. [2]	C	Centralized systems, Peer-to-peer systems	Anonymous credential systems, E-cash (bank), Blind signatures, Mixnets/ Onion Routing	If the underlying primitives (anonymous credential system, e-cash system, and blind signatures) are secure	$O(1)$
Kinateder and Pearson [37]	D	Peer-to-peer systems	Trusted platform, MIX cascades, Digital signatures	If the underlying primitives (trusted platform, MIX cascades, digital signatures) are secure	Not Provided
Steinbrecher [56]	C	E-commerce, Self-help forums, etc.	Pseudonym/ Identity management	If the provider (central server) is honest and secure	$O(1)$

that is, it may be used for many different applications. The protocol also does not rely on any TTPs or centralized constructs. The scheme 3 by Gudes et al. has similar properties. However, both these protocols have communication complexity upwards of $O(n^2)$, which is quite expensive.

The protocol by Hasan et al. builds on secure multi-party computation, trust awareness, and secret sharing to achieve a low complexity of $O(n)$ messages, where n is the number of feedback providers. The privacy of an agent a is preserved with high probability if it finds k trustworthy agents in the set of feedback provider agents S_t, such that $k \ll n$ and the probability that all k agents will collude to break agent a's privacy is low.

8.6.2   The Disruptive Malicious Adversarial Model

The reputation systems by Androulaki et al. and Steinbrecher are very efficient. They require a constant number of messages to be exchanged despite the number of feedback providers and the size of the system. However, each of these systems relies on a centralized construct. The reputation system by Androulaki et al. is based on the E-Cash system, which uses a centralized construct called the bank. Steinbrecher's reputation system has a central server as an integral part of its architecture. These centralized entities make these two systems unsuitable for fully decentralized environments.

Kinateder et al.'s reputation system provides anonymity in peer-to-peer systems under the disruptive malicious model. However, the system requires the presence of special hardware called Trusted Platform (TP) at each peer. Additionally, the system requires that messages be exchanged using MIX cascades. These requirements limit the reputation system to specialized networks where TPs are available at each peer and where MIX cascades are in use.

The protocol by Pavlov et al. (based on their second witness selection scheme) is secure under the disruptive malicious model. The protocol does not require centralized constructs or specialized networks. However, the issue with the protocol is that it needs $O(n^3)$ messages to be exchanged, which is very expensive.

References

1. Amirbekyan, A., Estivill-Castro, V.:   A new efficient privacy-preserving scalar product protocol. In: Proceedings of the Sixth Australasian Conference on Data Mining and Analytics, 2007
2. Androulaki, E., Choi, S.G., Bellovin, S.M., Malkin, T.:   Reputation systems for anonymous networks. In: Proc. of PETS'08, 2008
3. Atallah, M.J., Du, W.:   Secure multi-party computational geometry. In: Proceedings of the Seventh International Workshop on Algorithms and Data Structures (WADS 2001), 2001
4. Belenkiy, M., Chase, M., Erway, C.C., Jannotti, J., Kupcu, A., Lysyanskaya, A., Rachlin, E.: Making p2p accountable without losing privacy. In: Proceedings of the 2007 ACM Workshop on Privacy in Electronic Society, 2007
5. Belenkiy, M., Chase, M., Kohlweiss, M., Lysyanskaya, A.:   P-signatures and noninteractive anonymous credentials. In: Theory of Cryptography, 2008
6. Bertino, E., Ferrari, E., Squicciarini, A.C.:   Trust-x: A peer-to-peer framework for trust establishment. IEEE Trans. Knowl. Data Eng. **16**(7), 827–842 (2004)
7. Bigley, G.A., Pearce, J.L.:   Straining for shared meaning in organization science: Problems of trust and distrust. Acad. Manag. Rev. **23**(3), 405–421 (1998)
8. Bo, Y., Min, Z., Guohuan, L.:   A reputation system with privacy and incentive. In: Proceedings of the Eighth ACIS International Conference on Software Engineering, Artificial Intelligence, Networking, and Parallel/Distributed Computing (SNPD'07), 2007
9. Burt, R.:   Structural Holes: The Social Structure of Competition. Harvard University Press, Cambridge (1995)
10. Camenisch, J., Lysyanskaya, A.:   An efficient system for non-transferable anonymous credentials with optional anonymity revocation. In: EUROCRYPT 2001, 2001
11. Camenisch, J., Lysyanskaya, A., Meyerovich, M.:   Endorsed e-cash. In: Proceedings of the IEEE Symposium on Security and Privacy, 2007

12. Capra, L.: Engineering human trust in mobile system collaborations. In: Proceedings of the 12th ACM SIGSOFT International Symposium on Foundations of Software Engineering, Newport Beach, CA, USA, 2004

13. Chaum, D.: Untraceable electronic mail, return addresses, and digital pseudonyms. Comm. ACM **24**(2), 84–88 (1981)

14. Chaum, D.: Blind signatures for untraceable payments. In: Proc. Advances in Cryptology (CRYPTO '82), 1982

15. Chaum, D.: Blind signature systems. In: Advances in Cryptology (CRYPTO'83), 1983

16. Clifton, C., Kantarcioglu, M., Vaidya, J., Lin, X., Zhu, M.Y.: Tools for privacy preserving distributed data mining. SIGKDD Explor. **4**(2), 28–34 (2003)

17. Dingledine, R., Freedman, M.J., Hopwood, D., Molnar, D.: A reputation system to increase mix-net reliability. In: Proceedings of the 4th International Workshop on Information Hiding, 2001

18. Dingledine, R., Mathewson, N., Syverson, P.: Reputation in privacy enhancing technologies. In: Proceedings of the 12th Annual Conference on Computers, Freedom and Privacy, 2002

19. Dingledine, R., Mathewson, N., Syverson, P.: Reputation in p2p anonymity systems. In: Proceedings of the Workshop on Economics of Peer-to-Peer Systems, 2003

20. Dingledine, R., Mathewson, N., Syverson, P.F.: Tor: The second-generation onion router. In: Proceedings of the USENIX Security Symposium, 2004

21. Du, W.: A study of several specific secure two-party computation problems. PhD thesis, Purdue University, West Lafayette, IN (2001)

22. ElGamal, T.: A public-key cryptosystem and a signature scheme based on discrete logarithms. IEEE Trans. Inform. Theor **IT-31**(4), 469–472 (1985)

23. Gal-Oz, N., Gudes, E., Hendler, D.: A robust and knot-aware trust-based reputation model. In: Proceedings of the Joint iTrust and PST Conferences on Privacy, Trust Management and Security (IFIPTM 2008), 2008

24. Gambetta, D.: Trust: Making and Breaking Cooperative Relations, chapter Can We Trust Trust? pp. 213–237. University of Oxford, Oxford (2000)

25. Gilbert, E., Karahalios, K.: Predicting tie strength with social media. In: Proceedings of the Conferece on Human Factors in Computing Systems (CHI'09), 2009

26. Goldreich, O.: The Foundations of Crypto. - Vol. 2. Cambridge University Press, Cambridge (2004)

27. Granovetter, M.: The strength of weak ties. Am. J. Sociol. **78**, 1360–1380 (1973)

28. Gudes, E., Gal-Oz, N., Grubshtein, A.: Methods for computing trust and reputation while preserving privacy. In: Proc. of DBSec'09, 2009

29. Guha, R., Kumar, R., Raghavan, P., Tomkins, A.: Propagation of trust and distrust. In: Proceedings of the International World Wide Web Conference (WWW 2004), 2004

30. Hasan, O., Bertino, E., Brunie, L.: Efficient privacy preserving reputation protocols inspired by secure sum. In: Proceedings of the 8th International Conference on Privacy, Security and Trust (PST 2010), Ottawa, Canada, August 17–19 2010

31. Hasan, O., Brunie, L., Bertino, E.: k-shares: A privacy preserving reputation protocol for decentralized environments. In: Proceedings of the 25th IFIP International Information Security Conference (SEC 2010), pp. 253–264, Brisbane, Australia, September 20–23 2010

32. Hasan, O., Brunie, L., Bertino, E.: Preserving privacy of feedback providers in decentralized reputation systems. Comput. Secur. **31**(7), 816–826 (2012)

33. Hoffman, K., Zage, D., Nita-Rotaru, C.: A survey of attack and defense techniques for reputation systems. ACM Comput. Surv. **41**(4) (2009)

34. Homans, G.: The Human Group. Harcourt, Brace, & World, New York (1950)

35. Ioannidis, I., Grama, A., Atallah, M.: A secure protocol for computing dot-products in clustered and distributed environments. In: Proceedings of the 2002 International Conference on Parallel Processing, 2002

36. Josang, A., Ismail, R.: The beta reputation system. In: Proceedings of the 15th Bled Electronic Commerce Conference, Bled, Slovenia, 2002

37. Kinateder, M., Pearson, S.: A privacy-enhanced peer-to-peer reputation system. In: Proc. of the 4th Intl. Conf. on E-Commerce and Web Technologies, 2003

38. Kinateder, M., Terdic, R., Rothermel, K.: Strong pseudonymous communication for peer-to-peer reputation systems. In: Proceedings of the 2005 ACM Symposium on Applied computing, 2005

39. Kissner, L.: Privacy-preserving distributed information sharing. PhD thesis, Computer Science Department, Carnegie Mellon University, PA, USA, July 2006. CMU-CS-06-149

40. Lin, N., Ensel, W.M., Vaughn, J.C.: Social resources and strength of ties: Structural factors in occupational status attainment. Am. Socio. Rev. **46**(4), 393–405 (1981)

41. Marsden, P.V., Campbell, K.E.: Measuring tie-strength. Social Forces **63**, 482–501 (1984)

42. McKnight, D.H., Cummings, L.L., Chervany, N.L.: Initial trust formation in new organizational relationships. Acad. Manag. Rev. **23**(3), 473–490 (1998)

43. Mika, P., Gangemi, A.: Descriptions of social relations. Technical report, Department of Business Informatics, Free University Amsterdam, The Netherlands, Retrieved February 17, 2011

44. Mitchell, C. (ed.): Trusted computing. The Institution of Engineering and Technology, Stevenage, Herts., SG1 2AY, UK (2005)

45. Nin, J., Carminati, B., Ferrari, E., Torra, V.: Computing reputation for collaborative private networks. In: Proceedings of the 33rd Annual IEEE International Computer Software and Applications Conference, 2009

46. Pavlov, E., Rosenschein, J.S., Topol, Z.: Supporting privacy in decentralized additive reputation systems. In: Proceedings of the Second International Conference on Trust Management (iTrust 2004), Oxford, UK, 2004

47. Pearson, S., Balacheff, B. (eds.): Trusted Computing Platforms: TCPA Technology in Context. Prentice Hall, Upper Saddle River, NJ 07458, USA (2003)

48. Pedersen, T.P.: Non-interactive and information-theoretic secure verifiable secret sharing. In: Proceedings of the 11th Annual International Cryptology Conference on Advances in Cryptology, 1991

49. Pennebaker, J.W., Francis, M.E., Booth, R.: Linguistic Inquiry and Word Count: LIWC2001. Erlbaum Publishers, Mahwah, NJ (2001)

50. Petroczi, A., Nepusz, T., Bazso, F.: Measuring tie-strength in virtual social networks. Connections **27**(2), 39–52 (2007)

51. Pingel, F., Steinbrecher, S.: Multilateral secure cross-community reputation systems for internet communities. In: Proceedings of the Fifth International Conference on Trust and Privacy in Digital Business (TrustBus 2008), 2008

52. Pouwelse, J.A., Garbacki, P., Epema, D.H.J., Sips, H.J.: The bittorrent p2p file-sharing system: Measurements and analysis. In: Peer-to-Peer Systems IV, 2005

53. Resnick, P., Zeckhauser, R.: Trust among strangers in internet transactions: empirical analysis of ebay's reputation system. In: Volume 11 of Advances in Applied Microeconomics, pp. 127–157, 2002

54. Resnick, P., Zeckhauser, R., Friedman, E., Kuwabara, K.: Reputation systems. Comm. ACM **43**(12), 45–48 (2000)

55. Schiffner, S., Clauß, S., Steinbrecher, S.: Privacy and liveliness for reputation systems. In: Proc. of EuroPKI'09, pp. 209–224, 2009

56. Steinbrecher, S.: Design options for privacy-respecting reputation systems. In: Security and Privacy in Dynamic Environments, 2006

57. Voss, M., Heinemann, A., Muhlhauser, M.: A privacy preserving reputation system for mobile information dissemination networks. In: Proceedings of the First International Conference on Security and Privacy for Emerging Areas in Communications Networks (SECURECOMM), 2005

58. Wellman, B., Wortley, S.: Different strokes from different folks: community ties and social support. Am. J. Sociol. **96**(3), 558–588 (1990)

59. Yao, A.C.: Protocols for secure computations. In: Proceedings of the 23rd Annual Symposium on Foundations of Computer Science, 1982

Chapter 9
Security and Privacy Issues in Mobile Social Networks

Ariel Teles, Francisco José Silva, and Rômulo Batista

Abstract Mobile devices are becoming cheaper and resourceful, with more processing power and storage, multiple network interfaces, GPS and a variety of physical sensors, allowing the execution of sophisticated context-aware applications. Through the use of Mobile Social Networks (MSN), users may access, publish and share content generated by them at anytime and anywhere, enhancing their social interactions. MSN applications are characterized by the integration of context information to the social network content, enriching the existing applications and providing new services. On the other hand, the extensive use of context information leads to new privacy and security challenges, which is the scope of this chapter, which aims to describe the main concepts, research challenges and solutions for this area.

9.1 Introduction

The development and popularization of Information and Communication Technologies, in particular the Internet, has caused great impact in people lifestyle. Some of the most important changes are related to the way that people establish their social relations and interact with each other. Different ways for interaction are now currently available for individuals and organizations, ranging from e-mail and instant messages to online conferencing, collaborative editing tools and social media.

Social media are means of communication used for social interactions among its users. Social Networks are the most popular social media and have recently motivated great interest in both academia and industry. Conceptually, a social

A.S. Teles (✉) · F.J. da Silva e Silva · R. de Carvalho Batista
Distributed Systems Laboratory, Federal University of Maranhão, São Luís, MA, Brazil
e-mail: ariel@dee.ufma.br; fssilva@deinf.ufma.br; romulocbatista@gmail.com

R. Chbeir and B. Al Bouna (eds.), *Security and Privacy Preserving in Social Networks*,
Lecture Notes in Social Networks, DOI 10.1007/978-3-7091-0894-9_9,
© Springer-Verlag Wien 2013

network is a structure of entities connected to each other through one or more interdependencies [72]. These entities can be individuals, organizations or systems that are related in groups and whose interaction is done through Information and Communication Technologies.

In recent years, it is possible to notice a rapid growth of mobile computing[60], which has become increasingly part of the human society. Mobile devices can aggregate connectivity and allow the access, processing and sharing of information any time and anywhere, providing ubiquity. Also they are becoming cheaper and providing more resources (e.g., storage and processing power, multiple network interfaces, GPS receiver, and a variety of sensors, such as proximity, gyroscopes and accelerometers), allowing the execution of even more sophisticated applications.

Mobile Social Networks (MSN) are a subclass of Social Networks, in which users make use of mobile devices with wireless communication technologies to access social networks resources. In this way users can access, publish and share content in a mobile environment to explore social relationships. MSNs are well characterized by adding rich context information because mobile devices can sense, combine and infer a broad range of contextual information, providing them to applications. Thus, since mobile devices provide ubiquity, MSNs can greatly benefit human interactions by improving the efficiency and effectiveness of its services [43].

General privacy policies frequently applied in social networks are not sufficient for MSNs, since a variety of sensible information are usually made available in real time, such as the user's physical location, his/her preferences, relationships and interests. The use of contextual information gathered from mobile devices sensors, which in many cases occurs without the user's awareness, leads to new privacy issues that must be addressed. In addition, there are greater difficulties in providing security for mobile applications, once these environments have intermittent wireless connectivity and the bandwidth is typically low. The amount of users accessing resources, which can be of hundreds of thousands of nodes, also influence the design and implementation of mobile system security mechanisms. In this way, the provision of security and privacy in MSN environments is a challenging task [63], and several recent research initiatives aim to address information security issues, including privacy, trust and data security. In this chapter we survey the state of the art on security and privacy for MSNs.

This chapter aims to describe the main concepts, research challenges and solutions related to security and privacy in MSNs. It is organized as follows. Section 9.2 provides the necessary background to the understanding of the rest of this chapter by providing a detailed characterization of MSNs, highlighting the benefits of aggregating context information in social networks and describing basic concepts of security and privacy. Section 9.3 shows multiple security and privacy problems found on social networks that are also present in MSNs. Section 9.4 describes how the inclusion of context information may violate privacy and security issues in MSNs. In recent years, several middleware solutions for MSNs were presented, but only some of those projects have made proposals in the field of privacy and security, which are described in Sect. 9.5. Finally, Sect. 9.6 provides

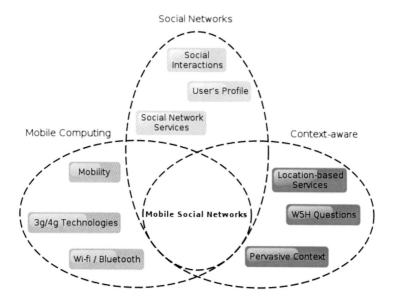

Fig. 9.1 Definition of mobile social networks

a critical analysis of the area, identifying future directions and open research issues.

9.2 Background

MSNs can be viewed as a combination of three threads of knowledge: Social Networks, Mobile Computing and Context-aware Systems—an extension of the concept provided by Kayastha et al. [43]. Social Networks provide functionalities for setting up profiles that represent entities which relate socially by exchanging information. Mobile computing brings users the ability to be "always on", due to the mobility support provided by portable devices and wireless network technologies. Context-aware computing enables to determine meta-information based on the sensors embedded in devices, such as the user's current location, climate and local temperature, people in neighbourhood, and also to infer information, such as the action being performed by the user and her/his intention to do it. This view is illustrated in Fig. 9.1.

In the literature, even though one can find several architectures for MSNs that allow different ways for establishing mobile users communities and promoting their interaction, MSN software can be broadly classified into two major groups: centralized and distributed, as illustrated in Fig. 9.2. The architectural choice used to build the mobile social networking software generates a large impact on its applicability and services that may be available. The architecture is also decisive for

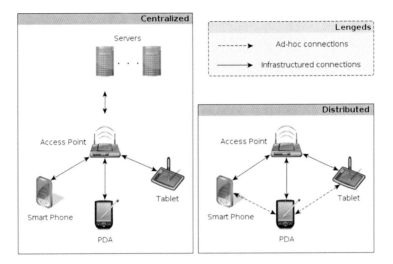

Fig. 9.2 Architectures of mobile social networks

choosing the algorithms to be adopted for the construction of software, including the mechanisms used for enforcing security and privacy.

In a *centralized architecture*, data are centralized in one or more servers, which are responsible for managing and delivering them to mobile users. These data correspond to information from user's profiles, groups, context information, dashboards, notifications and so on. Mobile users are represented by mobile applications hosted by devices such as smart phones, PDAs or tablets. In this architecture, all communication is mediated by the servers. In this way, content is necessarily accessed through servers, and only from them, through a established link with the mobile applications. Links are established through an access infrastructure using wireless communication technologies. In some cases, context information is gathered through a Body Sensor Network, such as in [11, 18]. Examples of popular MSNs with centralized architecture are Facebook,[1] Twitter[2] and Instagram.[3] CenceMe [56] and WhozThat [9] also adopt a centralized approach, differing by the way that users access the social network, which is done exclusively by mobile devices.

In a *Distributed Architecture*, mobile users communicate directly without the need of using servers to mediate their interactions. As seen in Fig. 9.2, connections between users devices can be established through an infra-structured network using access points or by means of ad hoc connections. In this MSN architecture, users can communicate and share content without Internet access, with minimal network

---

[1]http://m.facebook.com/.

[2]https://mobile.twitter.com/.

[3]http://instagr.am/.

infrastructure. Knowledge inferred from social relations between users can be used for building better routing and security protocols (e.g., considering the frequency in which users physically met) [1, 46, 61].

A final way that MSN components may be arranged is through a *Hybrid Architecture*. This architecture combines centralized and distributed approaches by allowing mobile users to access and share data across servers (centralized), as well as to establish direct connections with each other (distributed).

All those MSNs architectures can provide resources for allowing the dynamic establishment of associations between users, forming what is called Dynamic Communities (or dynamic groups) [15, 48, 75]. Dynamic communities correspond to groups of users that are automatically built based on (a) common interests derived from the users profiles, (b) inference of common interests based on historical user activity, (c) users physical vicinity or his/her location. Dynamic communities are formed spontaneously by physically close users and the performance of some joint activity based on their common interests (e.g., students of the same course exchanging information about a test before it is applied). In this case, the network is more dynamic, since users can get in and out of communities as they move or change their interests.

9.2.1 The Role of Context Information

Context is any information used to characterize a situation of an entity. An entity can be a person, place, or object that is considered relevant to the interaction between user and application, including themselves [24]. Context-aware systems are able to adapt their functionality to the user's current context without her/his explicit intervention. MSNs are well characterized by adding rich context information, since mobile devices can sense, combine and infer a broad range of contextual information and provide them to the applications.

Context information can be acquired by sensors of various types, including: light, visual (camera), audio, motion or acceleration, location, touch, temperature and physical (biosensors) [8]. Data from these sensors can be used individually or combined to infer a user situation (e.g., whether her/he is busy or free to receive notifications, connections or calls). This inference can provide a wealth of information that can be used by the MSN software to increase the level of collaboration between its users. Users can become more aware of their social network friends situation (such as their location or mood), leading to a greater integration between the real and virtual worlds. An example toward this integration is the project Touch Me Wear [12], where context information collected through a wearable device, such as vicinity and physical contact of friends, is used to sense physical interactions between people and display it on Facebook.

Location-aware systems are a subclass of context-aware systems. They focus on explore location as context information, providing added value to a user [64]. Several indoor and outdoor positioning systems have been developed providing location data with different characteristics, such as precision and coverage. Among the most used ones, we can highlight the GPS, a satellite-based global position system, cellular networks position systems, and wireless networks approaches, which are usually based on measuring the signal strengths received from nearby access points. The term Geo-social networks is used to denote networks which provide context-aware services that help to associate location with users and content [71]. A combination of geographic information technology and social network services can help people make social connections more convenient [37]. Location information is used by various MSN applications, such as Google Latitude[4] and Foursquare.[5] The term Situated Social Context denotes a set of people who share some common spatio-temporal relationship with an individual, which turn them into potential peers for information sharing or interaction in a specific situation [27].

Social computing embraces mobility and creates a trend of Pervasive Social Context. The Pervasive Social Context of an individual is the set of information that arises from direct or indirect interaction between people carrying sensor-equipped pervasive devices, connected through the same social network service. Schuster et al. [66] classify the variety set of information comprising a Pervasive Social Context based on the W5G questions:

- *Who*: express *who* are the participants involved in the consuming and/or production of context information;
- *What*: relates to *what* kind of context is used or is important for the application;
- *Where*: relates to *where* (physical location) social links and interactions are consumed and/or produced;
- *When*: relates to the characterization of interactions between peers and the context information they produce based on a temporal perspective;
- *Why*: relates to *why* a context information is to be used, determining the cause or reason of it being used by the application (i.e., related to the application goal);
- *How*: express *how* the context information (from the real world, virtual world or both) may influence or compose virtual applications.

9.2.2   Security and Privacy Principles

Information security comprises the set of measures aiming to preserve and protect information and information systems. The fundamental properties concerning information security are confidentiality, integrity, and availability, called CIA properties. Confidentiality ensures that information should be available only to entities duly authorized. Integrity denotes that information should not be destroyed or compromised (modified during its transmission). Availability ensures that information,

---

[4]https://www.google.com/latitude/.

[5]https://foursquare.com/.

services and system resources must be available whenever they are needed for legitimate users. Basic security requirements also include Authentication, which ensures that communication from one node to another is genuine (i.e., a malicious node cannot masquerade as a trusted network node), and Non-repudiation, the ability to ensure that a node cannot deny the sending of a message that it originated.

Some authors distinguish between privacy and confidentiality. Confidentiality refers to the ability of protect any given data from those who are not authorized to view it [3], while privacy is confidentiality applied to personal data. Oyomno et al. [58] define privacy based on three abilities: (a) solitude—the freedom from observation or surveillance (i.e., capacity to prevent current or past private data from being visible to others); (b) anonymity—the freedom from being identified in public and (c) reservation—the freedom to withdraw from communication (i.e., power to interrupt any communication at any time).

Another issue related to security is trust. We say that a given entity "trust" a second one when the first entity makes the assumption that the second one will behave exactly as he/she expects [13]. Trust and distrust guide the actual behaviours of an entity towards another entity. They are beliefs about the past and future behaviour of the other entity [20].

The need of reaching these properties varies according to the application requirements. To define the security requirements that an application must achieve, developers should consider several issues, such as the identification of the system's resources that must be protected, the computational environment where the application will be executed and the identification of its potential threats, what are application specific requirements and legislation related to the application domain (in health care, for instance, strict limitations on the manipulation of patient data are usually specified by law).

9.3   Security and Privacy Issues in Online Social Networks

An online social network can be defined as a collection of web-based services that allow individuals to (1) construct a public or semi-public profile within a bounded system, (2) articulate a list of other users with whom they share a connection and (3) view and traverse their list of connections and those made by others within the system. The nature and nomenclature of these connections may vary from site to site [17]. Mobile Social Networks (MSNs) comprise a subset of online social networks. This section highlights the main security and privacy issues found in online social networks and discuss which of them also applies to MSNs.

The text is organized as follows. Initially, privacy issues will be presented and organized aiming to show their origins. Several attacks or attempts to misuse the social network by malicious users (e.g., malware attacks, spam and sybill) are presented next, with the proposed solutions to them that can be found in the literature. Finally, anonymization and de-anonymization issues in social networks are also discussed.

Fig. 9.3 Social graph

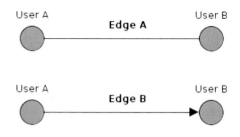

9.3.1 Privacy Issues in Social Networks

As observed by Zhang et al. in [74], there is a conflict in social network projects that must be addressed by the system developers. On the one hand, there is a demand for providing a variety of social network resources in order to improve usability and sociability, such as recommendation of friends, places and pages. On the other hand, as more resources are inserted, more security breaches arise that are used by the attackers to obtain improper access to personal information, which is called privacy leak.

In social networks, users usually generate a large amount of data (or content). Some of these data are strictly related to user's personal information, such as photos, videos, messages or posts, among others. Normally data reports to the user personal life activities and as she/he is usually almost always connected, mainly through mobile devices (aggregating the ability of accessing the social network "every time, everywhere"). All information can be shared and accessed by many entities, which represent the origins of privacy attacks to the social network users. To Gao et al. [31], these origins are considered breaches and may be: other users of the social network, third-party applications and the service provider itself.

Other Users of the Social Network

Other users represent a threat due to, mostly, the ease that malicious users have to create an account through the service provider and become social network genuine members. When this happens, they can do attempts of privacy invasion to obtain other users personal information. The protection offered by the provider is to promote access control mechanisms, so the user can set a configuration informing the users she/he trusts or not. However, the access control mechanism can be implemented in several ways, ranging from a fine granularity to a coarse grained one.

Social network can be represented by a graph, where the vertex express users and the edges express relationship between them. For example, when an user A has a friendship relationship with an user B, there is an edge linking the vertex A (user A) to the vertex B (user B), as illustrated in the Fig. 9.3 by the edge A. Depending of the social network, this edge may possess a guidance (direction), edge B on the figure.

In this scenario, privacy invasion attacks can occur via:

- Direct access or connected users: vertex A has an edge linking him to the vertex B;
- Indirect access or users connected indirectly: vertex A does not have an edge linking him to the vertex B, but both vertex are linked by a vertex C. So, there is an indirect link between them, with one hop. In this case, vertex A is "friend-of-friend" of vertex B;
- Public access: there is no edge or other vertex linking vertex A to vertex B or the amount of hops between them is high. Some access control mechanisms do not enable the addition of access restrictions to users in general. With this policy the access to content is public for every user.

The work presented in [68] suggests the use of privacy policies to restrict permission only to users who have direct access. So, only connected users can see published content between them. Nevertheless, the authors conclude that this is not a viable alternative, since nowadays few users may choose to use a social network with such a restrictive privacy policy. This happens because this approach would decrease the sociability and the possibility of making new friendships, restricting one of the most used resources in social networks, the discovery of friendship (or the forming of edges from the virtual world).

Normally it is possible to categorize friends according to some criteria or attribute in such way that privacy policies can be applied. Each category can have a set of access rules. Fang et al. [28] propose a mechanism for the automatic creation of categories of friends. The authors create a model that, through machine learning, allows the creation of users communities (forming clusters) and replicates access permissions for all the users that comprise these communities. To perform the replication, the user must define a permission to a given resource for at least one friend and the proposed mechanism automatically replicates the permission to all other friends comprising the user's communities. This model greatly decreases the users effort to grant access permissions even though, in some cases, friends can be erroneously categorized.

To exemplify the model presented in [28] consider the following example. Bob accepts a friendship invitation sent by Frank, his study buddy. In this moment, Bob informs which are the permissions of Frank for accessing his personal data. Later on, Bob receives more friendship invitations and accepts them. Bob does not need to explicitly inform for all those new friends what are the access control permissions for his personal data since the proposed model will automatically detect that they are part of the same community, due to common attributes that they share (e.g., where they study). This approach can also be applied in MSNs.

Rahman et al. presented in [62] a new privacy architecture focused on allowing a finer granularity for defining access permissions to published content in social networks called PCO (Privacy-sensitive architecture for Context Obfuscation). Their approach combines an obfuscation technique that generalizes the user's published content. For example, if an user publishes "I'm watching the Batman's movie", it makes available this content to some users as just "I'm watching a movie", according to some defined criteria. So, the information containing the name of the movie is obfuscated. Furthermore, the user can also set a time parameter that is added to the access restrictions defining the duration (period) in which the information will stay available.

These previous solutions are interesting from the privacy assurance viewpoint. However, when not properly used they can also lead to an imbalance between sociability and privacy and some issues can arise. For example, in [62], the PCO technique hides content in posts that could be important for the exchange of messages between users or to form new friendships. In the example above, users could have a common interest (the Batman's movie) that motivates a relationship.

In [14], Leyla Bilge et al. presented two attacks called *Same-site profile cloning* and *Cross-site profile cloning*. In the first one, the attacker creates an account which has an identical profile information of some user. Next, she/he sends requests for the establishment of edges (friendship relationship) to a list of friends of the legitimate user that was cloned. These friends, believing in the familiarity of the person (study buddy, employment colleague, family member), accept the request and, then, start to expose personal information to the attacker. The second attack is similar to the first one, differentiated by the fact that the creation of a fake profile is not done in the same social network where the legitimate user has a valid profile but in a different one. In [14], the authors proposed a method for the detection of these attacks through Captcha,[6] since the definition of fake profiles is usually done automatically through malicious code.

Regarding the indirect access in social networks, an issue found in literature is the sharing of publish content. The user profile can reveal personal information to friends such as gender, birth date, relationship status, e-mail address, phone number, home address, and even political affiliations. A user can share these personal information with friends through the user's profile, status updates, messages, and status replies. This puts an implicit responsibility on a user's social networking friends to keep shared information private and honour the implicit, and sometimes explicit, trust placed in friends by the user [51]. In Facebook, this resource is called *Share*, and in Twitter is called *Retweet* [52, 54]. Using these resources, a user can see published content by another user that does not have an edge established in the social graph. In other words, contents are accessed by other users (non-friends) without permission of their owners.

Third-Party Applications

Third-party applications are written codes that use an API (Application Programming Interface) made available by the service provider, which gives an open platform for the creation and addition of new applications inside the social network,

---

[6]A Captcha is a type of challenge-response test used in computing as an attempt to ensure that the response is generated by a person (human being), not a machine.

aiming to allow the extension of existing services. This kind of application is necessary to provide new social network services, demanded by the user themselves. Nevertheless, these applications are written by other entities and naturally cannot be always trusted. Moreover, they usually require to the user the permission to freely access her/his profile information. These third-party application can be categorized as in Facebook: games, fun, lifestyle, music, news, photos, videos, sports, places and so on.

One type of the third-party applications that offers more risks come from publicity agencies. These companies usually have access to the user's information authorized by the service provider in order to promote personalized advertisement according to inferred interests of users. These advertisements are considered context aware since they usually take into consideration context information extracted from the users profiles, such as frequented places, published messages, communities in which users are inserted, among others. In this way, the publicity agencies can better direct their announcements.

In the work [45], a solution is proposed to control the disclosure of personal information by third-party applications. The authors suggest the use of a component called FAITH that works as a firewall among third-party applications and the social network. FAITH is a server that controls all requests originated by third-party applications, controlling which applications can have access to which user's information. All calls from third-party applications to the social network API is done through FAITH, which creates a log that is made available to the user, informing all successful and unsuccessful attempts to access the user personal data. To prove the viability of the proposed approach, the authors developed an implementation used on Facebook.

Third-party applications that have users explicit permission to access their information can make the user data available to other entities. For example, a social game that requires a subset of the user's profile information (e.g., her/his name, last name, e-mail, date of birth and hometown) can distribute it to another entity. In [67], the authors propose a framework called *xBook* which controls the data flow trans-mission of third-party application. So, if a third-party application wants to re-pass a personal information acquired from the social network to another entity, an explicit permission issued by the user to do so is needed. Some framework components are responsible to control, alert and restrict unauthorized information flow.

Social Network Provider

The provider is the entity responsible for supplying necessary services to the social network users. It has access to all personal data inserted by users. However, users are ceasing to have confidence in this service provision and are increasingly concerned with how and by whom their personal data is being used.

Many works found in the literature have proposed encryption techniques to protect data transfer to and from the social network and its storage in the server provider [22, 34, 49, 50, 70] in a way that the latter does not have direct access to the users published data. While the cited proposals can be applied to MSNs, their implementation are based on extensions for desktop web browsers or are built in applications that uses the social network API for desktops (mostly the Facebook one).

Among these works, one stands by proposing an solution that is independent of the online social network to which it is applied, the *Persona* [7]. It embraces two encryption mechanisms: the first one uses an asymmetric key mechanism, where the user informs his public key to her/his friends. The second one, which differentiates the proposal from the other ones, is called ABE (Attribute-Based Encryption). In this mechanism, each user generates an ABE public key (APK) and a secret master key (AMSK). The user can add to the APK attributes used for controlling the access permissions to resources that are shared with groups of friends. In this way, the APK also works as an access control mechanism.

To exemplify the use of the Persona ABE mechanism, consider the following example: Bob sends his APK to Alice containing the attribute "family". Bob must organize his friends in groups (e.g., "family" and "friends"). When Bob publishes a content, he restricts the access to it to the groups that he wants. Alice will be able to access the published content if the attribute inserted in the APK received from Bob is for the group "family". The restriction defined for controlling the access to a given content also accepts logical operators. For example, when publishing a photo, Bob can restrict the access to this content to the groups "family" OR "friends", or "family" AND "friends" (meaning that an user must be in both the groups in order to have access to the content).

Other works found in the literature propose the utilization of a distributed social network architecture as solution to the privacy leak problem originated by the service provider. In this case, there is no centralized server that possesses all the user's personal information. In [2], the authors propose an approach where the service provider only acts as a name server. Communication performed between social network users is done directly and the user's data are stored locally by the client application. For instance, consider that Bob wants to communicate with Alice and that she is in his friends list. Bob knows only Alice's identifier, in this case, her name itself. Both users must be registered in the social network server that is used by Bob in order to obtain Alice IP address given her identifier. Communication and content transfers are done directly between Bob and Alice, without a server mediation.

9.3.2   Attacks Against Social Networks

In the literature, one can found several attacks aiming to disturb the social network correct operation, where the service provider, third-party applications or users may be damaged. Some attacks intend to undermine the users trust in the server provider that, consequently, will lose money due to the reduction of advertisements revenue. That will also have an impact on developers of third-party applications, which will decrease their product negotiations. Users can also be damaged in several ways, suffering privacy leak or being infected with malicious code.

An issue that is being increasingly common in social networks is spam. This term is used to reference the receipt of a message not requested, generally assuming a form of a product advertisement. Spam becomes a serious issue when combined to the dissemination of malicious code (malware). Grier et al. [33] analysed 25 millions URLS shared on Twitter and found out that 8 % of them pointed to web pages catalogued in black lists for containing malicious codes. The user, by clicking in a URL with malicious code, infects her/his computer and allows the malware owner to access her/his account. By analysing the accounts used for sending spams, the authors also found evidences that spam massages were originated by legitimate accounts infected by malicious code.

The dissemination of malicious code generates a large impact on social networks, as occurred in August 2008 when the *Koobface* malware caused great problems to several networks [73]. In this case, only the users (and not the service providers) were infected and were used for the dissemination of the malware. However, service providers can act by detecting, reducing or even stopping malware dissemination. In [73], Xu et al. propose a worm detection system for online social networks. The general idea is to take advantage of the worms propagation characteristics, capture evidences that probes it and identify when an infection occurs based on the network message flow. Thus, the approach is considered a detection system based on anomaly. An evaluation of the system was performed using the Flickr[7] social network and identified that in a universe of 1.8 million users 0.13 % of the users accounts were infected with some type of malicious code.

A Sybil attack [26] happens when an attacker adulterates the reputation system of the network by creating several accounts (some attacks require only 7 accounts, while others millions of them). The attacker influences sundry results using the controlled accounts that act collectively as legitimate users. In MSNs, unique identifiers associated with the mobile devices, such as the SIM card used in GSM networks—Global System for Mobile Communications, can be used to difficult the creation of simultaneous multiple accounts, a requirement for the effectiveness of this attack. In this case, one can limit the amount of the social network accounts that can be created for each mobile device equipment.

9.3.3   Anonymization in Social Networks

Service providers of social networks usually profit from their user's stored data by offering them to other companies and the academy. However, since the user's generated content are very sensitive, data should be provided in a way that it does not reveal the users identity. By failing to do so, users of the social network will

---

[7]http://www.flickr.com/.

be harmed by the leak of private information and this would also damage the confidence that they have on their service provider. The technique used to avoid users identification from data provided by social networks service providers is called anonymization. The reverse process is called de-anonymization.

The main goal of anonymization in social networks is to guarantee that a single vertex (user) will not be distinguished among the many other vertex, which is called l-anonymization. The data set is called k-anonymous if, given some vertex of the network, it is undistinguished of $k - 1$ other vertex.

MSNs bring new challenges to the construction of anonymization/de-anonymization algorithms, since the network graph is usually much more dynamic. For example, in dynamic communities vertex and edges change quickly, according to the user context.

9.4   Privacy Issues in Context-Aware MSNs

Context information provides a wealth of resources and the possibility to construct many kinds of applications and services for MSNs. However, they also impose new privacy issues that are discussed in this section.

As physical sensors and applications with the ability to infer context evolve, more and more context information becomes published on MSNs. Context information published on social networks may compromise the users privacy because they can contain (or from them one can infer) very sensitive personal information that a user would not want to become available to other social network members. One aggravating factor is the common practice of aggregating context information to ordinary data (such as adding to a photo the location, date and time it was taken), which is not always done with the user consciousness. In this way, context information can be exposed to users who only have permission to access the published content (and not it's attached context information), leading to a privacy leak of the content owner. Moreover, depending on the social network application domain, very sensitive information must be continuously transmitted, such as in the healthcare domain, where through sensors (biosensors) healthcare professionals can continuously monitor the patient's health status. Schuster et al. [66] argue that privacy management should be applied to different levels of content information, from single personal contextual information (single sensor) to context information inferred from multiple sensors.

Another issue is that malicious users can produce false context information and send through an application. This information can be used by attackers to gain access to some resource (called spoofing attack) or gain a temporary identity by acting as a legitimate user (called faking or fingerprinting attack) [25]. In other words, the improper handling of context information can seriously influence the reliability of MSNs applications. This becomes a severe problem when social network platforms are used in critical environments, as in the health domain, where

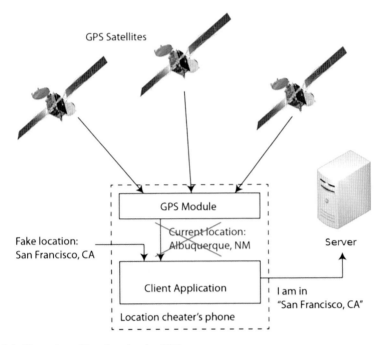

Fig. 9.4 Illustration of location cheating [36]

medical decisions can be taken based on manipulated context information, having a direct influence in the patient's treatment.

Another example related to the production of false context information is the location cheating attack [36], illustrated in Fig. 9.4. Here, when the GPS module of the mobile phone returns the current location information to the MSN application, the attacker blocks the genuine information and starts to feed fake location information, making the server provider believe that the mobile phone is in the fake location.

Another privacy problem that arises due to the use of context information is related to the tagging of published content. Several MSNs allow the user to associate data (usually context information) with any content to be published, which is called *user tagging*. If the associated data is a reference to a location, the tagging is called geotagging. Tagging can lead to a privacy problem, since a given user can tag a published content with data related to another social network user. For example, when publishing a photo of a bar meeting, Bob can tag the content with the date, time, location, and the identification of the meeting participants, such as Alice. Alice has no control over Bob acts, but she might not want to have revealed where and with whom she was at that date. In this way, a new challenge come into sight: how to avoid sensitive data revealing from users that post tagged contents? [30].

Considering the described privacy issues related to context-aware MSNs, we can highlight some requirements related to the design of the MSN architecture and its applications that developers must address:

- Applications should provide mechanisms that allow the user not only to define access control rules to published content but also to its associated context information in various granularity levels;
- In many cases, publishing of context information is made implicitly and hidden from the user. Applications should expose to the user any type of context information that is being published and allow him to disable that behaviour whenever she/he wishes;
- Applications should provide an easy to use interface (usability) that allows the user to set the desired permissions and privacy settings for his profile.

9.4.1 Preserving Location Privacy in MSNs

The protection of the user's location context is particularly important in MSNs due to the amount of applications that provides features based on the location of the mobile devices. Location privacy can be defined as the right of the users to decide how, when, and for which purposes their location information could be released to other counterparts [6]. The following categories of location privacy can then be identified [4]:

- *Identity privacy*, where the main goal is to protect users identities associated with or inferable from location information. Location data can than be provided to any entity, but the users identity must be preserved;
- *Position privacy*, where the main goal is to adulterate users locations in order to protect individual users positions;
- *Path privacy* that has the objective of not revealing the paths traversed by the users.

An identity privacy solution for MSNs, named *identity server*, is proposed in [10] and is illustrated in Fig. 9.5. The approach is to provide identifiers, named *anonymous identifier* (AID), to mobile phones users so their identity are not compromised. In this way, any published information will be anonymised, including context information. Furthermore, the solution also provides secure communication between the MSN components.

As seen in Fig. 9.5, the AID is a nonce generated by the Identity Server (IS). Before a mobile device advertises its presence to other nearby mobile or stationary devices, it securely contacts the IS to obtain an AID. The IS generates an AID using a cryptographic hash function and returns it to the mobile device (step 1 illustrates a mobile device A requesting an AID). Then, the mobile device A sends its location to the IS in order to obtain a list of nearby devices (step 2). The mobile device A shares the received AID with a nearby mobile B using a Bluetooth or WiFi AID

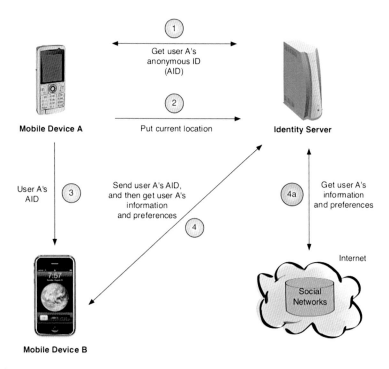

Fig. 9.5 Information sharing through the identity server [10]

sharing service (step 3). After that, the mobile device B queries the IS for the social network profile of the user associated with the received AID (steps 4 and 4a). Once the social network information for an AID has been retrieved from the IS, the IS removes this AID from the list of AIDs associated with the correspondent mobile device. In this way, a mobile device must obtain an AID for every device it wants to share its user information and preferences.

A privacy-aware proximity detection solution determines if two mobile users are close to each other without requiring them to disclose their exact locations. The *Longitude* protocol [53] is an example of this type of solution and also guarantees path privacy. In this protocol, each time a user location is sent to the server by the mobile application, it is first mapped to a two-dimensional area into a three-dimensional orthogonal coordinate system (called toroidal space). After, an encryption procedure (called solid transformation) is applied to the toroidal space and the result is then sent to the server. Each user shares a secret key with her/his friends that is used in the encryption and decryption of location messages. In this context a problem arises: if a user continuously sends her/his location to the server while moving using the same encryption key, the server can possibly learn the user movement (path). To avoid that, the encryption key is modified each time the user communicates her/his location to the server. Finally, the server computes the proximity between the user that is updating her/his location and all users from her/his friend list in the toroidal space and sends the result to each participating friend, that can then compute the proximity in the two-dimensional space based on the received toroidal space proximity.

The work [76] also addresses privacy-aware proximity detection. It proposes a service called *VicinityLocator* that notifies a user if any friend enter her/his specified area of interest (called vicinity region). This region can be of any specific shape and can be changed in real time by the user. Privacy policies are applied to the users locations and only friends in their contacts list have access permissions. These policies are based on vicinity region and only friends located in it receives the user location. In addition, the user specifies the maximum distance that his/her friends could know his/her location.

9.4.2 Trust in Distributed Mobile Social Networks

MSNs structured using a distributed architecture require specific security and privacy mechanisms, since they do not have a centralized server. Therefore, no central authority can take responsibility for tasks such as the generation of digital certificates, cryptographic key management, permissions and access control.

Trust is one of the most crucial concepts in distributed MSNs for improving the security of the establishment of new relationships. Trust can be a metric calculated from the user's encounters history (users finding other users and exchanging information between them) or through recommendations done by other social network users. Reputation-based systems use trust metrics to assist users in making new relationships. Many reputation systems have been recently proposed as an innovative solution to guarantee a minimum level of security to the interactions made between two entities belonging to a distributed system. Mármol and Pérez in [55] describe security threat scenarios that can be found in the area of trust and reputation for distributed systems.

In some reputation-based systems for distributed MSNs, a user has to be recommended by another one who is already part of the community in order to join it. Sometimes this recommendation is done by the application itself, based on context information, such as location or physical proximity. In this case the community is public and the application has authorization to add users into it. Nevertheless, without the existence of a central authority, revoking a community member becomes a challenging task.

In [61], the authors propose a decentralized framework for trusted information exchange, social interaction among users and membership management. Network nodes are labelled with a trust level indicator based on context information, like the degree of connectivity (number of connections), running out of battery, node being out of range, poor communication signal and so on. These labels are used to compute each individual node trust level and can also be used to calculate the trust level of the community/group. The goal is to create communities/groups of users with high trust ratings while identifying untrustworthy users and isolating them from the community, thus revoking their membership.

9.5 Security and Privacy Support in MSNs Middleware

Application development for MSN can be greatly simplified by using a middleware support. These systems aim to provide abstractions that reduce development effort, by providing mechanisms that hide the complexities arising from the distribution of components, programming paradigms that facilitate application development and promoting interoperability between systems/applications. The middleware is responsible for collect, organize, process and disseminate social data, providing a set of common services that can be reused in the construction of MSN applications. However, the design and development of a middleware for mobile social applications is not a trivial task, facing serious challenges [63], such as providing support for high scalability, heterogeneity and dynamic nature of mobile devices, the necessity of providing mechanisms for adapting to changing environmental and system conditions, and the provision of an adequate level of security and privacy.

In recent years, several middleware solutions for MSNs were presented [42], such as MobiSoft [44], SAMOA [16], MyNet [5, 41], MobiSoc [35], MobiClique [59], Mobilis [47] and MobileHealthNet [69]. However, only some of those projects have made proposals in the field of privacy and security, which are described in this section.

9.5.1 Mobilis

Mobilis is a middleware to support application development for MSN. This middleware architecture provides a good guideline for mobile collaborative applications developers [65]. It provides services to share context information between users, group management, a content storage service and a database of content meta-data, as well as collaborative editing of text and images. Its architecture is illustrated in Fig. 9.6.

Communication between the middleware components is based on the XMPP protocol.[8] All services have XMPP identifiers and are registered as XMMP clients in an XMPP server, such as Openfire.[9] Mobile clients also register themselves in the XMPP server and use the Smack XMPP client library[10] for exchanging messages with the Mobilis services. A component called *MobilisBeans* abstracts from the

---

[8]Extensible Messaging and Presence Protocol—http://xmpp.org/.

[9]http://www.igniterealtime.org/projects/openfire/.

[10]http://www.igniterealtime.org/projects/smack/.

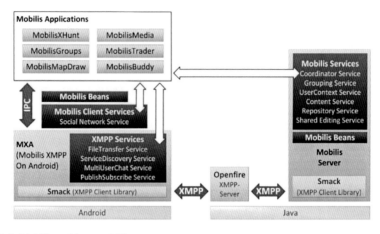

Fig. 9.6 Mobilis architecture [47]

programmer the creation of XMPP packets exchanged between client and server, simplifying the application development.

The *Coordinator Service* is responsible for managing all services offered by the *Mobilis Server* and also allows the instantiation of new services. It provides a service discovery function which returns the unique identifier of a requested service if it is already instantiated in the Mobilis Server.

The *Grouping Service* allows the creation of dynamic groups based on the users location. A group communication mechanism based on multicast is provided, which allows message exchange from simple chat communication to a more complex content exchange, such as spreading changes in a concurrent document editing process.

The *User Context Service* manages all user's context information that are collected and distributed through a publish/subscribe interface. Context information can include any kind of context, such as physical context (e.g., luminosity and temperature), technical context (e.g., bandwidth and error rate), personal context (e.g., the user address and name) and location.

The *Content Service* and *Repository Service* are responsible for the storage of user generated media content (photo or video, for instance) and their associated meta-data, respectively. The *Shared Editing Service*, which is based on the CEFX (Collaborative Editing Framework for XML) [32], allows the concurrent editing of XML structures by geographically distributed users. The current supported XML structures are formatted text, graphical files (SVG) or geographic information (KML).

The *MXA (Mobilis XMPP on Android)* encapsulates the functionality of the Smack XMPP client library and promotes an Android Interface Definition Language (AIDL) interface, used by Mobilis applications to communicate with the Mobilis server.

Mobilis Security

Part of the security mechanisms provided by the Mobilis middleware are inherited from the XMPP protocol. Communication security is done via the SSL (Secure Socket Layer) protocol, using an XMPP extension protocol (XEP) [57] implemented by Smack. Authentication is accomplished by Smack through the XMPP server. To access the Mobilis services, users need to register on the XMPP server, which will provide a unique identifier for each user called JID (Jabber Identifier) [39]. As stated before, Mobilis services also acquire a JID, which enables mutual authentication.

Furthermore, Mobilis provides a privacy management mechanism for context information shared by its users. Only users explicitly authorized by the owner of the context information have access to them. For this end, Mobilis uses a publish/subscribe mechanism with authorization that work as follows: if a user A wants to access context information of user B, he/she must issue a subscription request to the *Mobilis User Context Service*. User B must explicitly authorize the subscription before any update notification of context information be forwarded to user A.

Mobilis also has a group management mechanism, which allows users to create groups and invite other users to join them. There are two categories of groups in Mobilis: public, to which any user can join; and private, to which only users authorized by the group owner can join. In addition, just the group owner can invite users for joining a private group.

9.5.2 MobiSoc

MobiSoc provides a platform that enables the development of mobile social applications, allowing the capture, management and sharing of information concerning the social state of physical communities. This state consists of information regarding the user's profiles and places, and also contains data related to the affinity between people and that people have with places. A community state evolves continuously over time with the creation of new user profiles, social ties, information related to places or events. The MobiSoc also includes learning algorithms to discover geo-social patterns previously unknown (e.g., affinity between people and affinity that people have with places).

The *Data Collection* module consists of three sub-modules. The *People* sub-module allows applications to collect, store and modify information from user profiles (e.g., interests, preferences and social ratings). It has also mechanisms to introduce new groups and add new social contacts, and maintains a social network based on this information. The *Places* sub-module supports the collection of geographical data and maps about buildings, offices and outdoor locations. It also provides mechanisms to create social events associated with a place. The *Location* sub-module receives and stores updates of location updates from the user's mobile devices (Fig. 9.7).

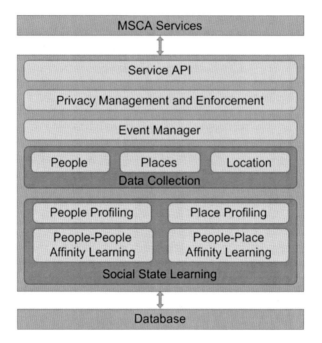

Fig. 9.7 MobiSoc architecture [35]

The *Social State Learning* is the social context learning module and is responsible for inferring new information concerning relationships between people and between people and places. This module consists of the following sub-modules:

- *People Profiling*: provides information about user's profiles, social links and social groups. This sub-module enhances the user profile based on newly discovered information regarding individual users;
- *Place Profiling*: shares place-centric information and enhances the semantics of the place with social information;
- *People-People Affinity Learning*: responsible for computing social affinities between pairs of users, based on similar personal interests, common friends, or common places that users usually go, even at different times;
- *People-Place Affinity Learning*: responsible for discovering what are the places of interest to users or groups. For this, it performs temporal synchronization on the mobility traces and uses clustering techniques to determine repeated user co-presence at a place.

The *Event Manager* is the module used for asynchronous communication with applications. The applications or middleware services can register events with the EventManager to receive notifications when a certain part of the social state changes (e.g., user A is notified when a user B accesses the MSN). Social state changes are detected via event triggers that include time, location and co-location (a positioning

Fig. 9.8 Structure of a
MobiSoc privacy statement
[35]

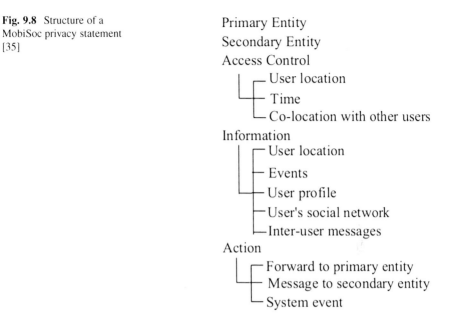

Primary Entity
Secondary Entity
Access Control
   ┌─ User location
  └┬─ Time
   └─ Co-location with other users
Information
   ┌─ User location
  │├─ Events
  └┼─ User profile
   ├─ User's social network
   └─ Inter-user messages
Action
   ┌─ Forward to primary entity
  └┼─ Message to secondary entity
   └─ System event

in relation to another location), as well as external triggers such as the generation of
a new social match (creation of a social edge).

MobiSoc Security: Privacy Management and Enforcement

The *Privacy Management and Enforcement* is the module that manages and enforces
privacy rules for the middleware entities (users or applications). Entities express
their privacy preferences using a mechanism called *privacy statement*. They have
a *primary entity* that issues the *statement* and a *secondary entity* to which the
statement is applied. The structure of a MobiSoc *privacy statement* is illustrated
in Fig. 9.8.

The *primary entity* is always an individual user and the *secondary entity* can
be individual users, groups of users, or applications. Whenever an application
requests an information, the middleware verifies if there is an applicable *statement*
(*Statements* are shared among all applications). If so, the middleware will perform
an action associated with the statement (Action field). The action can be to deny the
access or the send of an appropriate message to the secondary entity. If there is no
applicable statement, the middleware will forward the request to the primary entity,
which will decide the action to be taken (allow or disallow information access).

The *privacy statement* Access Control field defines the conditions under which
a statement is applied. These conditions are defined as user's location, co-location
with other users, time, or a combination of those parameters. The Information field
defines the information for which access restrictions are being set. Information can

be related to the user location, profile data, social network data, inter-user messages and events. The last one are spatial-temporal structures in which users can, for instance, schedule meetings.

Regarding secure communication, MobiSoc uses the RSA asymmetric encryption algorithm for providing integrity and authentication of the user's location updates. The RSA was chosen because it has low computational cost for signing messages, while presents a greater cost for verifying digital signatures. Since the signing of location update messages is performed in the mobile device and its verification is done in the server, RSA provides a good cost/benefit. For providing confidentiality, MobiSoc uses the AES symmetric encryption algorithm that can be combined with RSA by including the encrypted symmetric key in the header of the message and then using it (after RSA decryption) to decrypt the body of the message.

9.5.3 MyNet

MyNet is a middleware platform and a set of user interaction tools created using peer-to-peer (P2P) technologies, which enables users to establish ubiquitous connectivity and share services running on their devices, content, and social contacts without requiring a central repository or infrastructure. The main goal of this middleware is to simplify the deployment, management and use of distributed services in personal and social networks.

In order to show the use of this middleware, follows two use cases examples: (a) Alice has a personal network which consists of her mobile phone, laptop, home PC, and a wi-fi security camera at home. At work, she uses her laptop to retrieve the proposal she was working on over the weekend from her home computer. She also uses her phone to check on her new pet cat through the security camera. (b) In the same scenario, Alice wants to publish the proposal for some friends to review, using the social network to share the content.

This middleware uses the Unmanaged Internet Architecture (UIA) (for more details, see [29]) as its communication platform. The UIA architecture is a distributed name system and ad hoc routing infrastructure, which provides connectivity among the user's mobile devices without the use of centralized servers. The UIA allows users to share their profiles and personal data across the network.

Mynet Security: MyNetSec

MyNet provides a security framework called MyNetSec [40]. MyNetSec provides authentication, authorization, privacy and fine-grained access control to protect the resources (services and content) of a user's MyNet personal network, also allowing the secure sharing of the resources in the social network.

Each device has a unique, secure and authenticatable Endpoint IDentifier (EID). The device creates an asymmetric keypair and hashes the public key to create the EID. EIDs are used as the source and destination addresses in the routing overlay network and also for the encryption of messages exchanged between the hosts. In addition, there is a secure communication based on the SSL protocol.

MyNet access control mechanism is based on a structure called *Passlet*. *Passlets* include data regarding who is giving the permission, to whom the permission is destinated, the resources to which the permissions apply and for how long the defined permissions will last. Actually, *Passlet* is a mechanism to express high-level permissions provided by users. MyNetSec translates the user-level permissions carried in *Passlets* to the appropriate system-level settings, without exposing to the users concepts such as access control lists (ACLs), encryption keys and other security settings.

The middleware implements a dynamic firewall used to enforce fine-grained access control to resources. A firewall intercepts all MyNet overlay traffic before it reaches any service hosted in the device and it is controlled by the *MyNetSec Control Module*, which decides if the captured traffic should be allowed or rejected, based on the security policies expressed by the user through *Passlets*. Moreover, this module provides a mechanism for managing *Passlets* that allows to replicate, update and revoke permissions.

9.5.4 MobileHealthNet

MobileHealthNet is a software infrastructure developed for the creation of new services and applications of MSNs facing the health domain targeting off-side regions and under-served communities. The middleware architecture was developed based on well-defined requirements that include, among others, low resource consumption, so that it can run also on resource-poor and low-end equipment; provide mechanisms for real-time communication with QoS support, in order to enable high-priority notifications and remote monitoring of patients; a rich set of collaboration models and content sharing; a security and privacy model capable of protecting the sensitive data in accordance with the health care security legislation; intuitive user interfaces, considering age, education level and physical disability of potential users. The architecture of MobileHealthNet is organized into four layers, illustrated in Fig. 9.9 and described bellow.

Communication Layer—The MobileHealthNet system has its communication infrastructure based on the Scalable Data Distribution Layer (SDDL) [23] that employs two communication protocols: the Reliable UDP protocol (RUDP) for the inbound and outbound communication between the core network and the mobile nodes and the Data Distribution Service (DDS)[11] for the wired communication

---

[11]http://www.omg.org.

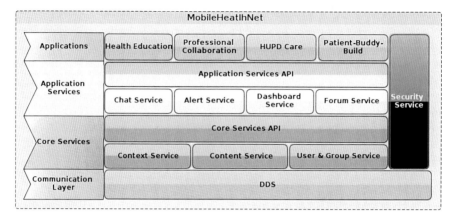

Fig. 9.9 MobileHealthNet architecture

within the MobileHealthNet core network. Mobile nodes are connected to the MobileHealthNet core network through a gateway, which is responsible for managing a separate RUDP connection with each node, forwarding any application-specific message or context information into the core network, and in the opposite direction, converting DDS messages to RUDP messages and delivering them reliably to the corresponding node(s). Gateways can be added to the infrastructure for achieving high scalability and a load balancing algorithm is provided that automatically handles the hand-off of mobile nodes between gateways. RUDP provides a publish/subscribe interaction with unicast communication between the mobile nodes and the gateway. It also has been customized to handle intermittent connectivity, firewall/NAT traversal and robustness to changes of IP addresses and network interfaces.

The MobileHealthNet core network is based on the DDS specification, a publish/subscribe communication standard defined by OMG (Object Management Group) aimed at high performance and real-time distribution of critical information in distributed systems, with Quality of Service (QoS) between consumers and producers of data (e.g., best effort or reliable communication, data persistence and several other message delivery optimizations). This specification was conceived around a Data-Centric Publish-Subscribe (DCPS) model based on topics. For the MobileHealthNet development, several topics have been created to provide communication between network users. The topics are structures that link publishers to subscribers and where is located the information that will be exchanged on the network. Each MobileHealthNet topic represents a set of information about an entity of the social network. For example, a set of information about a social network user, such as his/her name, age and address. So, every MobileHealthNet service is represented by an amount of topics that combined can facilitate the interaction between the mobile nodes and the Server.

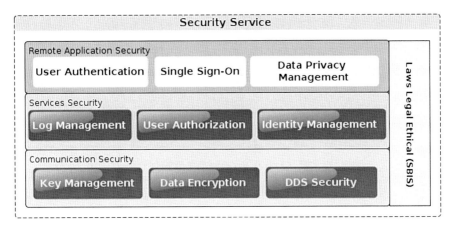

Fig. 9.10 MobileHealthNet security model

Core Services—The Core Services layer provides basic services that are shared by other services and applications. The Context Service is responsible for the storage and availability of context information. Several types of context can be exploited in MSNs applications focusing the health domain. The location information, for example, can be used to determine the members present in the vicinity of the network with which the user often interact. From the location, one can derive through a weather service the atmospheric conditions at the site, such as the moisture, that can affect patients with asthma.

The Content Service allows the sharing of media (text, photos, audio, videos, etc.) between the social network users. This service allows the tagging of each media element with meta-data that can be defined by the application in a flexible way, for example, tagging an element with the geographical position at which the media has been obtained or the model of a pump used for asthma treatment to which a tutorial video refers. The User and Group Service manages the creation of user accounts and groups, as well as the management of relationships (friendships) between network users.

Application Services—The Application Services layer provides typical social networks services, such as publishing messages on dashboards, forums and chat, as well as a notification (alert) service for mobile users. These services can be reused by several applications of the top layer, the Application Layer.

MobileHealthNet Security Model

An overview of the components that compound the MobileHealthNet security model can be viewed in Fig. 9.10.

The proposed model has the following components: User Authentication, responsible for authenticating users via a Single Sign-on mechanism, where a single

Table 9.1 Summary of the middleware analysed

System	Architecture	Authentication	Data privacy	Communication
Mobilis	Centralized	Provided by XMPP	Access request and authorization	SSL via XMPP
MobiSoc	Centralized	Exist, but not explicit	Statements	RSA/AES
MyNet	P2P	Provided by UIA	Passlets	SSL
MobileHealthNet	Centralized	Provided by JEE	Statements	DDS Security

authentication action allows a user to access all allowed services. Data Privacy Management, providing to users the means for specifying privacy policies to be applied to shared data and context information through specified statements. User Authorization that implements the mechanisms responsible for enforcing the defined policies and managing the access control to all shared resources.

The model also offers a component that records the access and manipulation of sensitive data and context information, the Log Management, allowing to audit the actions performed by the system users. This component is also responsible for monitoring the MSN usage, notifying the system administrator of unauthorized attempts to access data and intrusion attempts on the system. The DDS Security component is responsible for providing the establishment of secured communication channels in order to ensure the basic properties of authenticity, integrity and confidentiality in the context of the DDS publish/subscribe data distribution model.

9.5.5 Middleware Comparison

Table 9.1 summarizes a comparison of the analysed middleware, using the following metrics: system architecture, the techniques used for providing authentication and data privacy, and the security protocols and cryptographic algorithms used to ensure secure communication.

As observed in the Table 9.1, the unique middleware that uses a distributed architecture (P2P) is MyNet, the others have a centralized architecture. Considering authentication, Mobilis uses the security mechanisms provided by XMPP. In MyNet this is provided by the UIA via EIDs. MobiSoc offers some mechanism, but not explicit. MobileHealthNet provides a Single Sign-on mechanism provided by JEE.

Some middleware offer mechanisms to guarantee user privacy. The Mobilis implements a privacy mechanism for sharing context information through access request and authorization. Both MobiSoc and MobileHealthNet have a privacy management mechanism implemented through statements, through which each user can flexibly determine who will have access to her/his information. MyNet uses an access control mechanism based on Passlets.

To ensure secure communication, Mobilis and MyNet adopt SSL. MobiSoc use the RSA algorithm to negotiate a session key and the AES algorithm for encrypting messages. MobileHealthNet implements a component to provide the establishment of secure communication channels in the context of the DDS publish/subscribe data distribution model.

9.6 Conclusion

Several breakthroughs in mobile computing and social networks are being proposed in the last years. They involve new hardware resources in mobile devices (e.g., physical sensors and computing power), the development of middleware support for the development of applications and the creation of specific features focused on several domains (ranging form healthcare to entertainment) for which social mobile applications are now being destined. However, in spite of so many advances, there still exist several challenges to be faced in order to guarantee the security and privacy of MSN users [21].

The current social networks are in a security and privacy management pre-liminary stage. Most of them implement very basic access control systems that simply allow the user to decide which personal information are accessible by other members, which is usually done by marking a given item as public, private or accessible by direct contacts [19]. In this way, social network platforms do not provide open and flexible mechanisms that extend or adapt to the their users context.

Providing a balance between utility and security for social networks is a challenge [38]. As seen during this chapter, as more functionalities are inserted in the network (like location resources for a better usability or socialization), more difficult is to provide application security and privacy mechanisms. Thus, many issues deserve more research effort to give proper responses to questions like: How to avoid unauthorized access of personal information by the other users of the social network, social network provider and third-party applications? How to allow the publication of a specific content only for specific vertex on a user social graph? How to share context information in a safe way, without a privacy leak? Which security and privacy resources are necessaries in an MSNs middleware? How to design and provide them?

Acknowledgements The authors would like to thank FAPEMA (State of Maranhão Research Agency) for the support of this work, grant APP-00932/10, and Jesseildo Gonçalves for his review and suggestions specially concerning the text describing the middleware support for MSNs.

References

1. An, J., Ko, Y., Lee, D.: A social relation aware routing protocol for mobile ad hoc networks. In: Proceedings of the IEEE International Conference on Pervasive Computing and Communi-cations, PERCOM '09, pp. 1–6. IEEE Computer Society, Oakland, CA (2009)

2. Anderson, J., Diaz, C., Bonneau, J., Stajano, F.: Privacy-enabling social networking over untrusted networks. In: Proceedings of the 2nd ACM Workshop on Online Social Networks, WOSN '09, pp. 1–6. ACM, New York, NY (2009)

3. Andress, J.: The Basics of Information Security: Understanding the Fundamentals of InfoSec in Theory and Practice. Syngress Media. Elsevier (2011)

4. Anthony, D., Henderson, T., Kotz, D.: Privacy in location aware computing environments. IEEE Pervasive Comput. 6(4), 64–72 (2007)

5. Antoniou, Z., Kalofonos, D.: User-centered design of a secure p2p personal and social networking platform. In: Proceedings of the 3rd IASTED International Conference on Human Computer Interaction, HCI '08, pp. 186–191. ACTA Press, Anaheim, CA (2008)

6. Ardagna, C.A., Cremonini, M., Damiani, E., di Vimercati, S.D.C., Samarati, P.: Privacy-enhanced location services information. In: Digital Privacy: Theory, Technologies and Practices, pp. 307–326. Auerbach Publications (Taylor and Francis Group) (2007)

7. Baden, R., Bender, A., Spring, N., Bhattacharjee, B., Starin, D.: Persona: an online social network with user-defined privacy. In: Proceedings of the ACM SIGCOMM 2009 conference on Data communication, SIGCOMM '09, pp. 135–146. ACM, New York, NY (2009)

8. Baldauf, M., Dustdar, S., Rosenberg, F.: A survey on context-aware systems. Int. J. Ad Hoc Ubiquitous Comput. 2(4), 263–277 (2007)

9. Beach, A., Gartrell, M., Akkala, S., Elston, J., Kelley, J., Nishimoto, K., Ray, B., Razgulin, S., Sundaresan, K., Surendar, B., Terada, M., Han, R.: Whozthat? evolving an ecosystem for context-aware mobile social networks. IEEE Netw. 22(4), 50–55 (2008)

10. Beach, A., Gartrell, M., Han, R.: Solutions to security and privacy issues in mobile social networking. In: Proceedings of the International Conference on Computational Science and Engineering, CSE '09, vol. 4, pp. 1036–1042 (2009)

11. Beach, A., Gartrell, M., Xing, X., Han, R., Lv, Q., Mishra, S., Seada, K.: Fusing mobile, sensor, and social data to fully enable context-aware computing. In: Proceedings of the Eleventh Workshop on Mobile Computing Systems & Applications, HotMobile '10, pp. 60–65. ACM, New York, NY (2010)

12. Beach, A., Raz, B., Buechley, L.: Touch me wear: Getting physical with social networks. In: Proceedings of the 2009 International Conference on Computational Science and Engineering, CSE '09, vol. 4, pp. 960–965. IEEE Computer Society, Oakland, CA (2009)

13. Belapurkar, A., Chakrabarti, A., Ponnapalli, H., Varadarajan, N., Padmanabhuni, S., Sundarrajan, S.: Distributed Systems Security: Issues, Processes and Solutions. Wiley, (2009)

14. Bilge, L., Strufe, T., Balzarotti, D., Kirda, E.: All your contacts are belong to us: automated identity theft attacks on social networks. In: Proceedings of the 18th international conference on World wide web, WWW '09, pp. 551–560. ACM, New York, NY (2009)

15. Boix, E.G., Carreton, A.L., Scholliers, C., Van Cutsem, T., De Meuter, W., D'Hondt, T.: Flocks: enabling dynamic group interactions in mobile social networking applications. In: Proceedings of the 2011 ACM Symposium on Applied Computing, SAC '11, pp. 425–432. ACM, New York, NY (2011)

16. Bottazzi, D., Montanari, R., Toninelli, A.: Context-aware middleware for anytime, anywhere social networks. IEEE Intell. Syst. 22(5), 23–32 (2007)

17. Boyd, D., Ellison, N.B.: Social network sites: definition, history, and scholarship. J. Comput. Mediat. Comm. 13(1–2) (2007)

18. Breslin, J.G., Decker, S., Hauswirth, M., Hynes, G., Phuoc, D.L., Passant, A., Polleres, A., Rabsch, C., Reynolds, V.: Integrating social networks and sensor networks. In: Proceedings of the W3C Workshop on the Future of Social Networking (2009)

19. Carminati, B., Ferrari, E., Heatherly, R., Kantarcioglu, M., Thuraisingham, B.: Semantic web-based social network access control. Comput. Secur. 30(2–3), 108–115 (2011)

20. Chang, E., Thomson, P., Dillon, T., Hussain, F.: The fuzzy and dynamic nature of trust. In: Proceedings of the 2nd international Conference on Trust, Privacy, and Security in Digital Business, TrustBus'05, pp. 161–174. Springer, Berlin, Heidelberg (2005)

21. Chen, G., Rahman, F.: Analyzing privacy designs of mobile social networking applications. In: Proceedings of the IEEE/IFIP International Conference on Embedded and Ubiquitous Computing, EUC '08, vol. 2, pp. 83–88. IEEE Computer Society, Oakland, CA (2008)

22. Cutillo, L.A., Molva, R., Strufe, T.: Safebook: A privacy-preserving online social network leveraging on real-life trust. Comm. Mag. **47**(12), 94–101 (2009)

23. David, L., Vasconcelos, R., Alves, L., André, R., Baptista, G., Endler, M.: A communication middleware for scalable real-time mobile collaboration. In: WETICE, pp. 54–59 (2012)

24. Dey, A.K.: Understanding and using context. Pers. Ubiquitous Comput. **5**(1), 4–7 (2001)

25. Dong, W., Dave, V., Qiu, L., Zhang, Y.: Secure friend discovery in mobile social networks. In: Proceedings of 30th IEEE International Conference on Computer Communications, INFOCOM '11, pp. 1647–1655. IEEE Computer Society, Oakland, CA (2011)

26. Douceur, J.R.: The sybil attack. In: Revised Papers from the First International Workshop on Peer-to-Peer Systems, IPTPS '01, pp. 251–260. Springer, London (2002)

27. Endler, M., Skyrme, A., Schuster, D., Springer, T.: Defining situated social context for pervasive social computing. In: Proceedings of the IEEE International Conference on Pervasive Computing and Communications Workshops, PERCOM Workshops '11, pp. 519–524 (2011)

28. Fang, L., LeFevre, K.: Privacy wizards for social networking sites. In: Proceedings of the 19th International Conference on World Wide Web, WWW '10, pp. 351–360. ACM, New York, NY (2010)

29. Ford, B., Strauss, J., Lesniewski-Laas, C., Rhea, S., Kaashoek, F., Morris, R.: Persistent personal names for globally connected mobile devices. In: Proceedings of the 7th Symposium on Operating Systems Design and Implementation, OSDI '06, pp. 233–248. USENIX Association (2006)

30. Friedland, G., Sommer, R.: Cybercasing the joint: on the privacy implications of geo-tagging. In: Proceedings of the 5th USENIX Conference on Hot Topics in Security, HotSec'10, pp. 1–8. USENIX Association (2010)

31. Gao, H., Hu, J., Huang, T., Wang, J., Chen, Y.: Security issues in online social networks. IEEE Internet Comput. **15**(4), 56–63 (2011)

32. Gerlicher, A.R.S.: Developing collaborative xml editing systems. Ph.D. thesis, London, UK (2007)

33. Grier, C., Thomas, K., Paxson, V., Zhang, M.: @spam: the underground on 140 characters or less. In: Proceedings of the 17th ACM conference on Computer and communications security, CCS '10, pp. 27–37. ACM, New York, NY (2010)

34. Guha, S., Tang, K., Francis, P.: Noyb: privacy in online social networks. In: Proceedings of the first workshop on Online social networks, WOSN '08, pp. 49–54. ACM, New York, NY (2008)

35. Gupta, A., Kalra, A., Boston, D., Borcea, C.: Mobisoc: a middleware for mobile social computing applications. Mobile Network Appl. **14**(1), 35–52 (2009)

36. He, W., Liu, X., Ren, M.: Location cheating: A security challenge to location-based social network services. In: Proceedings of the 31st International Conference on Distributed Computing Systems, ICDCS '11, pp. 740–749. IEEE Computer Society, Oakland, CA (2011)

37. Huang, Q., Liu, Y.: On geo-social network services. In: Proceedings of the 17th International Conference on Geoinformatics, pp. 1–6 (2009)

38. Joshi, P., Kuo, C.C.J.: Security and privacy in online social networks: A survey. In: Proceedings of the IEEE International Conference on Multimedia and Expo, ICME '11, pp. 1–6. IEEE Computer Society, Oakland, CA (2011)

39. Kaes, C.: Xmpp extension protocol – definition of jabber identifiers (jids). Tech. Rep. XEP-0029, XMPP Standards Foundation (2003). URL http://xmpp.org/extensions/xep-0029.pdf

40. Kalofonos, D.N.: Mynetsec: Intuitive security for peer-to-peer (p2p) personal and social networking services. Tech. Rep. NRC-TR-2007-014, Nokia Research Center Cambridge (2007). URL http://research.nokia.com/files/tr/NRC-TR-2007-014.pdf

41. Kalofonos, D.N., Antoniou, Z., Reynolds, F.D., Van-Kleek, M., Strauss, J., Wisner, P.: Mynet: A platform for secure p2p personal and social networking services. In: Proceedings of the

6th Annual IEEE International Conference on Pervasive Computing and Communications, PerCom '08, pp. 135–146 (2008)

42. Karam, A., Mohamed, N.: Middleware for mobile social networks: A survey. In: Proceedings of the 45th Hawaii International Conference on System Sciences, pp. 1482–1490 (2012)

43. Kayastha, N., Niyato, D., Wang, P., Hossain, E.: Applications, architectures, and protocol design issues for mobile social networks: A survey. Proc. IEEE **99**(12), 2130–2158 (2011)

44. Kern, S., Braun, P., Rossak, W.: Mobisoft: an agent-based middleware for social-mobile applications. In: Lecture Notes in Computer Science including subseries Lecture Notes in Artificial Intelligence and Lecture Notes in Bioinformatics – Proceedings of the International Conference On the Move to Meaningful Internet Systems, OTM'06, pp. 984–993. Springer, Berlin, Heidelberg (2006)

45. Lee, R., Nia, R., Hsu, J., Levitt, K.N., Rowe, J., Wu, S.F., Ye, S.: Design and implementation of faith, an experimental system to intercept and manipulate online social informatics. In: Proceedings of the 2011 International Conference on Advances in Social Networks Analysis and Mining, ASONAM '11, pp. 195–202. IEEE Computer Society, Washington, DC (2011)

46. Li, J., Li, Q.: Decentralized self-management of trust for mobile ad hoc social networks. Int. J. Comput. Netw. Comm. **3**(6), 1–17 (2011)

47. Lübke, R.: Ein framework zur entwicklung mobiler social software auf basis von android. Ph.D. thesis, Dresden, Germany (2011)

48. Lubke, R., Schuster, D., Schill, A.: Mobilisgroups: Location-based group formation in mobile social networks. In: Proceedings of the 9th Annual IEEE International Conference on Pervasive Computing and Communications, PerCom 2011, 21–25 March 2011, Seattle, WA, USA, Workshop Proceedings, pp. 502–507. IEEE, Oakland, CA (2011)

49. Lucas, M.M., Borisov, N.: Flybynight: mitigating the privacy risks of social networking. In: Proceedings of the 7th ACM workshop on Privacy in the electronic society, WPES '08, pp. 1–8. ACM, New York, NY (2008)

50. Luo, W., Xie, Q., Hengartner, U.: Facecloak: An architecture for user privacy on social networking sites. In: Proceedings of the 2009 International Conference on Computational Science and Engineering - Volume 03, CSE '09, pp. 26–33. IEEE Computer Society, Washington, DC (2009)

51. Macropol, K., Singh, A.K.: Content-based modeling and prediction of information dissemination. In: ASONAM, pp. 21–28 (2011)

52. Mao, H., Shuai, X., Kapadia, A.: Loose tweets: an analysis of privacy leaks on twitter. In: Proceedings of the 10th annual ACM workshop on Privacy in the electronic society, WPES '11, pp. 1–12. ACM, New York, NY (2011)

53. Mascetti, S., Bettini, C., Freni, D.: Longitude: centralized privacy-preserving computation of users' proximity. In: Proceedings of the 6th VLDB Workshop on Secure Data Management, SDM '09, pp. 142–157. Springer, New York (2009)

54. Meeder, B., Kelley, P.G., Tam, J., Cranor, L.F.: Rt @iwantprivacy: Widespread violation of privacy settings in the twitter social network. In: Web 2.0 Security & Privacy, W2SP '10 (2010)

55. Mármol, F.G., Pérez, G.M.: Security threats scenarios in trust and reputation models for distributed systems. Comput. Secur. **28**(7), 545–556 (2009)

56. Miluzzo, E., Lane, N.D., Fodor, K., Peterson, R., Lu, H., Musolesi, M., Eisenman, S.B., Zheng, X., Campbell, A.T.: Sensing meets mobile social networks: The design, implementation and evaluation of the CenceMe application. In: Proceedings of the 6th ACM Conference on Embedded Network Sensor Systems (SenSys '08), pp. 337–350. ACM, Raleigh (2008)

57. Norris, R.: Xmpp extension protocol – ssl/tls integration. Tech. Rep. XEP-0035, XMPP Standards Foundation (2003). URL http://xmpp.org/extensions/xep-0035.pdf

58. Oyomno, W., Jäppinen, P., Kerttula, E.: Privacy implications of context-aware services. In: Proceedings of the 4th International ICST Conference on Communication System Software and Middleware, COMSWARE '09, pp. 17:1–17:9. ACM, New York, NY (2009)

59. Pietiläinen, A.K., Oliver, E., LeBrun, J., Varghese, G., Diot, C.: Mobiclique: middleware for mobile social networking. In: Proceedings of the 2nd ACM workshop on Online social networks, WOSN '09, pp. 49–54. ACM, New York, NY (2009)

60. Poslad, S.: Ubiquitous Computing: Smart Devices, Environments and Interactions. Wiley (2009)
61. Qureshi, B., Min, G., Kouvatsos, D.: A framework for building trust based communities in p2p mobile social networks. In: Proceedings of the 10th IEEE International Conference on Computer and Information Technology, CIT '10, pp. 567–574. IEEE Computer Society, Oakland, CA (2010)
62. Rahman, F., Hoque, M.E., Kawsar, F.A., Ahamed, S.I.: Preserve your privacy with pco: A privacy sensitive architecture for context obfuscation for pervasive e-community based applications. In: SocialCom/PASSAT, pp. 41–48 (2010)
63. Rana, J., Kristiansson, J., Hallberg, J., Synnes, K.: Challenges for mobile social networking applications. In: Proceedings of the International ICST Conference on Communications Infrastructure, Systems and Applications in Europe, Lecture Notes of the Institute for Computer Sciences, Social Informatics and Telecommunications Engineering, vol. 16, pp. 275–285. Springer, Berlin, Heidelberg (2009)
64. Schiller, J., Voisard, A.: Location-Based Services. Elsevier Morgan Kaufann Publishers (2004)
65. Schuster, D., Koren, I., Springer, T., Hering, D., Söllner, B., Endler, M., Schill, A.: Creating applications for real-time collaboration with XMPP and Android on Mobile devices. Handbook of Research on Mobile Software Engineering: Design, Implementation and Emergent Applications. IGI Global (2012)
66. Schuster, D., Rosi, A., Mamei, M., Springer, T., Endler, M., Zambonelli, F.: Pervasive social context - taxonomy and survey. ACM Trans. Intell. Syst. Tech. 9(4), 1–22 (2012)
67. Singh, K., Bhola, S., Lee, W.: xbook: redesigning privacy control in social networking platforms. In: Proceedings of the 18th Conference on USENIX Security Symposium, SSYM'09, pp. 249–266. USENIX Association, Berkeley, CA (2009)
68. Stutzman, F., Kramer-Duffield, J.: Friends only: examining a privacy-enhancing behavior in facebook. In: Proceedings of the 28th International Conference on Human Factors in Computing Systems, CHI '10, pp. 1553–1562. ACM, New York, NY (2010)
69. Teles, A., Pinheiro, D., Gonçalves, J., Batista, R., Silva, F., Pinheiro, V., Haeusler, E., , Endler, M.: Mobilehealthnet: A middleware for mobile social networks in m-health. In: Proceedings of the 3rd International Conference on Wireless Mobile Communication and Healthcare, MobiHealth '12 (2012)
70. Tootoonchian, A., Saroiu, S., Ganjali, Y., Wolman, A.: Lockr: better privacy for social networks. In: Proceedings of the 5th International Conference on Emerging Networking Experiments and Technologies, CoNEXT '09, pp. 169–180. ACM, New York, NY (2009)
71. Vicente, C.R., Freni, D., Bettini, C., Jensen, C.S.: Location-related privacy in geo-social networks. IEEE Internet Comput. 15(3), 20–27 (2011)
72. Wasserman, S., Faust, K.: Structural Analysis in the Social Sciences, 1st edn. Cambridge University Press, Cambridge (1994)
73. Xu, W., Zhang, F., Zhu, S.: Toward worm detection in online social networks. In: Proceedings of the 26th Annual Computer Security Applications Conference, ACSAC '10, pp. 11–20. ACM, New York, NY (2010)
74. Zhang, C., Sun, J., Zhu, X., Fang, Y.: Privacy and security for online social networks: challenges and opportunities. IEEE Netw. 24(4), 13–18 (2010)
75. Zhang, D., Wang, Z., Guo, B., Zhou, X., Raychoudhury, V.: A dynamic community creation mechanism in opportunistic mobile social networks. In: Proceedings of the IEEE 3rd International Conference on Social Computing, SocialCom/PASSAT '11, pp. 509–514 (2011)
76. Šikšnys, L., Thomsen, J.R., Šaltenis, S., Yiu, M.L.: Private and flexible proximity detection in mobile social networks. In: Proceedings of the 11th International Conference on Mobile Data Management, SDM '10, pp. 75–84 (2010)

Part IV
Multimedia-Based Authentication and Access Control Models for Social Networks

Chapter 10
Avatar Facial Biometric Authentication Using Wavelet Local Binary Patterns

Abdallah A. Mohamed and Roman V. Yampolskiy

Abstract Virtual worlds (e.g., Second Life) are populated by different types of people, businesses, and organizations. Users of virtual worlds, either individuals or organizations, might abuse the flexibility and adaptability offered by virtual environment by engaging in criminal activities. Even terrorist organizations have been active in virtual worlds. These organizations recently have used the virtual worlds for recruitment and to train their new members in an environment that is very similar to the real one. Since avatars are not just virtual creations as they have a great social and psychological correspondence with their creators, applying biometric techniques on avatars can give the law enforcement agencies and security experts the ability to identify who the actual users behind these avatars are. There is a mounting pressure to have techniques for verifying the real identities of the inhabitants of virtual worlds to secure cyberworld from incessant criminal activities (e.g., verbal harassment, fraud, money laundering, data or identity theft). In order to reduce the gap between our ability to recognizing human faces and avatar faces and to develop reliable tools for protecting virtual environments, we will discuss in this chapter how we can use different versions of local binary pattern (LBP) operators (traditional LBP, multi-scale LBP and hierarchical multi-scale LBP) to recognize avatar faces from two different virtual worlds (Second Life and Entropia Universe). This chapter includes a definition of discrete wavelet transform from a face recognition research perspective, a summary of previous work done on this topic, characteristics of the datasets used in the experiments as well as some suggestions for future work.

A.A. Mohamed (✉) • R.V. Yampolskiy
Department of Computer Engineering and Computer Science, University of Louisville,
Louisville, KY, USA
e-mail: aamoha04@louisville.edu; abdmoham@umail.iu.edu

R. Chbeir and B. Al Bouna (eds.), *Security and Privacy Preserving in Social Networks*,   317
Lecture Notes in Social Networks, DOI 10.1007/978-3-7091-0894-9_10,
© Springer-Verlag Wien 2013

10.1   Introduction

A virtual world can be defined as a 3D computer-based simulation of a real world. It consists of constellations of online communities connected over the Internet. Having the power to change society, virtual worlds have been attracting more and more attention recently [1, 2]. Some examples of popular virtual worlds are Second Life [3], Active Worlds [4], and Entropia Universe [5]. In a virtual world, any user would be able to experience a feeling of a private personalized digital space that mimics real life. In this way, the users can virtually tour the world with their avatars. An avatar is a user's chosen appearance or virtual identity. A user can change this identity as s/he wants. Users or visitors of these virtual worlds are thus provided with unlimited adaptability and flexibility. Virtual worlds have rules and instructions via which avatars can move around and teleport to different locations. Communication between users can be achieved via different media, i.e., texts, visual gestures, sound, and occasionally voice-commands. The currency used in virtual worlds for commercial activities is virtual money. Virtual money can be obtained through exchange of real money. Inhabitants of virtual worlds might use or abuse the flexibility and adaptability offered in these environments. Looking at the good side, we can see vibrant educational worlds where teaching, learning, and interactivity and collaboration between users flourish. However, looking at the downside, we can see daily reports of criminal activities (e.g., fraud, identity theft, tax evasion, illegal gambling, and terrorist activities) [6]. Quick investigation of Second Life reveals that it is inhabited by numerous terrorist organizations who train in these simulated contexts using weapons identical to their real world counterparts [7].

These illegal acts are a challenge to maintaining safe and peaceful virtual communities. They also constitute a serious problem for law enforcement agencies and forensic experts who seek to have an accurate and automatic tracking of real life users and their avatars [2]. Verifying humans' identities is indispensable. That is why there is a pressing need for a well-developed science to determine one's identity in today's modern society.

Profiling avatars is a new field and a challenging approach that contributes toward a new research direction in face detection and recognition. In response to the requests for an affordable, automatic, fast, secure, reliable, and accurate ways of profiling avatars, Yampolskiy and Gavrilova initiated a new direction in avatar recognition, *Artimetrics*, which will allow identifying, classifying, and authenticating robots, software, and virtual reality agents [1, 8].

Here, we examine four distinct situations that would require a face recognition algorithm to investigate criminal and terrorist activity in virtual worlds [6]:

(a) *Matching a Human face with an Avatar face*: Many users tend to connect their true identity with their online avatar which helps represent them well.
(b) *Matching one avatar face with another*: This approach is very useful as it allows for continuous tracking of an avatar through cyberspace at different places at different times.

(c) *Matching an Avatar's face from one virtual world with the same avatar in a different virtual world*: A recent improvement within the virtual communities is addition of connections between different virtual worlds. Consequently, this ability can help in identifying and tracking records of the avatars.

(d) *Matching an Avatar sketch with the Avatar face*: In a similar approach to the one followed in matching the forensic sketch of human faces provided by the victim or witness of a crime. A virtual criminal can be linked to his/her avatar identity by implementing this scheme within virtual worlds.

In this chapter, we implement different versions of local binary pattern (LBP) operators to recognize avatar faces from two different virtual worlds, Second Life and Entropia Universe, with a concentration on combining discrete wavelet transform with hierarchal multi-scale LBP (HMLBP) as a new technique to recognize avatar faces. The experimental results demonstrating the efficiency of the proposed algorithm are presented.

The remainder of this chapter is organized as follows: Sect. 10.2 provides a background on previous work related to this subject. Section 10.3 discusses the role played by in the field of image processing. Section 10.4 defines the LBP operator and histogram. Section 10.5 presents the concept of multi-scale local binary pattern. Section 10.6 explains the wavelet hierarchical multi-scale LBP (the proposed algorithm). Experiments implemented on avatar face datasets are presented in Sect. 10.6 as well. Comparisons of the wavelet hierarchical multi-scale LBP with various other methods are presented in Sect. 10.8. Finally, Sect. 10.8 provides concluding notes as well as suggestions for future research.

10.2   Related Work

The need for secure and affordable virtual worlds is attracting the attention of many researchers who seek to find fast, automatic, and reliable ways to identify virtual worlds' avatars [1, 6]. Most of the work done so far basically focuses on recognizing the identity of avatar facial images of gray scale or color using preexisting traditional face recognition techniques.

Yampolskiy et al. [1] introduced a new subfield of security research, *Artimetrics*. Artimetrics expands the research area of security to include security research for non-biological entities. Artimetrics is, thus, a sister subfield to security research for biological entities, Biometrics. The authors describe how visual and behavioral properties can be used to verify and recognize avatars. They also give a brief survey of the nature of non-biological entities, including avatars. They present some techniques for collecting and classifying datasets of avatars and introduce a multi-resolution system to enhance the performance of authenticating both biological (human being) and non-biological (avatars) entities. The authors also suggest potential directions for further research in Artimetrics.

One of the first contributions in this new subfield of security proposes Avatar Face Recognition (AFR) system that combines discrete wavelet transform (DWT) with support vector machine (SVM) technique [9]. DWT is used to extract the features while SVM is used to classify each facial image to its original subject. This technique follows the standard block diagram of a basic authentication system that has two main steps: training and testing or recognition. The authors used the web site MyWebFace to construct and enhance the tested dataset. For a dataset of 1,800 avatar facial images organized into 100 different avatar subjects, each of which has 18 different avatar facial images for the same avatar, the average recognition rate was better than the previous techniques.

One approach suggests combining discrete wavelet transform with LBP (WLBP) operator, as a descriptor, to extract features from each avatar facial image [10]. After applying the first and the second level of decomposition, a group of 581 avatar face images manually cropped from Second Life virtual world is organized into 83 different avatar subjects. Each of which has seven different avatar facial image for the same avatar with different frontal angle. This approach is good from the processing time point of view and gives better performance over similar approaches that only use gray-scale images. However, since it is a single scale technique mainly based on the original LBP, applying it to larger or noisy datasets may lead to a decrease in the recognition rate.

Another approach uses a set of algorithms to identify avatars within Second Life virtual world, following an avatar-to-avatar matching scenario [6]. Here, the technique tracks a person in both real and virtual worlds. In this approach, the authors implement a scripting technique to collect avatar faces automatically from Second Life virtual world while using the state-of-the-art picture-to-avatar conversion software, AvMaker, to convert FERET images to 3D avatars. AvMaker is used to generate an avatar face image for every photo in the FERET database. The proposed avatar-to-avatar matching algorithm achieved high true acceptance rates either in the case of using manual eye detection or in a fully automated mode. In case of the FERET-to-avatar face dataset, FaceVACS algorithm achieved high recognition rate. A combination of the proposed matching algorithm with FaceVACS algorithm increases the improvement in inter-reality-based scenario.

10.3   Wavelet Decomposition of an Image

Discrete wavelet transform (DWT) is a widespread instrument for image decomposition and texture description as it has the capability to perform multi-resolution decomposition analysis as well as the ability to extract essential features necessary for avatar face recognition [11]. DWT helps to view and process digital images at multiple resolutions. Its mathematical background as well as its merits has been discussed in several studies. Here, the main benefits of using wavelet transforms are summarized as follows:

- It decomposes an image by decreasing the resolutions of its sub-images and thus assists in reducing the computational complexity of the system. Harmon explains that images with resolution 16×16 pixels are sufficient to recognize human faces [12]. Comparing to original images of size 256×256 resolution, using sub-images of resolution 16×16 reducing the size of original images by 256 times and hence reduces the computational complexity of the system.
- It decomposes images into sub-bands corresponding to different frequency ranges [13]. They straightforwardly achieve the input requirements for the next major step, thus minimizing the computational overhead in the proposed system.
- Wavelet decomposition provides local information in both spatial and frequency domains. This makes it more advantageous than Fourier decomposition which supports only global information in the frequency domain [13].

One of the main qualities of wavelets is that they make available multi-resolution analysis of the image in the form of coefficient matrices. Arguments in favor of the use of this multi-resolution decomposition in psychovisual research support the suggestion that humans process images in a multi-scale way.

The computational complexity of wavelets is linear [13], with the number (N) of computed coefficients [$O(N)$], while other transformations in their fast implementation have $N \times \log_{-2}(N)$ complexity. Thus, wavelets are good candidates for implementation on dedicated hardware designs.

With respect to images, we have to apply WT in two directions or dimensions (row or horizontal direction and column or vertical direction) using four different filters [12, 13]:

$$
\begin{aligned}
\phi(n_1, n_2) &= \phi(n_1)\,\phi(n_2) \\
\psi^H(n_1, n_2) &= \psi(n_1)\,\phi(n_2) \\
\psi^V(n_1, n_2) &= \phi(n_1)\,\psi(n_2) \\
\psi^D(n_1, n_2) &= \psi(n_1)\,\psi(n_2)
\end{aligned}
\tag{10.1}
$$

where n_1 is the horizontal direction and n_2 is the vertical direction, φ is the scaling function which is essentially a low-pass filter, ψ is the wavelet function which is essentially a high-pass filter, the product $\varphi(n_1)\,\psi(n_2)$ means applying the low-pass filter in the horizontal direction and applying the high-pass filter in the vertical direction, by the same way we can understand the meanings of all the four filters. In the second filter there is a superscript H since there is a high-pass filter applied on the horizontal direction, by the same way we can understand the superscripts V and D.

Once these four filters are implemented, an image would be decomposed into four sub-bands LL (low-pass filter on the horizontal direction and low filter on the vertical one), HL (high-pass filter on the horizontal direction and low-pass filter on the vertical one), LH and HH (see Fig. 10.1) [14, 15]. The band LL is an approximation to the original image while bands LH and HL represent the changes

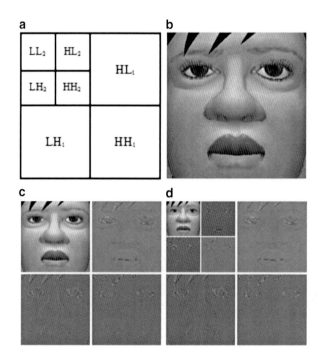

Fig. 10.1 (**a**) Wavelet coefficient structure [14, 15]. (**b**) A sample image of one of the avatar face images in the dataset. (**c**) One level Daubechies wavelet decomposition for the avatar face image in (**b**). (**d**) Two levels Daubechies wavelet decomposition for the face image in (**b**)

of the image along the vertical and horizontal directions, respectively. The band HH registers the high frequency component of the image.

To get a higher level of decomposition any one of the previous four sub-bands can be analyzed [14, 15]. However, since images are generally very rich (maximum information of the image) in the low frequency contents, so we have to decompose the LL sub-band of the previous decomposition level using four different filters as we did before to obtain the next level of decomposition. For instance, to reach the second level of decomposition we have to decompose the LL_1 sub-band.

The decomposition must be carried out for the LL_2 to achieve the third level of decomposition and so on. So, we can say that wavelet decomposition of an image provides an approximation image, which is used to acquire the next decomposition level, and three detailed images in horizontal, vertical, and diagonal directions.

Figure 10.1 provides an illustration of the structure of the wavelet coefficient and the first and the second level of wavelet decomposition for one of the avatar face images used in the experiments. Decomposing an image with two scales will give us seven sub-bands: LL_2, HL_2, LH_2, HH_2, HL_1, LH_1, and HH_1.

10.4 Local Binary Pattern

The LBP operator presented by Ojala et al. [16] is a powerful local descriptor for determining image texture. It has been used in many applications such as industrial visual inspection, image retrieval, automatic face recognition, and detection.

The LBP operator labels the pixels of an image by thresholding the value of the central pixel against its surrounding eight pixels (for a given size of 3×3 neighborhood of each pixel) and considering the result as a binary value [17]. The binary value is then converted to the decimal value to get the LBP value. The output value of the LBP operator can be defined as follows [17, 18]:

$$LBP(x_c, y_c) = \sum_{i=0}^{7} 2^i S(g_i - g_c) \qquad (10.2)$$

where g_c corresponds to the gray value of the central pixel, (x_c, y_c) are its coordinates, g_i $(i = 0, 1, 2, \ldots, 7)$ are the gray values of its surrounding eight pixels and $S(g_i - g_c)$ can be defined as follows:

$$S(g_i - g_c) = \begin{cases} 1, & g_i \geq g_c \\ 0, & \text{otherwise} \end{cases} \qquad (10.3)$$

So we can describe LBP as a sequenced set of binary comparisons between the central pixel value and the values of its neighborhood pixels [18]. Figure 10.2 gives an illustration to the basic LBP operator and how to compute the LBP value.

The LBP operator can be extended to use pixels from neighborhoods of different sizes. Figure 10.3 gives us some examples of different LBP operators, where R is the radius of the neighborhood and P is the number of pixels in the neighborhood.

The neighborhood can be either in a circular or square order. Importantly, it has been proven that using the circular order neighborhood gives better results. Using the circular order neighborhood allows any radius and number of the pixels in the neighborhood [20].

One of the most rewarding improvements is an extension of the basic LBP operator is called uniform LBP (ULBP). An LBP is called uniform when it contains at most two different conversions from 0 to 1 or 1 to 0 when the binary string is viewed as a circular bit string [19, 20]. For example, 11111111, 00011000, and 11110011 are uniform patterns. Ojala notes that with $P = 8$ and $R = 1$ neighborhood, uniform patterns account for around 90 % of all patterns and with $P = 16$ and $R = 2$ neighborhood, uniform patterns account for around 70 % of all patterns [19]. So only a little amount of information will be lost [18] when using uniform patterns.

After labeling an image using the LBP operator, the histogram of the labeled image has to be defined as follows [18]:

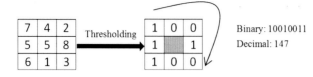

Fig. 10.2 The basic LBP operator [17–19]

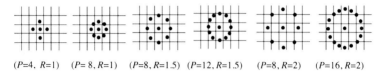

(P=4, R=1)   (P= 8, R=1)   (P=8, R=1.5)   (P=12, R=1.5)   (P=8, R=2)   (P=16, R=2)

Fig. 10.3 Different LBP operators [17–19]

$$H_i = \sum_{x,y} I(f(x, y) = i), \quad i = 0, 1, \dots, n - 1 \tag{10.4}$$

where $n = 2^P$ is the number of different labels produced by the LBP operator and $I(A)$ is a decision function with value 1 if the event A is true and 0 otherwise.

LBP histogram attains very important information about the distribution of the local microstructures, such as spots and edges, over the whole image. Therefore, it can be used to describe and represent the global characteristics of images [18, 21].

10.5 Multi-scale LBP

One of the main weaknesses of the original LBP (single scale LBP) is that it does not provide a complete image representation [22]. Features obtained by using a local 3×3 neighborhood around a central pixel can only capture small scale structures (microstructures). Hence, the LBP operator is not robust enough against any local changes in the image texture. To overcome this limitation of the original LBP and to capture large-scale structures that may have useful features of the faces, new representation of the image, multi-scale LBP, was presented as a solution.

There are many versions for multi-scale analysis of an image. Mäenpää and Pietikäinen [23] propose two novel ways to extend the LBP operator to be able to handle multiple scales. In the first one the authors use exponentially growing circular neighborhoods with Gaussian low-pass filters to collect information from a large texture area. In this study, both the filters and the sampling positions are planned in a way that makes them able to handle the neighborhood as much as possible and in the meantime be able to reduce repeated information. Additionally, the authors suggest an alternative way to encode arbitrary large neighborhood. The method was used successfully in compactly encoding even 12-scale LBP operators. Here, a feature vectoring, which is characterized by having marginal distributions of LBP codes

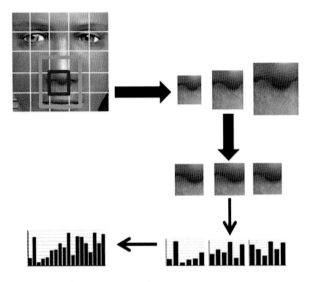

Fig. 10.4 Multi-scale avatar image representation

and cellular automation rules, was employed as a texture descriptor. However, it is important to note that in these experiments no significant progress could be achieved when performance was compared to the basic multi-scale approach.

Another improvement was done to the multi-scale LBP operator. It was extended to become a multi-scale block local binary pattern (MB-LBP) [22]. The main advantage that MB-LBP offers is to enable comparing average pixels values found within the small blocks in lieu of comparing pixel values. Here, the operators always use eight neighbors producing labels from 0 till 255. For example, if the block size is 3×3 pixels, the parallel MB-LBP operator performs a comparison of the average gray value of the center block to the average gray values of the eight neighboring blocks of the same size. The effective of the operator is 9×9 pixels. The MB-LBP was introduced to replace the fixed uniform pattern mapping and to be used with a mapping which is dynamically obtained from a training data. Here, the mapping works as follows: the N recurring MB-LBP patterns take labels $0, \ldots, N-1$. The rest of the patterns take a single label. Here, the number of labels and the length of the MB-LBP histogram are parameterized so that the user can set them.

Generally, the direct way for multi-scale analysis of an image can be performed by obtaining the input image computed at different scales and then concatenating the LBP histogram computed at each scale after resizing each image patch to the same size (see Fig. 10.4) [24]. The main problem existing in this approach is the high dimensionality of the final histogram which contains redundant information.

To overcome this problem Chan et al. [25] develop this approach which combines the multi-scale LBP representation with linear discriminant analysis (LDA) and the principal component analysis (PCA). In their approach, they first applied the uniform LBP operators at R scales on a face image. Then, they crop the resulting LBP images to the same size and divide these images into non-overlapping sub-regions. The set of histograms computed at different scales for the same sub-region are concatenated into a single histogram. To reduce the dimensionality of the descriptor they applied PCA before LDA. So, to derive discriminative facial features using LDA they applied PCA first to extract statistical independent information.

10.6 Wavelet Hierarchical Multi-scale LBP

Uniform patterns play an important role in classifying textures and recognizing faces [26]. By applying the idea of uniformity, many patterns which are not uniform have to be isolated into non-uniform patterns or clusters. So, we may have some discriminant but non-uniform patterns. These patterns fail to share in providing useful features. Recently some methods were proposed to deal with this issue [22, 23]. These methods require the use of some training samples in order to differentiate between useful nonuniform patterns and non-useful ones. Therefore, these methods are learning-based methods and their performance may be affected by the available number of training samples. Guo et al. [26] propose another version for multi-scale analysis of an image, hierarchical multi-scale LBP. The proposed algorithm is totally training free. In this approach, an LBP operator with the biggest radius has to be extracted first and then the extraction is done to operators with smaller radius.

The main issue with this approach is that the feature dimension of the multi-scale LBP is high. To address this problem, we combine discrete wavelet transform with hierarchical multi-scale LBP [27]. We present this as a new technique to recognize avatar faces from different virtual worlds. This technique has three steps: preprocessing, feature extraction, and recognition or classification.

10.6.1 Preprocessing Face Images

In order to improve the efficiency of extracting the face features we have to apply a set of preprocessing operations. First, we manually cropped the input images to pure face images by removing the background which is not useful in recognition. Second, these pure face images have to be normalized and then decomposed using different levels of wavelet decomposition to obtain pure facial expression images (see Fig. 10.5). Here, we focus on decomposing our datasets of images to a certain level of decomposition. This level will be determined after applying different families of wavelets. This part is discussed in Sect. 10.7.1.

Detailed images resulting from applying wavelet decomposition contain changes. These changes represent the differences between face images [13–15]. Here, considering only approximation images will enhance the common features of the same class of images and at the same time the differences between face images will be

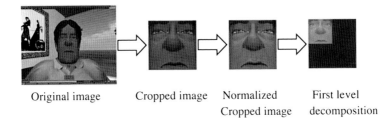

Original image   Cropped image   Normalized   First level
Cropped image   decomposition

Fig. 10.5 Face image preprocessing

reduced. For this reason, our experiments are concerned only with approximation images resulting from applying different levels of wavelet decomposition. That is, up to this stage, it is not determined whether the approximation images obtained from first level of wavelet transform are the ones that would be used to test and to evaluate the performance of the proposed algorithm or if we have to run a second level of wavelet decomposition.

10.6.2 HMLBP Feature Extraction

Many studies had shown that the performance of the multi-scale or multi-resolution LBP operator is better than that of a single scale LBP operator for many reasons:

(a) Extracting the features of an image using a limited size neighborhood as in a single scale LBP encodes only the microstructures of image patterns. So, calculating features based on this way can result in the loss of useful information of an image [26, 27]. On the other hand, using multi-scale operators can help to encode not only the microstructure patterns but also the macrostructures ones under different settings.

(b) Applying the single scale LBP operator can cause a substantial loss of data as "nonuniform" patterns are clustered into one nonuniform pattern. Specifically the cause of data loss would be in that the radius of the LBP increases, subsequently the cluster size of the "nonuniform" patterns increases, leading eventually to information loss [26, 27].

In HMLBP algorithm the LBPs for the biggest radius is extracted first. The new LBPs of "nonuniform" patterns have to be extracted further using a smaller radius in order to extract "uniform" patterns. This procedure continues until the smallest radius is processed. This hierarchical scheme does not have a training step and thus it is insensitive to training samples [26, 27].

Figure 10.6 displays an example of the hierarchical multi-scale LBP scheme. The LBP histogram for $R = 3$ is first built. For those "nonuniform" patterns of the $R = 3$ operator, a new histogram is built by the $R = 2$ operator. Then, the

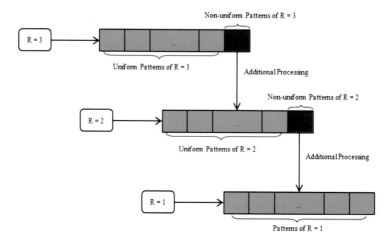

Fig. 10.6 An example of hierarchical multi-scale LBP scheme [26, 27]

"nonuniform" patterns of $R = 2$ lead to the histogram building process for the $R = 1$ operator. Finally, the three histograms are concatenated into one multi-scale histogram to form the feature histogram of an image [26, 27].

10.6.3 Dissimilarity Measure

The last stage of our proposed algorithm is to classify each facial image to its class by computing the dissimilarity between training samples and a test (input) sample. To do that we apply chi-square distance as follows [19]:

$$D(X, Y) = \sum_{n=1}^{N} \frac{(X_n - Y_n)^2}{X_n + Y_n} \tag{10.5}$$

where X is the tested image (sample), Y is the training sample(s) or image(s), and N is the sum dimension.

10.7 Experimental Results and Analysis

In this section, we present an overview of the collected datasets and we expose the different tests and the obtained results.

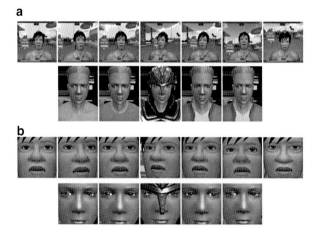

Fig. 10.7 (**a**) Two classes of unprocessed avatar images [3, 5]. (**b**) The same two classes after cropping avatar faces

10.7.1 Experimental Setup

In order to evaluate the robustness and the reliability of our proposed technique we apply it to two datasets of virtual characters aiming at dealing with problems of disguise or usurpation of identity in the virtual worlds.

In fact, our gallery contains two datasets of facial images belonging to two different virtual worlds, Second Life (SL) and Entropia Universe (ENT). The SL dataset contains 581 facial images belonging to 83 classes. Each class has seven facial images for the same avatar with different angles (front, far left, mid left, far right, mid right, top and bottom). All images of SL dataset are manually cropped to a fixed size of 260×260 pixels. The SL dataset is devised into two independent groups: the first contains 332 avatar faces (four facial images for each class) considered for training phase, whereas the second group is reserved for testing and contains the rest of 249 images (three images from each class).

The second dataset, ENT, contains 490 facial images belonging to 98 subjects. Each subject has five different images for the same avatar with different facial expressions and eye angles. All images in ENT dataset are manually cropped and then resized to a fixed size of 180×180 pixels. The ENT dataset is divided into two independent groups: the first contains 294 (three images from each subject) facial images considered for training and the second group contains the rest of 196 images (two images from each subject) reserved for testing.

All images in both datasets are in gray scale format taken in frontal position. Some of these images were taken with different angles (front, far left, mid left, far right, mid right, top and bottom). All chosen images either for training or testing are chosen randomly. Figure 10.7 displays an example of two classes of avatars (one from each dataset) before and after cropping.

Table 10.1 Recognition rate and processing time for SL dataset using tested wavelet families

Wavelet family		Recognition rate (%)	Processing time (s)
Haar		93.1713	8.45
BiorSplines	Bior1.1	92.7723	9.34
	Bior2.2	93.9822	9.78
	Bior6.8	94.3861	10.46
Daubechies	Db2	93.9854	8.34
	Db4	94.38	8.78
	Db5	94.4312	8.34
	Db7	93.9809	7.48
	Db10	93.7834	8.19
Coiflets	Coif1	94.385	11.34
	Coif3	93.1713	12.31
	Coif5	92.37	11.78

Table 10.2 Recognition rate and processing time for ENT dataset using tested wavelet families

Wavelet family		Recognition rate (%)	Processing time (s)
Haar		92.86	6.85
BiorSplines	Bior1.1	91.84	6.98
	Bior2.2	93.348	6.88
	Bior6.8	93.3237	7.59
Daubechies	Db2	93.5621	8.57
	Db4	94.425	8.19
	Db5	93.8183	7.89
	Db7	93.375	7.69
	Db10	92.325	7.89
Coiflets	Coif1	94.3917	9.56
	Coif3	93.375	9.28
	Coif5	91.84	8.95

Also, applying wavelet transform will reduce the size of each image based on the decomposition level. The following tables summarize many tests carried out to determine the best wavelet family, its best level of decomposition for each one of the tested datasets, and the processing time required by each wavelet family.

After examining, various test results of applying different discrete wavelet families (provided in Tables 10.1 and 10.2), we figured out that the recognition rates for SL and ENT datasets are very similar. The rates range from 92 to 95 %. Importantly, the best recognition rate was obtained by the same wavelet family for both datasets but with different index, Daubechies5 for SL dataset and Daubechies4 for ENT dataset.

Also, these results showed that the average processing time required by both datasets with different wavelet families ranged between 6 and 12 s. The best processing time required by SL dataset was recorded with the wavelet family "Db7." This

Table 10.3 Recognition rate for SL and ENT datasets using different decomposition levels of the retained wavelet family

Decomposition level	SL recognition rate (%)	ENT recognition rate (%)
Level 1	92.37	92.35
Level 2	93.57	93.88
Level 3	93.57	95.92
Level 4	95.58	95.41
Level 5	95.98	95.41
Level 6	94.78	94.90
Level 7	94.37	94.39

time is 7.48 s. The difference between the best processing time and the time required by "Db4" wavelet family which provides the best recognition rate is too little, less than a second. The best processing time required by ENT dataset was recorded with the wavelet family "Haar" and this time was 6.85 s. The difference between the best recoded time and the time required by the wavelet family with the best recognition rate was less than 20 % increase in the time. Since the difference between the best required time and the time required by the wavelet family with the best recognition rate is little, we considered the best wavelet family is the one with the highest recognition rate.

Once we selected the wavelet family via which the recognition rate could be obtained in the highest recognition rates, we have to decide the best decomposition level for the performance rate within this family (as in Table 10.3).

10.7.2   Comparing WHMLBP with HMLBP and Other Algorithms

In order to gain better understanding on whether using wavelet transform with HMLBP is advantageous or not, we compared WHMLBP with HMLBP, MLBP, WLBP, and LBP with several experiments. First we obtain the performance of WHMLBP with different block sizes with $R = [3, 2, 1]$ and $P = [16, 16, 16]$ as we can see in Fig. 10.8. Changing the block size affects the result of the recognition rate. In Fig. 10.8, the recognition rate increases as the block size gets larger. However, the performance drops down when the block size is larger than 44×44 in SL dataset and 40×40 in ENT dataset.

Next, we compare the performance of WHMLBP and HMLBP using 44×44 block size for SL dataset and 40×40 block size for ENT dataset with the same radius $R = [3, 2, 1]$ and with different neighborhood sizes for the two datasets as in Fig. 10.9. The experimental results show that the recognition rate of WHMLBP increases about 5 % in average for SL dataset. The greatest accuracy is 97.59 % which occurs when the neighborhood size is $24 \times 24 \times 24$. In case of ENT dataset, almost all the results obtained where the WHMLBP was applied are better than

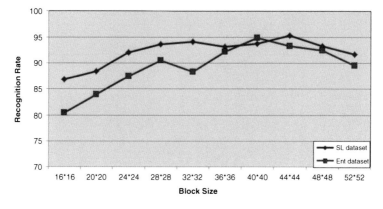

Fig. 10.8 Performance of WHMLBP with different block size

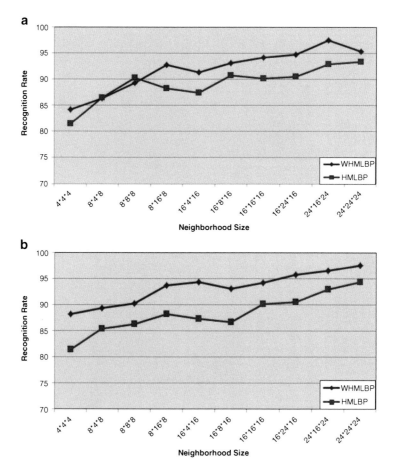

Fig. 10.9 The recognition rate for WHMLBP and HMLBP on: (**a**) Entropia dataset; (**b**) Second Life dataset

Table 10.4 Average recognition rate for different algorithms

Dataset	LBP (%)	WLBP (%)	MLBP (%)	HMLBP (%)	WHMLBP (%)
SL	84.45	82.12	87.34	88.32	93.29
ENT	81.98	85.31	86.88	89.11	91.87

when the HMLBP is applied. The accuracy rate increases about 3 % in average. The greatest accuracy rate is 97.45 % and is obtained when the neighborhood size is $24 \times 16 \times 24$. The average of the recognition rate of the two methods for both datasets using different neighborhood sizes is presented in Table 10.4.

We compared the performance of WHMLBP method with other methods, MLBP, WLBP, and LBP, on the same two datasets. We applied these methods with $R = 1, 2, 3$ and $P = 8, 16, 24$. The average of the recognition rate for both datasets is presented in Table 10.4.

The results we obtained demonstrate the effectiveness of our proposed technique in comparison with other methods.

10.8   Conclusion

In this chapter, we propose the application of wavelet transform to normalized manually cropped images to improve the efficiency of the HMLBP in extracting useful features from images. We applied different wavelet families to determine the best wavelet family and wavelet decomposition level fitting for each tested dataset. The effectiveness of this proposed method is shown in two avatar face datasets. Compared with HMLBP method, the proposed method gets more than 4 % improvement rate in the first dataset and approximately 3 % improvement rate in the second one. Additionally, if compared with other well-known methods, the proposed method gets higher recognition rate.

For future work, we suggest that moving from LBP to local ternary pattern, multi-scale local ternary pattern and hierarchical multi-scale local ternary pattern (for extracting features) and uBoost (for classifying) may lead to better accuracy rates in recognizing avatar faces. This is what we intend to work on in the future. We also plan to use larger datasets from different virtual worlds. The final goal is to build a complete automatic system for avatar face detection and recognition which will be applied on different virtual worlds to increase the level of security and reduce the effect of criminal activities.

References

1. Gavrilova, M.L., Yampolskiy, R.V.: Applying biometric principles to avatar recognition. In: Proceedings of the International Conference on Cyberworlds, pp. 179–186, 2010

2. Ajina, S., Yampolskiy, R.V., Amara, N.E.B.: Avatar facial biometric authentication. In: 2nd International Conference on Image Processing Theory, Tools and Applications, Paris, pp. 2–5, 2010
3. Second Life. Available from: www.secondlife.com
4. Active Worlds. Available from: www.activeworlds.com
5. Entropia Universe. Available from: www.entropiauniverse.com
6. Yampolskiy, R.V., Klare, B., Jain, A.K.: Face recognition in the virtual world: recognizing avatar faces. In: Proceedings of the 11th IEEE International Conference on Machine Learning and Applications, Boca Raton, FL, pp. 40–45 (2012)
7. O'Brien, N.: Spies watch rise of virtual terrorists. Available from: http://www.news.com.au/top-stories/spies-watch-rise-of-virtual-terrorists/story-e6frfkp9-1111114075761
8. Yampolskiy, R.V., Govindaraju, V.: Behavioral biometrics for verification and recognition of malicious software agents. In: Sensors and Command Control Communications and Intelligence (C3I) Technologies for Homeland Security and Homeland Defense VII. SPIE Defense and Security Symposium, Orlando, 2008
9. Ajina, S., Yampolskiy, R.V., Amara, N.E.B.: Evaluation of SVM classification of avatar facial recognition. In: Proceedings of the 8th International Conference on Advances in Neural Networks, pp. 132–142, 2011
10. Mohamed, A.A., Yampolskiy, R.V.: An improved LBP algorithm for avatar face recognition. In: 23th International Symposium on Information, Communication and Automation Technologies, Sarajevo, pp. 1–5, 2011
11. Garcia, C., Zikos, G., Tziritas, G.: A wavelet-based framework for face recognition. In: Proceedings of the 5th European Conference on Computer Vision, Freiburg, Allemagne, pp. 84–92, 1998
12. Harmon, L.: The recognition of faces. Sci. Am. **229**(5), 71–82 (1973)
13. Mazloom, M., Ayat, S.: Combinational method for face recognition: wavelet, PCA and ANN. In: International Conference on Digital Image Computing: Techniques and Applications, Canberra, pp. 90–95, 2008
14. Wang, H., Yang, S., Liao, W.: An improved PCA face recognition algorithm based on the discrete wavelet transform and the support vector machines. In: International Conference on Computational Intelligence and Security Workshops, Harbin, pp. 308–311, 2007
15. Luo, B., Zhang, Y., Pan, Y.-H.: Face recognition based on wavelet transform and SVM. In: IEEE International Conference on Information Acquisition, Hong Kong and Macau, pp. 373–377, 2005
16. Ojala, T., Pietikäinen, M., Harwood, D.: A comparative study of texture measures with classification based on feature distributions. Pattern Recognit. **29**(1), 51–59 (1996)
17. Meng, J., Gao, Y., Wang, X., Lin, T., Zhang, J.: Face recognition based on local binary patterns with threshold. In: IEEE International Conference on Granular Computing, San Jose, pp. 352–356, 2010
18. Wang, W., Chang, F., Zhao, J., Chen, Z.: Automatic facial expression recognition using local binary pattern. In: 8th World Congress on Intelligent Control and Automation, Jinan, pp. 6375–6378, 2010
19. Ahonen, T., Hadid, A., Pietikäinen, M.: Face description with local binary patterns: application to face recognition. IEEE Trans. Pattern Anal. Mach. Intell. **28**(12), 2037–2041 (2006)
20. Tan, N., Huang, L., Liu, C.: Face recognition based on a single image and local binary pattern. In: 2nd International Symposium on Intelligent Information Technology Application, Shanghai, pp. 366–370, 2008
21. Liu, X., Du, M., Jin, L.: Face features extraction based on multi-scale LBP. In: 2nd International Conference on Signal Processing Systems, Dalian, pp. v2 438–v2 441, 2010
22. Liao, S., Zhu, X., Lei, Z., Zhang, L., Li, S.: Learning multi-scale local binary patterns for face recognition. In: International Conference on Biometrics. Lecture Notes in Computer Science, vol. 4642, pp. 828–837. Springer, Berlin (2007)
23. Mäenpää, T., Pietikäinen, M.: Multi-scale binary patterns for texture analysis. In: Proceedings of the 13th Scandinavian Conference on Image Analysis, pp. 885–892, 2003

24. Turtinen, M., Pietikäinen, M.: Contextual analysis of textured scene images. In: Proceedings of the British Machine Vision Conference, pp. 849–858, 2006
25. Chan, C., Kittler, J., Messer, K.: Multi-scale binary pattern histogram for face recognition. In: International Conference on Biometrics. Lecture Notes in Computer Science, vol. 4642, pp. 809–818. Springer, Berlin (2007)
26. Guo, Z., Zhang, L., Zhang, D., Xuanqin, M.: Hierarchical multi-scale LBP for face and palmprint recognition. In: Proceedings of the IEEE International Conference on Image Processing, Hong Kong, pp. 4521–4524 (2010)
27. Mohamed, A.A., D'Souza, D., Baili, N., Yampolskiy, R.V.: Avatar face recognition using wavelet transform and hierarchical multi-scale LBP. In: Proceedings of the 10th IEEE International Conference on Machine Learning and Applications, Honolulu, pp. 194–199, 2011

Chapter 11
A Flexible Image-Based Access Control Model for Social Networks

Bechara Al Bouna, Richard Chbeir, Alban Gabillon, and Patrick Capolsini

Abstract Nowadays, social network are tremendously spreading their tentacles over the web community providing appropriate and well adapted tools for sharing images. A fundamental glitch to consider is their ability to provide suitable techniques to preserve individuals' privacy. Indeed, there is an urgent need to guarantee privacy by making available to end-users, tools to enforce their privacy constraints. This cannot be done over images as simple as it has been designed so far for textual data. In fact, images, as all other multimedia objects are of complex structure due to the gap between their raw data and their actual semantic descriptions. Without these descriptions, protecting their content is a difficult matter which outlines the premise of our work. In this chapter, we present a novel security model for image content protection. In our model, we provide dynamic security rules based on first order logic to express constraints that can be applied to contextual information as well as low level features of images. We finally discuss a set of experiments and studies carried out to evaluate the proposed approach.

B. Al Bouna (✉)
Ticket Labs, Antonine University, Baabda Beirut, Lebanon
e-mail: bechara.albouna@upa.edu.lb

R. Chbeir
University of Pau and Adour Countries, LIUPPA Laboratory, 64200 Anglet Cedex, France
e-mail: richard.chbeir@univ-pau.fr

A. Gabillon · P. Capolsini
GePaSud Laboratory, University of French Polynesia, Punaauia, France
e-mail: alban.gabillon@upf.pf; patrick.capolsini@upf.pf

R. Chbeir and B. Al Bouna (eds.), *Security and Privacy Preserving in Social Networks*,
Lecture Notes in Social Networks, DOI 10.1007/978-3-7091-0894-9_11,
© Springer-Verlag Wien 2013

11.1 Introduction

In the last decades, there has been a rapid evolution of Web technologies and mobile devices (PDAs, cell phones, etc.) enabling users to produce and easily share large amounts of data. Multimedia objects such as pictures, videos, and audio files operate across networks and are used mainly for personal purposes but also in a multitude of areas and information systems (medical, military, education, etc.). This massive use of multimedia objects has raised several problems including privacy and confidentiality preserving. In fact, providing safe browsing and publishing of multimedia objects' contents particularly, masking out objects of interests is considered of great importance in several privacy scenarios (e.g., hiding the face of a popular person in a photo, hiding violent scenes in a TV show, etc.). This pushed the research community to extend and/or explore dedicated access control models for multimedia database [2, 3, 8, 9, 21, 22, 25]. While these models have proved effective in some contexts, their scalability is constrained by several drawbacks that can be summarized as follows:

- *Lack of proper multimedia description model.* Some of the proposed approaches such as [21, 22] adopt multimedia description models that are XML-Schema based which has been considered cumbersome for the Web. In fact, the wide range of vocabularies included in these multimedia models that should be processed by an access control mechanism limit direct machine processing of low level features and focus mainly on their semantic descriptions.
- *Lack of proper decision-making under uncertainty.* Multimedia objects processing leads eventually to uncertain decisions due to multimedia objects' complex structure. For instance, computing a suitable similarity score between two distinct multimedia objects depends on (1) the choice of the similarity function, (2) the content of multimedia objects (e.g., person's face and body), and (3) their semantic descriptions. This can be explicitly important when protecting multimedia objects resulting from similarity-based queries where uncertain decisions must be interpreted.

To point out these drawbacks, we provide the following scenario.

11.1.1 Motivating Scenario

Let us consider, for example, the case of the company SOURA[1] provides, for her public image, a photo database accessible by all employees, visitors, students, and clients through terminals in the waiting rooms. Some of the images contained in

---

[1] Works in photo editing, montage, and publishing.

Fig. 11.1 A photo captured by the surveillance camera at the main entrance

the database are stored without textual description and they are categorized as follows:

- *Social photos*: includes all photos taken in the social gatherings of employees with their relatives
- *Meeting photos*: contains all photos captured during important official meetings and general assemblies of the company
- *Promotional photos*: gathers photos of heads and directors who have participated in promotional events (conferences, workshops, etc.).

To use and manage the photo database, a set of dedicated processing functions (face detection, blob detection, blur, etc.) are used along with a search engine allowing users to search for photos using different techniques (i.e., categories, queries by examples, keywords, colors, etc.).

For security reasons (privacy preserving, access controlling, surveillance, etc.), the company decided to install in most of the rooms inside the company building several surveillance cameras and adopted the following authorization rules that allow to specify the users (employees, trainees, visitors, etc.) who have the right to view the database content:

- **Rule 1**: All users must be authenticated using their identity and picture (trainees and employees via their photos in their personal profile form, and visitors via surveillance cameras capturing their faces at the entrance gate)
- **Rule 2**: Allow all visitors and trainees to view social and meeting photos
- **Rule 3**: Deny a visitor, who identified himself as *Bob* when capturing his photo (Fig. 11.1), to view social and meeting photos
- **Rule 4**: Deny users to view social and meeting photos when the visitor *Bob* is nearby
- **Rule 5**: Deny trainees and visitors to identify the existence of Mr. *Dupont* (head of the company) in all photo sets.

Consequently, the access control model to be adopted and used by the company must be able to:

- Define a user/subject via a photo/picture
- Define accessible/authorized photos using a textual and low-level criteria (e.g., so to detect Mr. *Dupont* since some photos are annotated while others are not)
- Manage uncertain-decision making
- Specify the context of a user.

To meet all these requirements, the need for a new solution capable to adequately controlling the access to image database becomes a must.

11.1.2   Contributions

In this chapter, we propose a new model, called FOL-IBAC, for access controlling of Image data to cope with traditional access control models drawbacks. In FOL-IBAC, we use first order logic which has been adopted in several approaches [14, 15, 19, 24] and [16], to express complex policies and provide suitable rule specification formalism. Our contributions can be summarized as follows:

- We provide a simple data representation model to properly describe the content of images and their context. We also present a set of methods and functions to process images data at different levels.
- We propose a framework for specifying the security policy. In our model the security policy consists of contextual security rules which are based on constraints applying to subjects, objects, and the environment, as it is done in the ABAC model and the ORBAC model.
- We provide a second framework for specifying access control mechanisms where several multimedia-based predicates are defined to handle content processing and uncertain decisions. In fact, this framework allows us to express the security mechanism that should be enforced whenever a prohibition is derived from the security policy and reduce as well possible uncertainty that comes across multimedia processing. Indeed, in image databases, access controls can be more complicated than simply granting or denying access to resources. For example, one may prefer to apply a blurring function on a sensitive object contained in a picture rather than denying access to the whole picture.

The same language based on first-order logic is used to specify the security policy and the access control mechanisms.

The organization of the remainder of this paper is as follows. In Sect. 11.2, we review existing access control models for image and multimedia database, and we discuss their drawbacks. In Sect. 11.3, we present several definitions needed to fully understand our approach detailed in Sect. 11.4 where we also discuss our security model. Section 11.6 is dedicated to present the set of experiments conducted to validate our approach. The last section concludes this work and pinpoints several future directions.

11.2 Related Work

Several studies [2,8,9,13,18] were dedicated and proposed in the literature to protect multimedia objects of different types. In the following, we will give a brief overview of these approaches and discuss their drawbacks.

Models proposed by Bertino et al. [8] and Thurainsigham et al. [25, 26] create an abstraction between the content of the video and their description. This can actually facilitate the specification of permissions based on hierarchical concepts. Thus, access control is actually implemented on the description of objects instead of their content. However, specifying multimedia objects content using expressions based on the concepts and/or objects' identifiers is not always sufficient especially when multimedia objects lack description and annotation. In addition, classification techniques are still inefficient to reduce the gap between the semantic description of multimedia objects and their content. In other words, these techniques often require a preprocessing phase which can be less effective with few trained multimedia objects.

In [2], Nabil R. Adam et al. propose an access control model dedicated to the protection of digital libraries' content. Although the proposed model handles fine granularities (including digital libraries content and links to object at different levels), the ability to specify flexible permissions remains limited due to the abstraction that the authors wished to create through their model descriptions of electronic libraries which is limited to concepts and objects specified by their IDs. However, the structure of multimedia objects cannot be limited to the model description as multimedia objects are characterized by attributes describing their semantics (annotation, keywords, concepts, etc.), content (URI, Blobs, etc.), and relations of different types (semantic, spatial, etc.) which may be the target of an authorization or restriction.

In [7], the authors present an access control model for geo-spatial data. They specify objects using expressions that take into consideration raw data (images, satellite image, vector, etc.), regions, object resolution, timestamps, and data type(s). Regardless of permission definition flexibility which is based on low level attributes and description, the approach holds some limitations related to: (1) the complexity of defining permissions based on attributes with missing semantics such as coordinates and region types (rectangle, circle, etc.) and (2) the lack of tolerance for error due to uncertain values (geo-spatial coordinates, longitude, latitude, etc.) that may be directly related to the used equipments.

In the following, we present our Image Access Control Model that cope with previously discussed drawbacks.

11.3 Definitions and Terminologies

In this section, we will present several definitions needed to fully understand the proposed approach.

11.3.1 Data Representation

Below, we will explain our proposal for representing images. The proposed definitions are built on relational-object paradigm in order to be able to consider both relational, object-oriented DBMS, and XML-Based DBMS.

Image

An image is an object that contains embedded attributes with descriptions that illustrate and give important information about the captured scene. More formally, we represent an image in the following way:

$$IM :< iid, O, meta, desc, f, SOs, rel >$$ (11.1)

where:

- *iid* is a unique identifier associated with an image
- *O* is a reference to the raw data of the object (or the file) which can be stored as BLOB or URI
- *meta* is the set of technical descriptions and metadata embedded in the image that takes the form of data about data. Here, metadata can be defined as information not related to the semantic content of the image (such as author, location, and date/time). Adding metadata at the creation time requires no effort. Actually, recent digital devices provide such functionalities. This is considered as a best practice if all the information related to the date, the place, and other technical characteristics are available when capturing an image/photo. Using Exif standard,[2] captured images are described using a predefined set of attributes in terms of the context, the environment, and the technical characteristics (exposure time, resolution, focal length, shutter speed, etc.). The value of the Location metadata is extracted from the following attributes: exif:gpsLatitude and exif:gpsLongitude. The values of the Date and the Time metadata are extracted from the following attribute: exif:dateTime
- *desc* is the set of textual descriptions, keywords, or annotations provided by the main user as an image and/or album/folder[3] caption(s). Usually, an image description gives some relevant information about the image in terms of places, objects, people, etc. In this study, the description of an image is represented using the Dublin Core (DC) standard,[4] which is a compact standard used for cross-domain information resource description. It is simple to use and easy

---

[2]http://www.exif.org.

[3]When an image belongs to an album or a folder.

[4]DC is an XML/RDF-based syntax used to describe textual as well as audiovisual documents (see http://dublincore.org/).

to implement thus allowing the non-specialist to use it (librarians, researchers, museum curators, music collectors, etc.)

- f is the set of features describing the physical and spatial content of an image such as color and texture. The physical features (e.g., colors) can be described via several descriptors (such as color distribution, histograms, and dominant color(s)) where each descriptor is obtained by assembling a set of values with respect to a color space model. Similarly, spatial features (e.g., shape) can have several descriptors such as MBR (Minimum Bounding Rectangle), bounding circle, surface, and volume.
- SOs is a set of salient objects contained in the image which will be detailed in the next section
- rel is a set of triplets where each contains two sets of objects (either image or salient object) as well as their relationship(s) (semantic, spatial, and spatiotemporal).

Salient Object

A salient object represents an object of interest in the image such as a person (face), a logo, and a building. Formally, a salient object is defined as:

$$SO :\prec soid, loc, f, desc \succ \qquad (11.2)$$

where:

- $soid$ is a unique identifier associated with the salient object
- loc is the region containing the salient object in the image. The salient object is not assigned a fixed width and height. Its coordinates determine the location and the size of the salient object region. We represent a salient object as a region in the image using the Image vocabulary[5]
- f describes the physical and visual content of a salient object (such as color histogram(s), color distribution, and texture histogram, etc.)
- $desc$ is the set of textual tags and annotations assigned to this salient object.

Relations

Images and salient objects can share several types of relations between them.

Spatial relations as described in [12] may exist either between two salient objects or a salient object and its image. Using our representation model, they can be automatically computed using the *location* and images and salient objects identifiers.

---

[5]http://www.bnowack.de/w3photo/pages/image_vocabs.

Table 11.1 Sample object predicates

Image-related predicates	Description
Type/2	Type(o, Image): reads "o is an Image"
Description/2	Description(o, NewYear10): reads "o contains in its descriptions the text NewYear10"
Date/2	Date(o, 27/07/2009): reads "o is taken on 27/07/2009"
Keyword/2	Keyword(o, Party): reads "o is associated to a keyword Party"
Owner/2	Owner(o, s_1): reads "o has an owner s_1"
R/2	$R(o_1, o_2)$: reads "o_1 is related to o_2 with the relation R, i.e., R(o_1, o_2)"
Sim/3	$Sim(o_1, o_2, 0.7)$: reads "o_1 is similar o_2" within a threshold 0.7.
Belong/3	$Belong(o_1, o_2, 0.7)$: reads "o is similar to at least one of the objects contained in o_2" within a threshold 0.7.

Furthermore, *semantic relations* are of high importance and can be represented as well. In several domains, users need to describe and search images using familiar terms like invading, attacking, and shifting. In this manner, the user will have access to a comprehensive and intelligent description of image and its related data.

We note that some of the semantic relations can be described using spatial relations and derived according to domain application requirements. For instance, to define the relation *MarriedTo*, one can use the following rule:

$$\forall o_1 \forall o_2 \ (MarriedTo(o_1, o_2) \rightarrow Co - occur(o_1, o_2))$$

That is, if in a given situation, the semantic relation *MarriedTo* is not explicitly defined between two distinct objects, the $co - occur$ relation holds.

11.3.2 Data Methods and Predicates

Each attribute in our representation model can be associated with a set of simple functions/methods/predicates. Table 11.1 shows a sample nonexhaustive list of predicates. In addition, similarity and derived functions/predicates can be defined as well as detailed in the following subsections.

Distance and Similarity

Since in our representation model, we consider two kinds of attributes, comparing images and/or salient objects comes down to comparing atomic values A (e.g., tag and caption etc.), multimedia values M (e.g., low-level features), or both.

On the one hand, when images come to play, comparing two atomic attribute values using standard operators ($<$, $>$, Like, $\leqslant$, $=$, $\geqslant$, etc.) can be inefficient and inaccurate. More particularly, when multimedia objects (images and/or salient objects) are associated with textual or spatial attributes (e.g., genre and place), the use of such operators becomes inappropriate since they do not take into account the related semantics (for example, comparing the strings Paris to France is not relevant).

On the other hand, similarity between two multimedia values in $\mathcal{M}$ can be computed using various distance measures defined on feature spaces (color, texture, etc.) [6] which eventually gives rise to uncertain decisions depending on the various scores returned.

In our approach, we take into account these important features of atomic and multimedia attributes by assuming that, given an attribute A in $U = \mathcal{A} \cup \mathcal{M}$, either atomic or multimedia, several distances can be adopted. Consequently, the similarity of two values is defined according to a function that aggregates the results of related distances which provide reduced uncertainty and more precise results as we will show in Sect. 11.6. Let A be an attribute in U (atomic or multimedia) over which n distance functions $d_A^1, \ldots, d_A^n$ are used. We associate A with a *global distance function*, denoted by ω_a, such that for all a_1 and a_2 in its domain $dom(A)$:

$$\omega_A(a_1, a_2) = g_A(d_A^1(a_1, a_2), \ldots, d_A^n(a_1, a_2)) \tag{11.3}$$

where g_A is an aggregation function assumed to satisfy the following triangular co-norm properties:

- $g_A(0, \ldots, 0, x_i, 0, \ldots, 0) = x_i$ (zero identity),
- if for every $i = 1, \ldots, n$, $x_i^1 \leq x_i^2$, then $g_A(x_1^1, \ldots, x_n^1) \leq g_A(x_1^2, \ldots, x_n^2)$ (monotonicity),
- for every permutation ϖ of $\{1, \ldots, n\}$, $g_A(x_1, \ldots, x_n) = g_A(x_{\varpi(1)}, \ldots, x_{\varpi(n)})$ (commutativity), and
- $g_A(x_1^1, \ldots, x_{i-1}^1, g_A(x_1^2, \ldots, x_n^2), x_{i+1}^1, \ldots, x_n^1) =$
  $g_A(x_1^1, \ldots, g_A(x_{i-1}^1, x_1^2, \ldots, x_{n-1}^2), x_n^2, x_{i+1}^1, \ldots, x_n^1)$ (associativity).

We note that if $n = 1$, then g_A is the identity function and so, $\omega_A(a_1, a_2) = d_A^1(a_1, a_2)$, for all a_1 and a_2 in $dom(A)$.

In what follows, given an attribute A in U and a real number ϵ, when writing expressions such as $\omega_A(a_1, a_2) \leq \epsilon$, ϵ refers to the radius of the range operator or to the number of k neighbors to be returned, depending on the similarity operator being considered.

Let A be an attribute in U (atomic or multimedia) and ϵ a positive real number. For all a_1 and a_2 in $dom(A)$, a_1 and a_2 are said to be *similar within ϵ*, denoted as $Sim(a_1, a_2, \epsilon)$:

$$\forall a_1, \forall a_2, \forall \epsilon, Sim(a_1, a_2, \epsilon) \equiv \omega_A(a_1, a_2) \leq \epsilon \tag{11.4}$$

For instance, the distance between two photo places p1 and p2, denoted as $\omega_{place}(p1, p2)$, can be computed as follows. According to the predefined KB

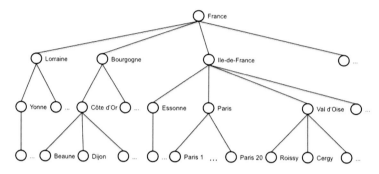

Fig. 11.2 A sample KB with one relationship *belongs*

shown in Fig. 11.2, $\omega_{\texttt{place}}(\texttt{p1},\texttt{p2})$ is the height of the minimal subtree of KB whose leaves are p1 and p2. In this case, it is easy to see that $\omega_{\texttt{place}}$ is a distance for which $\omega_{\texttt{place}}(\texttt{Dijon},\texttt{Beaune}) = \omega_{\texttt{place}}(\texttt{Cergy},\texttt{Roissy}) = 1$ and $\omega_{\texttt{place}}(\texttt{Paris},\texttt{Bourgogne}) = 2$. Therefore, *Sim*(Dijon, Beaune, 1) and *Sim*(Cergy, Roissy, 2) both hold, whereas *Sim*(Paris, Bourgogne, 1) does not.

Derived Methods

In addition to the similarity method(s) and simple functions/predicates to be assigned to each component of our data model, several derived methods and predicates can be defined as well. To illustrate this, let's consider the following two examples. The next definition says that any picture taken on 31st of December 2010 and associated with a textual keyword party should be described by "New year's eve 2010"

$$\forall o \ (TypeO(o,\texttt{Image}) \wedge Keyword(o,\texttt{Party}) \wedge Date(o,31/12/10) \rightarrow$$

$$Description(o,\texttt{NewYear10}))$$

The function *Belong* can be defined as follows:
$\forall o_1, \forall o_2,$ *and* $\forall o_3$

$$Contain(o_1,o_2) \vee (Contain(o_1,o_3) \wedge Sim(o_2,o_3,\epsilon))$$

$$\rightarrow Belong(o_1,o_2,\epsilon)$$

This means that one object o_2 is contained in o_1 or is similar to one of its salient objects o_3.

Table 11.2 Sample subject predicates

Subject predicates	Description
Identity/2	Identity (s, 1): reads "subject s has an identity 1"
Role/2	Role (s, Director): reads "subject s plays the role Director"
Username/2	Username (s, Bob): reads "subject s has a username Bob"
Password/2	Password (s, pass): reads "subject s has a pass"
Age/2	Age (s, 18): reads "subject s has 18"
TrustLevel/2	TrustLevel (s, Trusted): reads "subject s has a Trusted Level"
Organization/2	Organization (s, UPF): reads "subject s is affiliated to UPF organization"
PictureOf/2	PictureOf (s, pic1.jpg): reads "subject s has the picture pic1.jpg"
Sim/3	Sim(s, Bob, 0.7): reads "subject s has a predefined set of textual attributes similar to the word Bob within a threshold 0.7"

11.4 Security Model

The security model defined in this section is based on first-order logic and on the model provided in [14] and [15]. As already mentioned in the introduction, it allows us to express the security policy and the access control mechanisms that should be enforced whenever prohibitions are derived from the security policy. First, we define predicates of our language i.e., predicates applying to subject, objects, and actions. We also specify predicates related to the environment since we aim at specifying dynamic contextual security policies. Second, we define the authorization rules and show how to solve the conflicts that may arise between positive and negative authorization rules. Finally, we define rules specifying access control mechanisms which should be enforced.

11.4.1 Subject

A subject is a user or a process run on behalf of a user. The predicate $Subject(s)$ reads "s is a subject". We assume here that each subject has associated atomic and complex attributes A in $U = \mathcal{A} \cup \mathcal{M}$ to describe its identity, role, age, location, photo, and so on. Each attribute can be associated with a predicate in our language. Table 11.2 shows a sample list of predicates representing some subject attributes. Of course, this is not an exhaustive list and can be extended depending on the application. Instances of some predicates (e.g., identity/2 and age/2) have to be specified one by one, whereas instances of other predicates (e.g., role/2) may be partially or totally specified by logical conditions. To illustrate this, let's consider the following examples. The first example says that any subject which belongs to the University of French Polynesia (UPF) should be considered as trusted.

$$\forall s \ (Organization(s, \text{UPF}) \rightarrow TrustLevel(s, \text{Trusted}))$$

The second example says any subject whose identity is unknown (i.e., any unregistered user) should belong to role Guest:

$$\forall s \ (\neg \exists i \ Identity(s, i) \rightarrow Role(s, \text{Guest}))$$

The third definition indicates that every guest whose photo is similar to (Fig. 11.1) must not be trusted:

$$\forall s \ (Role(s, \text{Guest}) \ \wedge \ Sim(s, \text{Suspect.jpg}, 0.7) \rightarrow$$
$$TrustLevel(s, \text{Untrusted}))$$

Note that some predicates should respect some integrity constraints. We did not include these integrity constraints into Table 11.2, since they may depend on the application. For example, the following integrity constraint requires all subjects to provide a photo so to be authenticated:

$$\forall s \ (Subject(s) \rightarrow \exists i \exists p \ (Identity(s, i) \ \wedge \ PictureOf(s, p)))$$

The following integrity constraint requires users to have only one identity:

$$\forall s \forall i_1 \forall i_2 \ (Identity(s, i_1) \wedge Identity(s, i_2) \rightarrow i_1 = i_2)$$

Note that predicates for which a similar constraint applies represent a mono-valued attribute whereas predicate for which there is no similar constraint represent a multi-valued attribute.

11.4.2 Object

In our model, an object is mainly an image represented using our data representation previously detailed to describe its owner, content, and features (colors, textures, etc.), semantic (keywords), as well as its relationships with other objects. As for subjects, some integrity constraints can be associated with object predicates depending on the application at hand. For instance, the following integrity constraint requires objects to have only one owner:

$$\forall o, \forall s_1, \forall s_2 \ (Owner(o, s_1) \wedge Owner(o, s_2) \rightarrow s_1 = s_2)$$

11.4.3 Action

An action corresponds to a composition of (sequential or parallel) a number of operations to visualize, create, edit, delete images. Each action is associated with a

Table 11.3 Sample action predicates

Action predicates	Description
TypeA/2	TypeA(a, view) : reads "a is view action"
IsA/2	IsA(a_1, a_2): reads 'a_1 is hierarchically linked to a_2"

set of atomic attributes allowing to define its type, category, and various parameters attached to the action. Predicate Action(a) reads "a is an action" and each attribute corresponds to a predicate. Table 11.3 shows a list of sample predicates representing some action attributes. Potentially, instances of some action predicates may be partially or totally defined by logical conditions and should also respect some integrity constraints. These integrity constraints may depend on the application. For example, the following constraint says that all actions should be typed.

$$\forall a \; (Action(a) \rightarrow \exists t \; TypeA(a,t))$$

The second following domain constraint says that any action should be of type Read or Edit.

$$\forall a \forall t \; (TypeA(a,t) \rightarrow t = \text{Read} \; \vee \; t = \text{Edit})$$

The third following constraint says that the type of an action is unique.

$$\forall a \forall t_1 \forall t_2 \; (TypeA(a,t_1) \wedge TypeA(a,t_2) \rightarrow t_1 = t_2)$$

11.4.4 Environment Signature

In order to enforce access control mechanisms and to consider the context information sensed from multimedia devices and sensors, we include in our language the notion of environment signature so to represent user environment properties. Each *signature* is: (1) composed of atomic and complex attributes A in $U = \mathcal{A} \cup \mathcal{M}$ such as user location, user clothes, temperature, etc. and (2) associated with some predicates. For instance, InArea(s, pic.jpg, 0.7) reads "subject s is in the area described by the picture pic.jpg", i.e.,$\exists \; o_1 \; PictureOf(o_1,s) \wedge Belong(o_1, \text{pic.jpg}, 0.7)$.

11.4.5 Security Rule

Security rules specify how (and in which context) subjects can execute actions on objects. Our model includes permissions (positive rules) and prohibitions (negative rules). We define a positive authorization rule as a logical rule having the following form:

$$\forall s \forall a \forall o \ (Condition \rightarrow Permit(s, a, o))$$

where *Condition* is a Boolean formula and *Permit*(s, a, o) reads "s is permitted to execute action a on object o."

Similarly, we define a negative authorization rule as a logical rule having the following form:

$$\forall s \forall a \forall o \ (Condition \rightarrow Deny(s, a, o))$$

where *Deny*(s, a, o) reads "s is forbidden to execute action a on object o."

Authorizations propagate to sub-objects i.e., the following entailments should be considered as part of the security policy:

E1:

$$\forall s \forall a \forall o_1 \forall o_2 \ (Permit(s, a, o_1) \wedge Contain(o_1, o_2) \rightarrow Permit(s, a, o_2))$$

E2:

$$\forall s \forall a \forall o_1 \forall o_2 \ (Deny(s, a, o_1) \wedge Belong(o_2, o_1, \epsilon) \rightarrow Deny(s, a, o_2))$$

Authorizations propagate to sub-actions i.e. the following entailments should be considered as part of the security policy:

E3:

$$\forall s \forall a_1 \forall a_2 \forall o \ (Permit(s, a_1, o) \wedge Contain(a_1, a_2) \rightarrow Permit(s, a_2, o))$$

E4:

$$\forall s \forall a_1 \forall a_2 \forall o \ (Deny(s, a_1, o) \wedge Contain(a_1, a_2) \rightarrow Deny(s, a_2, o))$$

The default policy of our model is closed. This means that, given a subject s requesting to execute action a on object o, if neither *Permit*(s, a, o) nor *Deny*(s, a, o) can be derived from the security policy then the subject s should be denied to execute action a on object o.

Let us consider the following example of security policy which consists of the aforementioned propagation rules and the three following rules:

Rule 1: This statement grants to people who are on *PicUsers.jpg* the right to see pictures from the New Year's Eve party:

$$\forall s \forall a \forall o_1 \forall o_2 \forall o_3$$

$$(PictureOf(o_1, s) \wedge Belong(o_2, PicUsers.jpg, 0.7) \wedge Sim(o_1, o_2, 0.8)$$

$$\land\ TypeA(a, Read)$$

$$\land\ Album(o_3, NewYear10)$$

$$\rightarrow\ Permit(s, a, o_3))$$

The first line is for authenticating subject s. Subject s should exhibit a picture of himself (o_1) which looks like somebody (o_2) appearing on PicUsers.jpg. The second line says that action a should be a read action. The third line says that security object o_3 should belong to the NewYear10 album.

Rule 2: This statement denies to Bob the right to see pictures from the New Year's Eve party

$$\forall s \forall a \forall o_1 \forall o_2 \forall o_3 \forall o_4 \forall t$$

$$(PictureOf(o_1, s) \land Sim(o_1, Bob.jpg, 0.8) \land TypeA(a, Read)$$

$$\land\ Album(o_2, NewYear10)$$

$$\rightarrow\ Deny(s, a, o_2))$$

The first line is for authenticating Bob. Subject s should exhibit a picture of himself (o_1) which looks like a reference picture of Bob (*Bob.jpg*). The second line says that action a should be a read action. The third line says that security object o_2 should belong to the NewYear10 album.

Rule 3: This statement denies to people who were on *PicUsers.jpg* the right to see bob on the party images

$$\forall s \forall a \forall o_1 \forall o_2 \forall o_3 (PictureOf(o_1, s) \quad \land \quad Belong(o_2, PicUsers.jpg)$$

$$\land\ Sim(o_1, o_2, 0.8)$$

$$\land\ TypeA(a, Read)$$

$$\land\ Album(o_3, NewYear10) \land Belong(o_4, o_3, 0.7)$$

$$\land\ Sim(o_4, Bob.jpg, 0.8)$$

$$\rightarrow\ Deny(s, a, o_4))$$

The first line is for authenticating subject s. Subject s should exhibit a picture of himself (o_1) which looks like somebody (o_2) appearing on PicUsers.jpg. The second line says that action a should be a read action. The third line says that security object o_4 should belong to a picture (o_3) from the NewYear10 album. The fourth line says that o_4 should look like a reference picture of Bob (Bob.jpg).

11.4.6 Conflicts

A positive rule may conflict with a negative rule. In the security policy specified in the previous section, we may have the following two conflicts:

- If Bob appears on `PicUsers.jpg` and attempts to see pictures from the NewYear10 album, then Rule 1 conflicts with Rule 2.
- If somebody who is on `PicUsers.jpg` attempts to see a New Year's Eve picture where Bob appears, then Rule 3 conflicts with propagation rule E1 (Rule 1 says that people who are on `PicUsers.jpg` have the right to see pictures from the New Year's Eve party; E1 grants to these people the right to see all sub-objects (including Bob) belonging to these pictures, but Rule 3 denies to these people the right to see Bob).

There exist various strategies for solving such conflicts such as the most specific takes precedence, the negative rule takes precedence, the first match, or a policy based on priority levels which would be assigned to rules.

We prefer discarding the most specific takes precedence strategy. Indeed, it is sometimes impossible to decide on which rule is more specific than the other (for instance, consider the case where the first rule has a more specific subject whereas the second rule has a more specific object). The negative rule takes precedence strategy is acceptable but only if we assume that the global policy can always be expressed by means of positive rules and that negative rules are used only for specifying exceptions to the global policy. Unfortunately, there are some policies which are easier to express by means of negative rules specifying the global policy and positive rules specifying exceptions. The first match strategy[6] can in some scenarios be adopted but it requires being very careful when ordering the rules since the risk of having rules which never apply is very high.

For all theses reasons, our model requires that the security administrator assigns priority levels to rules. If given an access request, a positive rule conflicts with a negative rule then:

- If the two rules have the same priority, then the "first match" strategy is used i.e., the rule which is first read applies
- otherwise, the one with the highest priority applies.

Propagation rules E1 to E4 are also assigned with a priority level i.e., authorizations propagate downward to sub-objects (or sub-actions) unless overridden by higher priority rules.

---

[6]The first match strategy is particularly used in firewalls where security rules are read sequentially. The first rule which matches the packet applies whether it is a positive or a negative rule.

Considering these principles, the two potential conflicts which may arise between Rule 1 & Rule 2 and E1 & Rule 3 can be solved as follows:

- by assigning a priority level to Rule 1 and Rule 2 such that the priority level of Rule 2 is higher than the priority level of Rule 1
- by assigning a priority level to Rule 3 which is higher than the priority level of propagation rule E1.

Note that, in a model enabling contextual security rules, it is not always easy to predict the conflicts which may arise between rules and consequently the security administrator may find difficult to assign priority levels to rules. The Or-BAC model [1] includes an algorithm for a priori detecting potential conflicts between security rules. Once these potential conflicts have been identified, the security administrator may specify extra information saying that some of these potential conflicts will never occur and therefore should be ignored. For example, if the security administrator indicates that Bob is not on PicUsers.jpg then, according to the Or-BAC model, there is no need to assign priority levels to Rule 1 and Rule 2 since these two rules will never conflict.

Now, if Bob is on PicUsers.jpg, then priority levels should be assigned to Rule 1 and Rule 2. We believe we can easily adapt the Or-BAC potential conflicts detection algorithm to our model although this remains work to be done.

11.4.7 Protection Mechanism

Let us consider Rule 3 and assume that the priority level of Rule 3 is higher than the priority level of E1 (i.e., Rule 3 applies). Rule 3 denies to people who are on PicUsers.jpg the right to see Bob on the New Year's Eve party images although these people have the right to see these pictures. Therefore, we may wonder what should happen if a subject who is on PicUsers.jpg requests viewing a picture from NewYear10 where Bob appears. Should the subject be denied to see the whole picture? Should Bob be hidden in the picture which is shown to the subject? Should the object representing Bob be blurred? etc.

To answer this question, we extend our security framework in order to enforce the access control mechanism when a given subject is forbidden to execute a given action on a given object. The idea is to hide and/or disseminate an information by transforming an object o into a secured/protected one o_s. Several techniques are provided in the literature to protect textual and image data by encrypting them [13], adding some noise [10], blurring a face [10], etc.

Consequently, a negative rule is extended as follows:

$$\forall s \forall a \forall o \ (Condition \rightarrow Deny(s,a,o) \land Protect(o,m))$$

Fig. 11.3 Sample image where a sensitive object *Bob* appears

Fig. 11.4 Two protection mechanisms

where *Protect(o,m)* reads "o should be protected with a mechanism *m*" and should respect the following integrity constraint:

$$\forall o \forall m \ (Protect(o, m) \rightarrow m = \texttt{Reject} \lor m = \texttt{Blur} \lor m = \texttt{Avatar} \lor m$$
$$= \texttt{Downgrade} \lor \dots)$$

If the conclusion does not include the Protect predicate, then a default protection mechanism applies. For instance, if the default protection mechanism is *Reject* i.e., if *Deny(s,a,o)* is derived from the security policy, then request *(s,a,o)* is rejected.

To illustrate this, let us consider the following two rules which consists of applying protection functions on a sensitive object *Bob* in Fig. 11.3.

The first rule consists of blurring the object, while the second one allows substituting *Bob*'s face by an avatar (Fig. 11.4).

Rule 1e: $\forall s \forall a \forall o_1 \forall o_2$

$$(Type(o_1, \texttt{Image}) \land Description(o_2, \texttt{Bob}) \land Belong(o_2, o_1, 0.7) \land IsA(a, \texttt{Read})$$
$$\rightarrow Deny(s, a, o_2) \land Protect(o_2, \texttt{Blur}))$$

Rule 2e: $\forall s \forall a \forall o_1 \forall o_2$

$$(Type(o_1, \text{Image}) \wedge Description(o_2, \text{Bob}) \wedge Belong(o_2, o_1, 0.7) \wedge IsA(a, \text{Read})$$
$$\rightarrow Deny(s, a, o_2) \wedge Protect(o_2, \text{Avatar}))$$

An even more restrictive option would be to deny access to pictures containing sensitive objects. However, this cannot be done using protection mechanisms. If pictures containing sensitive objects should not be displayed, then it means that such pictures should not be authorized at all. This should be expressed at the security policy level as follows:

$$\forall s \forall a \forall o_1 \forall o_2 \; (Type(o_1, \text{Image}) \wedge Belong(o_2, o_1, 0.7) \wedge IsA(a, \text{Read}) \wedge$$
$$Deny(s, a, o_2) \rightarrow Deny(s, a, o_1))$$

This rule propagates authorizations upward in the image aggregation hierarchy. If stated, it should have a higher priority than Rule 1.

It is important to note that protection mechanisms can also be extended to target some attributes of images and salient objects instead of main objects' raw data. However, we do not consider this issue in this study since this process is application dependent and depends on how the protection mechanism is implemented. For instance, if one intends to protect the *annotation* attribute of the image rather than the image itself, (s)he could develop/use a protection function which encrypts the textual annotation of an image.

11.5 Prototype

In this section, we present our java-based prototype called ImageProtector[7] dedicated to protect images to be published on Facebook social network [4]. Through its *BackOffice* module, ImageProtector allows one to define a set of security rules so to specify which contact(s) is entitled to visualize images and/or part of their content. Accessing and visualizing photos is done using *FrontOffice* module. For more details concerning the *Image Protector*, we invite the reader to consult the demo paper published in [4]. The architecture of ImageProtector is provided in Fig. 11.5.

Before we detail the prototype architecture and related modules, we will first present in the following subsection the core ontology used to describe the security model and store photos related data.

---

[7] An online demonstration of the prototype is provided at http://www.upa.edu.lb/ImageProtector/.

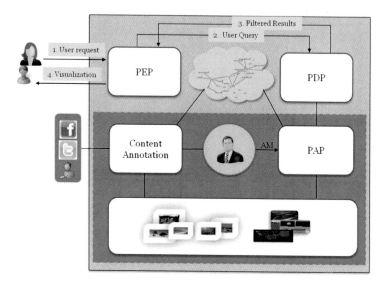

Fig. 11.5 ImageProtector architecture

11.5.1 Ontology-Based Storage Model

When images and photos come to play, the use of ontologies becomes required in order to reduce the semantic gap between users vocabulary and descriptions associated with those data. This has been proven in many situations [11, 20]. To handle this and also to provide extensible framework, we adopt in our prototype an ontology-based storage model. This latter can be formally defined as follows:

$$O := (C, \leq_C, R, \leq_R, I, \tau, \sigma)$$

where:

- C is a non-empty set of concepts
- $\leq_C$ is a partial order over the set C
- R is a set of relations (unary or binary)
- $\leq_R$ is a partial order over R
- I is a set of instances
- $\tau : R \to C^2$ is a mapping of a relation r to a couple of concepts
- $\sigma : C \to I$ is an instantiation of a concept c

Image features and descriptions are stored in the core ontology (MO class) with a link (URI) indicating the location of its content (i.e., raw data). Figure 11.6 shows a subset of the core ontology and more precisely how subject and object instances related to our access control model are interconnected.

In the following, we will describe the main two modules in ImageProtector.

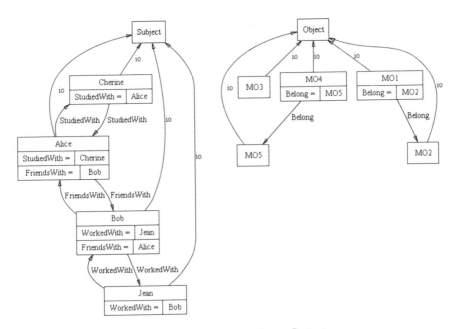

Fig. 11.6 Snapshot of a subset of the core ontology of ImageProtector

11.5.2 BackOffice Module

This module helps the main user[8] to specify who, among his/her contacts, have the right to manipulate a multimedia object and/or its content. In its current version, ImageProtector allows only to associate *view* action. Using SWRL specification module [17], the main user can define security rules, written in SWRL, so to:

1. map the multimedia objects to the core ontology, and map semantic descriptions to their content through the *content annotation* component,
2. define subjects (users who are able or not to visualize images) on the basis of semantically similar profiles [23] through the *user specification* component, and
3. specify actions and needed protection mechanisms (in the current version, we provide three main mechanisms: pixelize, blur, and spiral) (see Fig. 11.7).

11.5.3 FrontOffice Module

The *FrontOffice* module is used to manage (query, view, etc.) stored images and photos. In order to access this module and use provided functionalities, users should

---

[8]The user who logs into FaceBook.

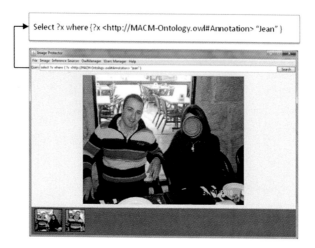

Fig. 11.7 Screenshot of the ImageViewer interface in ImageProtector

be authenticated. Before displaying requested images, a filtering engine is used to remove and/or protect confidential content according to the predefined authorization rules.

11.6   Experiments

In order to validate our approach, we conducted two sets of experiments using our prototype described in [4]. In the first one, we evaluated decision making and studied the effects of false negatives and positives when using multimedia and textual predicates in the policy enforcement. In the second one, we evaluated the utility of using FOL in multimedia access control. In the following, we detail each of the two experiments and discuss obtained results.

11.6.1   Decision Making Evaluation

In order to assess and evaluate decisions made when similarity predicates are used, we carried out various tests as shown in the following subsections. To do that, we used a computer with 2.2 GHZ and 4GB RAM, a webcam with 1.3 mega pixels in video mode and more than 5 mega pixels in image mode. We also used a set of 100 images with more than 200 tagged persons (a tag represents an object of interest in a Minimum Bounding Rectangle).

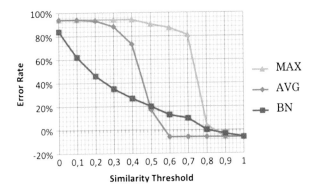

Fig. 11.8 Error rate vs. similarity threshold using a textual similarity predicate

Test 1: Evaluation of False Positives and False Negatives When Using a Textual Similarity Predicate

In this test, we used 100 images randomly tagged with 4 tags per image. We prefixed 6 tags representing the tags that should be protected. We used in this case, a predicate of the form $Sim(o_1, o_2, \beta)$ to search for all tags whose textual description is similar to the textual description of supplied multimedia object o_1. The similarity score is computed using three different aggregation functions (Average, BayesianNetwork,[9] and Max). After varying the similarity threshold from 0 to 1, we obtained the result shown in Fig. 11.8.

Please note that, on the one hand, the rates of false positives (objects of interest which should not be targeted by the predicate) are located between 0 and 100 % of the error rate axis, while, on the other hand, the false negatives (objects of interest which should be targeted but they are not) are identified when the error rates are less than 0 %. This shows in reality the trade-off in this test between privacy and availability of information. Please note also:

- MAX gives the highest (with respect to the other functions) error rate when similarity threshold ≤ 0.8,
- AVG returns a false positive error rate lower than the others when similarity threshold ≤ 0.5. Nevertheless, it returns a relatively high false negative error rate (with respect to the other functions) when similarity threshold > 0.55.
- BN reaches, when similarity threshold $= 0.8$, an error rate of 0. In addition, when threshold $\in [0.5, 0.8]$, the false negative error rate is the lowest.

---

[9] We invite the reader to consult [5] for more details regarding aggregation functions.

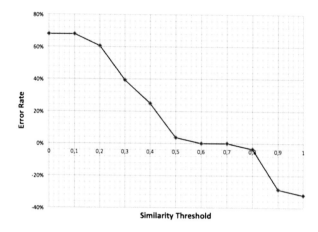

Fig. 11.9 Error rate vs. similarity threshold using a multimedia similarity predicate

Test 2: Evaluation of False Positives and False Negatives When Using a Multimedia Similarity Predicate

The objective of this test was to evaluate the error rate related to the protection of multimedia objects using their raw content and without referring to their textual descriptions. Here, we used Oracle Intermedia similarity function to protect all tags *similar* to an image specified by its URI. The used predicate was of the form of $Sim(o_1, shape1.jpg, \beta)$. Moreover, we tagged several shapes in the photos to specify the tags to be protected. More precisely, we specified 9 tags to protect in order to estimate the corresponding error rate which appears in Fig. 11.9. The graph above shows that the similarity function used has an error rate close to 0 when threshold $\in [0.5, 0.8]$. However, we noticed that the false negative error rate is relatively too high when threshold > 0.8, and it decreases quickly until reaching -38%.

Discussion

According to the tests conducted in the first experiment, protecting multimedia objects using textual and multimedia predicates is appropriate to a certain extent. In fact, it is crucial to take into consideration several aspects in order to obtain satisfactory results:

- The choice of similarity functions must be done according to the application domain and the types of images to be protected,
- The profile of the authorization manager cannot be only security-based anymore and must include some background in image processing,
- The value of the similarity threshold must be chosen carefully and its tuning remains a complex task.

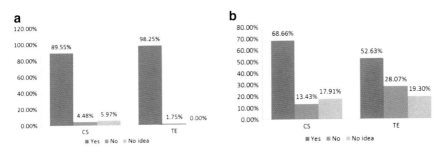

Fig. 11.10 Usefulness and Relevance of expressing security rules. (**a**) Evaluating the usefulness of FOL for security rules (**b**) Evaluating the relevance of FOL for security rules

11.6.2   Utility Evaluation

As mentioned earlier, the goal of this experiment was to evaluate the utility of using our approach for access control. To do so, we designed an online questionnaire[10] that has been answered by 124 graduate students. Sixty-seven students were from Computer Science (CS) department while the others were from Telecommunication's (TE). The first section of the questionnaire contained the same scenario defined in the motivation section as a working example for which the students had to write several security rules using FOL. A set of sample predicates were supplied to help students write their own rules and avoid potential errors since all chosen students were discovering FOL for the first time. The rest of the questionnaire was divided into three main sections as we present in the following.

Evaluating Security Rule Expressiveness

This was computed using two measures: *relevance* and *usefulness* of using FOL to express security rules. The graphs displayed in Fig. 11.10 show students' feedback. One can notice that the majority of students agreed on the fact that FOL is both useful and relevant to express security rules (although students in TE were less convinced about FOL relevance obviously due to their background).

Evaluation the Difficulty of Expressing Security Rules

The graph given in Fig. 11.11 shows the feedback of students regarding the difficulty of using FOL to express security rules. Most of the students agreed on the fact that expressing security rules is a moderate task, with a slight difference between CS and TE students who think that FOL might complicate the use of security rules.

---

[10]http://sigappfr.acm.org/rulessurvey/.

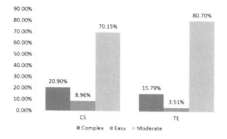

Fig. 11.11 Difficulty of expressing security rules

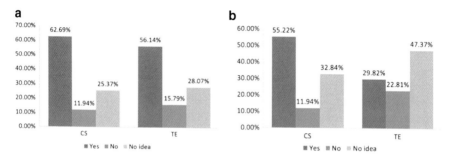

Fig. 11.12 Usefulness and Relevance of multimedia predicates. (**a**) Evaluating the usefulness of multimedia predicates (**b**) Evaluating the relevance of multimedia predicates

Evaluating the Utility of Using Multimedia Predicates

The aim here was to evaluating multimedia predicate utility when handling multi-media objects security. The results of the test are presented in Fig. 11.12. A relevant number of CS students agree on the fact that multimedia predicates are useful and relevant to include in security rules. However, TE students were less convinced.

11.6.3   Accuracy of the Study

Unlike common experiments including computational tests and computer analysis, human-based experiments are often subjective and involve considerable amount of ambiguity. Hence, the results presented here, regardless of their modesty, remain valid even for large margin of errors. In fact, the goal of these studies was limited to demonstrate the usability of our approach and its ability to express security rules to address at the same time the low-level features and textual descriptions of multimedia objects. To deal with complexity of rule specification, adopting a user-friendly GUI for users in order to specify their requirements easily can be considered in future work.

11.7   Conclusion

In this chapter, we presented the FOL-IBAC, a fine-grained access control model for images described based on their semantic description as well as their low level features. We used an FOL-based language to define the model's related information and specify security rules addressing images content. We also provided a new mechanism to deal with uncertain decision-making at policy enforcement time which has been tested in a set of experiments with textual and multimedia predicates.

In a future work, we intend to extend the proposed model to make it suitable for videos taking into consideration the complex aspects of video segments. We are also exploring some studies to automate the value of the similarity threshold so to ease the work of the end user.

References

1. Abou El Kalam, A., Baida, R.E., Balbiani, P., Benferhat, S., Cuppens, F., Deswarte, Y., Miège, A., Saurel, C., Trouessin, G.: Organization Based Access Control. In: 4th IEEE International Workshop on Policies for Distributed Systems and Networks (Policy'03), June 2003
2. Adam, N.R., Atluri, V., Bertino, E., Ferrari, E.: Content-based authorization model for digital libraries. IEEE Trans. Knowl. Data Eng. 14(2), 296–315 (2002)
3. Al Bouna, B., Chbeir, R.: Content-based policy specification for multimedia authorization and access control model. In: Cyber Warfare and Cyber Terrorism, pp. 345–357. Idea Group Reference, 2007
4. Al Bouna, B., Chbeir, R., Gabillon, A.: The image protector - a flexible security rule specification toolkit. In: SECRYPT, pp. 345–350, 2011
5. Al Bouna, B., Chbeir, R., Miteran, J.: Mca2cm: Multimedia context-aware access control model. In: Intelligence and Security Informatics, pp. 115–123. IEEE, New Brunswick, New Jersey (2007)
6. Androutsos, D., Plataniotis, K.N., Venetsanopoulos, A.N.: A novel vector-based approach to color image retrieval using a vector angular-based distance measure. Comput. Vis. Image Understand. 75, 46–58 (1999)
7. Atluri, V., Ae Chun, S.: An authorization model for geospatial data. IEEE Trans. Dependable Sec. Comput. 1(4), 238–254 (2004)
8. Bertino, E., Fan, J., Ferrari, E., Hacid, M.-S., Elmagarmid, K.A., Zhu, X.: A hierarchical access control model for video database systems. ACM Trans. Inf. Syst. 155–191 (2003)
9. Bertino, E., Hammad, M.A., Aref, W.G., Elmagarmid, A.K.: Access control model for video databases. In: 9th International Conference on Information Knowledge Management, CIKM, pp. 336–343, 2000
10. Boyle, M., Edwards, C., Greenberg, S.: The effects of filtered video on awareness and privacy. In: CSCW, pp. 1–10. ACM, Philadelphia, Pennsylvania (2000)
11. Chen, T.-Y.: Knowledge sharing in virtual enterprises via an ontology-based access control approach. Comput. Ind. 59(5), 502–519 (2008)
12. Egenhofer, M.J., Frank, A.U., Jackson, J.P.: A topological data model for spatial databases. In: SSD, pp. 271–286, 1989
13. Fan, J., Luo, H., Hacid, M.-S., Bertino, E.: A novel approach for privacy-preserving video sharing. In: CIKM, pp. 609–616. ACM, Bremen, Germany (2005)

14. Gabillon, A., Capolsini, P.: Rule-based policy enforcement point for map services. In: ACM SIGSPATIAL International Workshop on Security and Privacy in GIS and LBS. ACM, San Jose, CA (2010)
15. Gabillon, A., Capolsini, P.: Security mechanisms for geographic data. In: International Conference on Management of Emergent Digital EcoSystems (MEDES), pp. 297–302, Bangkok, Thailand (2010)
16. Gabillon, A., Capolsini, P.: Enforcing protection mechanisms for geographic data (to appear). In: International Symposium on Web and Wireless Geographical Information Systems (W2GIS). Naples, Italy (2012)
17. Horrocks, I., Patel-Schneider, P.F., Boley, H., Tabet, S., Grosof, B., Dean, M.: Swrl: A semantic web rule language combining owl and ruleml, 07 2010
18. Joshi, J.B.D., Li, K., Fahmi, H., Shafiq, B., Ghafoor, A.: A model for secure multimedia document database system in a distributed environment. IEEE Trans. Multimed. Special Issue Multimedia Databases 215–234 (2002)
19. Li, N., Mitchell, J.C.: Understanding spki/sdsi using first-order logic. In: CSFW, pp. 89, 2003
20. Masoumzadeh, A., Joshi, J.B.D.: Osnac: An ontology-based access control model for social networking systems. In: SocialCom/PASSAT, pp. 751–759, 2010
21. Pan, L., Zhang, C.N.: A web-based multilayer access control model for multimedia applications in mpeg-7. Int. J. Netw. Secur. 4(2), 155–165 (2007)
22. Pan, L., Zhang, C.N.: A criterion-based multilayer access control approach for multimedia applications and the implementation considerations. ACM Trans. Multimed. Comput. Comm. Appl. 5 17:1–17:29 (2008)
23. Raad, E., Chbeir, R., Dipanda, A.: User profile matching in social networks. In: NBiS, pp. 297–304, 2010
24. Sharma, L.K., Vyas, O.P., Tiwary, U.S., Vyas, R.: A novel approach of multilevel positive and negative association rule mining for spatial databases. In: MLDM, pp. 620–629, 2005
25. Thuraisingham, B.M.: Security and privacy for multimedia database management systems. Multimed. Tool Appl. (33), 13–29 (2007)
26. Thuraisingham, B.M., Lavee, G., Bertino, E., Fan, J., Khan, L.: Access control, confidentiality and privacy for video surveillance databases. In: SACMAT, pp. 1–10. ACM, California, USA (2006)

Problems and Questions

1 Social Networks: Definitions, Properties and Analysis

1. What are the most common ways to represent social networks? What are the characteristics of each of these representations?
2. What is the difference between implicit and explicit social network data?
3. What are the indicators of the strength of a social relationship?
4. When an actor is considered to be the most central? Is it according to the degree centrality metric? the closeness centrality metric? or the betweenness centrality metric?
5. How is a Web Based Social Network (WBSN) defined?
6. For which purpose context information are used in MSNs?
7. What are the main characteristics of Peer to Peer (P2P) networks?
8. What is the most appropriate link mining task to apply in order to classify a large number of users that are members of the same group on a social network? to discover the type of relationship between two users? to rank users based on their measured structural importance? to find correspondence between same real-world users across two social networks? to predict the existence of a relationship between two users?

2 Privacy and Security in Social Networks

1. Identify the origins of privacy threats in Social Networks that are also present in MSNs.
2. What is the difference between anonymization in Non-mobile Social Networks and MSNs?
3. When context information are added in MSNs, several challenges are raised regarding the information privacy. Name some of them.

R. Chbeir and B. Al Bouna (eds.), *Security and Privacy Preserving in Social Networks*,
Lecture Notes in Social Networks, DOI 10.1007/978-3-7091-0894-9,
© Springer-Verlag Wien 2013

4. Compare the security and privacy resources provided by the Mobilis, MobiSoc, MyNet, and MobileHealthNet middleware.
5. Define location privacy.
6. What is the difference between data privacy and user privacy in P2P systems?
7. What is the main advantage of P2P versus cloud regarding data privacy?
8. Why current privacy settings are not enough for social network users?
9. Many conventional data protection exist, name some of these techniques and explain their advantages as well as their drawbacks.
10. Give an example of a privacy threat that can exploit the structure of a social network. Give another privacy threat that can exploit information that a user previously published online.
11. What are the most common privacy threats that an attacker can exploit by recoding and analyzing the habits of a social network user?
12. As a social network user, what are your current expectations in terms of online privacy? Try to map each of your these needs to its closest threat based on the social network analysis and link mining threat classification presented in this chapter.

3 Access Control Models for Social Networks

1. What are the features required to achieve fine-grained access control management?
2. Differentiate between data security and data integrity.
3. What are the main contributions of the Clark–Wilson model to the integrity models?
4. Experiment with Alloy by formalizing the Role Based Access Control Policy.
5. Compare and contrast between the Biba Model and its predecessors, the Low-Water-Mark Model and the Ring Model.
6. State two various strategies for conflict resolution in access control models.
7. How does the mechanism to achieve interoperability works?
8. What is the general mechanism to manage data exposure minimization?
9. What are the requirements needed to achieve a successful access control management system in WBSNs?
10. How is an S4AC-PRISSMA Access Condition implemented?
11. Describe the S4AC authorization procedure.
12. What are the advantages of using Semantic Web lightweight vocabularies for defining access policies?
13. What is the role of the S4AC Access Evaluation Context?
14. What does a privacy policy contain according to PriMod?

4 Trust and Reputation in Social Networks

1. What is trust in social networks?
2. What are the various characteristics of trust?
3. What are the techniques for inferring trust?
4. How trust can be used in Distributed MSNs as a metric?
5. Why should the privacy of feedback providers be preserved in reputation systems?
6. List the key sets of security building blocks used by current privacy preserving reputation systems secure under the semi-honest model.
7. List the key sets of security building blocks used by current privacy preserving reputation systems secure under the disruptive malicious model.

5 Social Multimedia Content Protection and Manipulation

1. Why image content protection differs from other techniques offered for textual data?
2. How FOL is used in the scope of preserving privacy in images?
3. Express the utility of environmental predicates used for context-based protection?
4. Enumerate various scenarios that require face recognition in virtual worlds?
5. What is the difference between original Local Binary pattern and Multi-scale Local Binary Pattern?
6. Define Artimetrics.
7. What is the difference between applying Hierarchical Multi-scale Local Binary Pattern and normal Multi-scale Locale Binary Pattern on an image?